CONTOURING
A Guide To The
Analysis And Display
Of Spatial Data

COMPUTER METHODS IN THE GEOSCIENCES

Daniel F. Merriam, Series Editor

Volumes published by Van Nostrand Reinhold Co. Inc.:

Computer Applications in Petroleum Geology: J. E. Robinson
Graphic Display of Two- and Three-Dimensional Markov Computer Models in Geology: C. Lin and J. W. Harbaugh
Image Processing of Geological Data: A. G. Fabbri
Contouring Geologic Surfaces with the Computer: T. A. Jones, D. E. Hamilton, and C. R. Johnson
Exploration-Geochemical Data Analysis with the IBM PC: G. S. Koch, Jr.(with programs on diskettes)
Geostatistics and Petroleum Geology: M. E. Hohn
Simulating Clastic Sedimentation: D. M. Tetzlaff and J. W. Harbaugh

* Orders to: Van Nostrand Reinhold Co. Inc, 7625, Empire Drive, Florence, KY 41042, USA.

Volumes in the series published by Pergamon:

Geological Problem Solving with Lotus 1-2-3 for Exploration and Mining Geology: G. S. Koch Jr. (with program on diskette)
Exploration with a Computer: Geoscience Data Analysis Applications: W. R. Green
Management of Geological Databases: J. Frizado (Editor)

Related Pergamon Publications

Books

GAAL & MERRIAM (Editors): Computer Applications in Resource Estimation: Prediction and Assessment for Metals and Petroleum

HANLEY & MERRIAM (Editors): Microcomputer Applications in Geology II

MALING: Coordinate Systems and Map Projections, 2nd edition

MALING: Measurements from Maps

Journals

Computers & Geosciences

Computer Languages

Information Processing & Management

International Journal of Rock Mechanics and Mining Sciences (& Geomechanics Abstracts)

Minerals and Engineering

Full details of all Pergamon publications/free specimen copy of any Pergamon journal available on request from your nearest Pergamon office.

CONTOURING
A Guide To The Analysis And Display Of Spatial Data

(with programs on diskette)

David F. Watson

PERGAMON PRESS

OXFORD · NEW YORK · SEOUL · TOKYO

UK	Pergamon Press Ltd, Headington Hill Hall, Oxford OX3 0BW, England
USA	Pergamon Press Inc., 660 White Plains Road, Tarrytown, New York 10591-5153, USA
KOREA	Pergamon Press Korea, KPO Box 315, Seoul 110-603, Korea
JAPAN	Pergamon Press Japan, Tsunashima Building Annex, 3-20-12 Yushima, Bunkyo-ku, Tokyo 113, Japan

Copyright © 1992 D. F. Watson

All Rights Reserved. No part of this publication may be reproduced, stored in a retrieval system or transmitted in any form or by any means: electronic, electrostatic, magnetic tape, mechanical, photocopying, recording or otherwise, without permission in writing from the publishers.

First edition 1992

Library of Congress Cataloging in Publication Data
A catalogue record for this book is available from the Library of Congress.

British Library Cataloguing in Publication Data
A catalogue record for this book is available from the British Library.

ISBN 0 08 040286 0

Printed in Great Britain by B.P.P.C. Wheatons Ltd, Exeter

Contents

Figures and Tables xi

Series Editor's Foreword xiii

Preface xv

PART 1 - Practical Contouring 1

1.0 Getting a Picture 3

1.1 Data Preparation 7
Data Set A:\HILL.DAT 8

1.2 Setting Contour Levels 11

1.3 Data Windows 13

1.4 Adjusting Surface Tautness 17

1.5 Display Types 21
Gridding 22
Planar and Isometric Contours 23

v

Isometric Orthogonal Profiles 23
Stereograms 23

1.6 Changing Color-fill Colors 25

1.7 Computing Volumes 27
Volume of a Prism 29
Volume Under a Surface 30
Numerical Integration 31
Volume of a Parallelepiped 34

1.8 Trouble Shooting 37

1.9 Program Adaptations 41
Histograms 42
Ogives 42
Block Averages 43

PART 2 - Principles of Contouring 45

2.0 General Concepts of Contouring 47
Topographical Data 48
History of Contouring 48
Isoline Maps 49
Dot Density Maps 50
Isochor Maps 50
Computing Contours 53
Major Difficulties of Contouring 54

2.1 Data Sorting 57
Proximal Order 57
 Arbitrary Proximal Order 58
 Optimal Proximal Order 58
 Greedy Proximal Order 58
 Natural Neighbor Proximal Order 59
Expressing Proximal Order 63
Natural Neighbor Sorting Algorithms 64

2.2 Subset Selection 69
Selection Criteria 71
 Fixed Number Subsets 71
 Fixed Distance Subsets 71
 Natural Neighbor Subsets 72

2.3 Local Coordinates   75
Binary Local Coordinates   76
Barycentric Coordinates   76
Rectangular Local Coordinates   78
Parallelogram Coordinates   79
Natural Neighbor Coordinates   81
    Computing Natural Neighbor Coordinates   82

2.4 Gradient Estimation   85
Surface Types   86
Gradient Subsets   87
Tangent Planes   87
    Least Squares Gradients   88
    Least Squares Quadratics   90
    Spline Gradients   92
    Cross Product Gradients   94
    Neighborhood-Based Gradients   97

2.5 Interpolation   101
Ideal Interpolation   103
Difficult Aspects   104
Fitted Function Interpolation   105
Weighted Average Interpolation   105
Manual Methods   106
    Statistical Method   106
    Proximal Polygons   107
    Natural Neighbor Proximal Polygons   108
    Triangulations   108
    Rectangular Grids   111
    Inverse Distance Weighting   112
    Linear Regression   113
Computer Methods   113
    Distance-Based Methods   113
        Inverse Distance Weighting   114
        Inverse Distance Weighted Gradients   116
    Fitted Function Methods   118
        Lagrange Interpolation   119
        Collocation   120
        Minimum Curvature Splines   122
        Kriging   123
        Relaxation Surfaces   124
        Approximation Surfaces   126

   Triangle-Based Methods 130
    Linear Facets 130
    Nonlinear Triangular Patches 131
   Rectangle-Based Methods 137
    Bilinear Patches 139
     Data Set **A:\SHARPG.DAT** 142
    Hermite Patches 143
    Bezier Patches 147
    B-Spline Patches 149
    Taylor Interpolants 151
    Tension Patches 152
    Fourier Surfaces 154
   Neighborhood-Based Interpolation 155
    Linear Interpolation 155
    Nonlinear Interpolation 157
  Synoptic Tabulation 159
  Discussion 159

2.6 Blending Functions 163
  Global Blending Functions 164
   Power Functions 164
   Trigometric Functions 165
   Nonfinite Kernels 165
  Local Blending Functions 167
   Explicit Functions 167
   Parametric Functions 169
   Compound Exponential Function 170
    Roughness Index 171
    Outlier Index 173

2.7 Output Techniques 177
  Display Media 177
  Output Database 178
  Triangular Grids 179
  Following Contours 180
  Isometric Views 181
  Interval Filling 182
  Color-filled Isometric Views 183
  Orthogonal Profiles 184
  Stereograms 184
  Numerical Output 185
   Precise Volumes 185
   Volume Products 186
   Isted's Formula 186

2.8 Execution Efficiency 189
   Spatial Sorting   189
   Gradient Estimation   190
   Interpolation Efficiency   191
   Display Efficiency   192

Some Published Programs   195

Glossary of Contouring Terms   197

References to Contouring Literature   201

Appendix: Included Software   223
   BASIC Program Listings   226
      CON2R.BAS   226, 273
      NEIGHBOR.BAS   234, 280
      TRIANGLE.BAS   244, 289
      DISTANCE.BAS   253, 297
      FITFUNCT.BAS   261, 305
      RECTANGL.BAS   267, 310

Index   317

Figures And Tables

1.0.1	Isolines of example data	6
1.2.1	Isometric contour levels	12
1.3.1	Input and display windows	14
1.4.1	Surface tautness	18
1.5.1	Interpolation grid nodes	22
1.5.2	Stereogram	24
1.7.1	Area of a parallelogram	28
1.7.2	Triangular prism volume	30
1.9.1	Histogram and cumulative distribution	42
1.9.2	Array of block averages	43
2.1.1	Systematic triangulations	60
2.1.2	Natural neighbor circumcircles	62
2.1.3	Delaunay and Voronoi tessellations	62
2.3.1	Binary local coordinates	76
2.3.2	Barycentric local coordinates	77
2.3.3	Rectangular local coordinates	79
2.3.4	Parallelogram local coordinates	80
2.3.5	Natural neighbor local coordinates	81
2.3.6	Computing natural neighbor coordinates	83

Figures and Tables

2.4.1 Tangent to a surface 87
2.4.2 Least squares plane gradients 89
2.4.3 Minimum curvature spline gradients 93
2.4.4 Cross product gradients #1 96
2.4.5 Cross product gradients #2 96
2.4.6 Natural neighbor gradients 98
2.5.1 Triangle-based interpolation 110
2.5.2 Inverse distance parameters 115
2.5.3 Rectangular hyperboloid interpolation 139
2.5.4 Hermite cubic polynomials 145
2.5.5 Bernstein cubic polynomials 148
2.5.6 B-spline cubic basis functions 150
2.5.7 Neighborhood-based linear interpolation 156
2.5.8 Neighborhood-based blended interpolation 158
2.5.9 Comparison of four surfaces 161
2.6.1 Exponential blending functions 170
2.6.2 Roughness index 172
2.6.3 Outlier index 174
2.7.1 Isted's volume-product 187

Table 1 Listing of Data Set A:\HILL.DAT 8
Table 2 Interior Angles 61
Table 3 Proximal Order Lists 64
Table 4 Listing of Data Set A:\SHARPG.DAT 142
Table 5 Synoptic Tabulation of Computer Methods 159

Series Editor's Foreword

Contouring, a method used by many scientists, especially earth scientists, is a basic research tool, yet probably little understood by most users. This book on the subject by Dave Watson explores in detail the practice and principles of contouring using a personal computer. Contouring allows a three dimensional view in two dimensions and thus is a fundamental technique to represent spatial data. All aspects of this type of representation are covered here including data preparation, selecting contour intervals, interpolation and gridding, computing volumes, and output and display.

A set of twelve computer programs in BASIC is included with the book (on the diskette and also listed in the appendix). A compilation of published programs is appended along with a glossary and references to pertinent literature. Everything you want or need to know about contouring is included between these two covers.

The book is well organized, well illustrated, and well written. Both the novice and the seasoned veteran will find the book easy to use and informative. Watson has tabulated many techniques that have been used in contouring, comparing them, and noting their advantages and disadvantages. The first part of the book is a step-by-step explanation of the PC software and how to use the programs with the reader's own data. The second part presents the rationale and concepts for contouring using the computer.

There are two other books in this series that would be of interest to those involved with contouring. J. E. Robinson's book (*Computer Applications in Petroleum Geology*) contains several chapters on contouring and map analysis.

Series Editor's Foreword

The other book (*Contouring Geologic Surfaces with the Computer*) by T. A. Jones, D. E. Hamilton, and C. R. Johnson, discusses in detail contouring and mapping as applied to geologic data. Both of these are good ancillary reading.

Graphics, including contouring, is especially important as evident by the coverage of the subject in recent years. This book by Dave Watson is an important addition to that literature.

D. F. Merriam

Preface

WHAT IS CONTOURING?

Computer contouring is a practical tool that generates and displays the form and shape of a representative surface. Such displays are useful to illustrate and evaluate bivariate data sets, as well as to explore functional concepts. Contouring can provide a profound insight into spatial phenomena, allows a precise estimate of volume under the representative surface, and can be done efficiently on microcomputers.

WHY PUBLISH A BOOK ON CONTOURING?

Many computer contouring programs are available, but usually offer little information about the algorithms that generate their display. A visually impressive display of the data, however, only is one-half of the picture; to make the contour display credible, the reasoning and logic underlying the contouring technique must be understood.

The rationale and technical details of particular computer contouring methods have been published in scientific journals during the last quarter century or so, but the general principles behind computer contouring methods are complex, and not widely appreciated. Although their application underpins many commercial and technical conclusions derived from topographical data, at this time there is no book devoted to providing a general and overall treatment of contouring methods.

Preface

This book collects, summarizes, and categorizes the broad range of concepts and mechanisms that determine the effectiveness of contouring programs, and it expresses and demonstrates these principles, in specific but practical terms, for technical readers. Procedural options, pertinent to displaying the surface contours implied by a topographical data set, are discussed in terms of their merits, insufficiencies, and possible alternatives, with references to the original literature.

To enable readers to evaluate effectively this information, as applied to their own data, a suite of computer programs has been included. The twelve BASIC programs, usually referred to as "the programs" where this is unambiguous, are intended to facilitate testing and comparison of contouring methods. Any of eighteen interpolation methods, five gradient estimation methods, and seven forms of display, can be used while all other aspects remain constant.

The generated surface is integrated numerically over arbitrary regions, and this provides a powerful check on the quality of the interpolation. Interpolated surfaces can be saved to a scratch file, and so a library of surfaces, in perfect register, can be accumulated.

In addition to the survey of contouring as a general subject, the keynote of the book is the presentation of neighborhood-based interpolation, using neighborhood-based gradient estimation, and blending the gradients with the linear interpolation. This method is shown to be the most general approach to topographical data, and its relationships to other contouring methods are discussed. The programs use a previously unpublished, computationally convenient, method for determining the intersection area of overlapping Voronoi polygons, and it is described for the first time in this book.

WHO WOULD USE THIS BOOK?

Because it combines practical and theoretical information, this book fulfills two requirements. In the first situation, this book shows you how to get a contour display of topographical data on your screen or plotter, if you have minimal computing equipment---64 Kbytes of memory and a BASIC Interpreter---using the BASIC programs on the DOS diskette at the back of this book. Their operation is described in practical terms of what to do, what not to do and, possibly, how to determine the cause when the program fails. Then the plausibility of the display may be established from the theoretical discussions.

In the second situation, if you are interested in the theory underlying contouring algorithms, then you will perceive this book to be useful. The many techniques that have been used to contour topographic data are organized in a manner that reveals the development and evolution of computer contouring theory over the last several decades. The advantages and disadvantages of each method are cataloged, illustrated, and referenced to technical articles listed in the bibliography. The algorithms may then be tested and compared by using the included programs.

Preface

To meet these two requirements, the book is organized into two parts. **Part 1** introduces you to practical contouring, beginning with basic details. How to set up your data, set contour intervals, select data windows, adjust surface tautness, and set output scales. Also, in this part, is a description of volume computations under the contoured surface, trouble shooting tips, and custom adaptations.

Part 2 specifies and explains the underlying concepts and reasoning behind the algorithmic procedures such as spatial sorting, subset selection, local coordinates, estimating gradients, methods of interpolation, blending functions, display techniques, and estimates of computational efficiency.

There also is a glossary of contouring terms and a list of programs published by other authors. Over 500 references to technical articles are included, and listings of the twelve BASIC programs, on the DOS diskette, are given.

HOW TO USE THIS BOOK

To obtain a contour display of your data, insert the DOS diskette, and see the next pages for step-by-step instructions. To obtain information on any particular aspect or topic of contouring, locate the topic in the Table of Contents or the Index. The book may be read from front to back, or you may follow the extensive cross referencing, as you require.

D.F.W.

PART 1
Practical
Contouring

OVERVIEW OF PART 1

This part presents an explanation of the practical details of displaying a surface using a personal computer and the programs that come with this book. As an introductory step, a picture can be generated from the example data included on the diskette. This is described in Section 1.0. To contour your own data, several options and requirements need to be considered. Initial data preparation is discussed in Sections 1.1, 1.2, and 1.3, leading to output design in Sections 1.4, 1.5, and 1.6, and on to computing volumes, trouble-shooting, and special adaptations for your own purposes, in the last three sections of this part.

SECTION 1.0

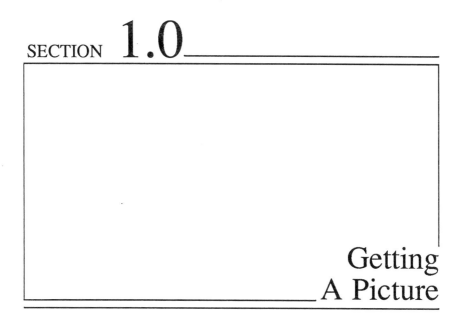

Getting A Picture

SECTION SUMMARY

This section discusses selecting and loading a program from the diskette, setting screen resolution, and altering graphics statements.

FIRST

It is recommended that the first thing you do is make backup copies of the programs on the diskette. The reason is that the rest of this book discusses changes that you can make to these programs, and so you may need a copy of the original code for future reference.

Then, when backup copies have been made, check the compatibility of your system with the programs by generating a display of the example data set included on the diskette. This will insure that any problems arising later will not be caused by system misalignment. Now load a program according to the following steps.

LOADING A PROGRAM

Two programs are suitable for the initial run. If you have only 64 Kbytes of memory, SCRN\CON2R.BAS uses 23 Kbytes excluding data. However, SCRN\NEIGHBOR.BAS, using 29 Kbytes, is more versatile and useful.

Getting A Picture

Step 1

Insert the diskette into drive **A:**, load the program, and get a listing of line **1100** on the screen. The instruction and path is

LOAD "A:\SCRN\NEIGHBOR.BAS"
or
LOAD "A:\SCRN\CON2R.BAS"

Step 2

These two programs produce output to the screen so you will need to know your screen resolution. If it is 640 pixels wide by 350 pixels high, you can skip to Step 3. If it is, say, 320 by 200, or 720 by 348, these values need to replace the values for **WD** and **HI** in line **1100**. For example, change the specification of screen size from

1100 . . . . :WD=640:HI=350: . . . .
to
1100 . . . . :WD=320:HI=200: . . . .

On the same line, **SZ** is the width, in pixels, of the border for the generated display. The height of the border is set automatically at .71*SZ, in line **5100**, to allow for pixel elongation. Therefore, you need to select a value for **SZ** that is less than about 1.4*(HI-10).

Step 3

Next get a listing of line **5080**. The SCREEN instruction should be set to suit your graphics card. If you have a monochrome screen then the COLOR statement should be deleted.

 5080 SCREEN 9:COLOR 2,0: . . . .   *suits the EGA graphics card*

If your system is compatible with the IBM CGA or EGA graphics calls, the program should now run; go to step 4. Otherwise you will need to substitute your system's graphics statements to draw a line and plot a point on the screen. BASIC code at lines **5130**, **5527**, **5531**, **5710**, **5740**, **6130**, **6580**, and **6600**, must be changed.

5710 . . . . . :LINE(x,y)-(u,v),9: . . . .
and
5740   draws a line from screen coordinates (x,y) to (u,v) in color no. 9. Retype these instructions to suit your system, using the same variable names in place of x,y,u, and v.

5130.....	:LINE(x,y)-(u,v),14: ....
6580	draws a line from screen coordinates (x,y) to
and	(u,v) in color no. 14. Retype these
6600	instructions to suit your system, using the same variable names in place of x, y, u, and v.
5527.....	:PSET(x,y),IC: ....
5531	sets pixel (x,y) to color no. **IC**.
and	Retype these instructions to suit your
6130	system, using the same variable names in place of x, y, and **IC**.

Step 4

Run, or compile and run. The program is set up to read the example data from **HILL.DAT** on the diskette, and draw an isoline map on the screen. Figure 1.0.1 shows the example isoline display as it appears on a plotter. The data, used in this example, are listed on p. 8.

PLOTTER OUTPUT

Programs PLOT\CON2R.BAS and PLOT\NEIGHBOR.BAS are equivalent to the programs discussed above, but send the output to a plotter. In this situation, the values for the width and height of the plot, in plotter units, need to be set. Change **WD** and **HI** in line **1100**. Also set **SZ**, in the same line, to the desired width for the border of the plotted display. Additionally, you may have to substitute the plotter instructions in lines **5170 - 5190** to initialize the plotter and draw the border, in line **6570** (two statements) to draw line segments, in lines **5540** and **5550** to plot the data points, and in lines **5570** and **5577** to write a title and close the plot.

Getting A Picture

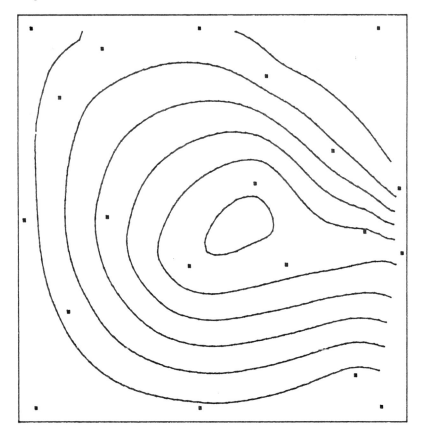

Figure 1.0.1 Isolines of example data

SECTION 1.1

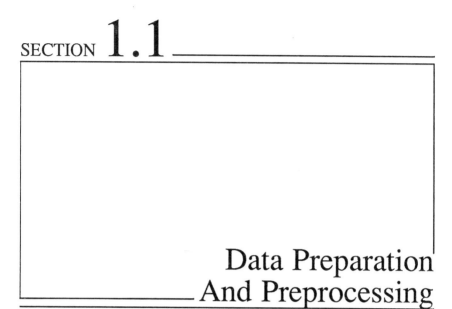

Data Preparation And Preprocessing

SECTION SUMMARY

This Section describes the structure of the data set and contouring information.

INTRODUCTION

Topographical data are measurements of height, for example, at various locations on the landscape. Each measurement involves three numbers: two geographical coordinates and the elevation at the location defined by the coordinates. Such data often are written as a three-column list, and each row of three numbers is a single datum representing a particular point on the topography. Except for RECTANGL.BAS, which requires the data to be on a rectangular grid, it does not matter whether the data have a regular pattern or are scattered.

DATA SET STRUCTURE

The programs are set up to read the contouring instructions and the data, in a fixed sequence. Section 1.9, p. 41, explains how to make changes in this sequence when rearranging the presentation order is desirable. To suit the programs as they are, your contouring instructions and data need to be arranged in the following order.

Data Preparation and Preprocessing

First of all, for each datum, the locational coordinates and height have a fixed order. Conventionally, the first number in the triplet is the x-coordinate, termed *easting*. The second number is the y-coordinate, termed *northing*, and the measured *height* is the third number. The programs expect the x values to increase toward the right or east, and the y values are expected to increase upwards or north. The contour map will be inverted if either the x or y coordinate axes are directed oppositely.

For these programs, the topographical data are preceded by the contouring instructions. An example topographical data set, **A:\HILL.DAT** plus labels on the right that indicate the contents of each row, is shown below in TABLE 1.

TABLE 1   Listing of Data Set A:\HILL.DAT

20	6		Number of data, number of contours
7	11	15	Six contour levels
19	23	27	.
0	75	750	Lower left corner of window and width
172	451	16.4	Twenty data triplets, $(x, y, f(x,y))$
81	671	8.7	.
163	762	8.1	.
350	800	7.4	.
480	711	8.9	.
608	573	12.1	.
668	423	23.5	.
650	160	5.8	.
350	100	6.3	.
97	276	8.2	.
330	360	25.6	.
458	513	26.0	.
518	363	24.6	.
34	100	0.5	.
12	445	5.0	.
26	800	3.2	.
695	800	0.1	.
735	504	8.9	.
740	384	24.1	.
700	102	2.3	.

Setting the information for contour levels is discussed in Section 1.2, p. 11. For now, just notice the structure of the data set; the computer programs expect the data to be in this order. The first number on the first row is the number of data points and, in this data set, there are 20 location pairs with their respective height measurements.

If there were more than 20 data, then the first number in the first row must be changed to suit, because otherwise the computer will ignore the extra data. If there were less, then again this number must be changed to suit, because the computer will look for, and fail to locate, the specified amount of data.

The topographical data are in the last 20 rows of TABLE 1. Column 1 has the x values, column 2 has the y values and the *heights* are in the third column. The units of measurement are not included because these are not involved in the calculations. Of course, they are necessary for an understanding of the contour display.

Figure 1.0.1, on p. 6, shows a plan view of the data locations listed in TABLE 1. The lower left corner has coordinates (0,75), and the square border is scaled to 750 units on a side, as specified by the fourth row of **A:\HILL.DAT** on the previous page.

The geographical configuration of the data, in conjunction with the heights at each datum, are restraints that influence the form and shape of a computer-generated surface representing the spatial variation in the data. There are many ways to display the variation in that surface.

Therefore, in addition to the topographical data, and the amount of data, contouring programs require information on the number and heights of the contour levels, and the position and size of the window or drawing frame around the data. The complete input form for a data set, such as that of TABLE 1, also includes the information discussed in Section 1.2 on assigning contour levels and in Section 1.3 on setting the size of the data window.

SECTION 1.2

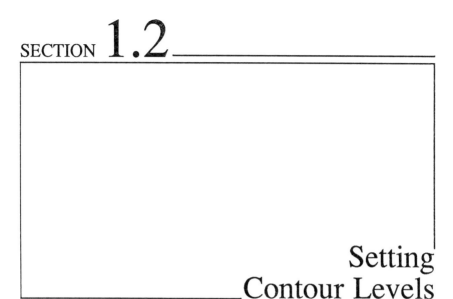

Setting Contour Levels

SECTION SUMMARY

How to determine and specify the number, and heights, of contour lines.

CONTOURS

A contour line (more precisely, an isoline) is a horizontal curve that marks the intersection of a sloping surface with a level plane. Contour levels are the heights where level planes cut the surface and give each isoline a constant elevation. The isolines formed by a set of level planes, at regular intervals, can illustrate and represent the shape and contour of a surface by the way they vary in relation to each other. For example, if several horizontal slices were made through a hill, the curved lines in Figure 1.2.1 on the next page are isolines that show where the blade would cut the surface of the ground.

SETTING CONTOUR LEVELS

The heights of the contour levels must be specified as additional information to the programs. By looking at the range of heights in your data set, you will get an idea of the levels that would suit your data. In TABLE 1, p. 8, the heights (third column, last 20 lines) range from 0.1 to 26.0.

Setting Contour Levels

The computer constructed surface through these data may extend above or below these values so, for example, six contour levels, from 7.0 to 27.0 at intervals of 4.0, are specified. The second number on the first row in TABLE 1 is the number of contour levels and, in the following two rows, those six contours levels are specified.

A constant vertical spacing between contour levels illustrates the surface shape in the most general way. However, the programs do not require that the interval be constant so irregular intervals may be used, if desired. Put the desired number of contour lines in second position on the first line of your data set. Then place the contour levels that you wish to use in the data set immediately after the number of contours. It is important to include the exact number of contours specified, no more and no less, or the programs will go wrong.

As well as irregular intervals, contour levels may be specified in any order for an isoline display. However, color-filling between contour levels requires the contour levels to be ordered from lowest to highest. The color-fill procedure skips all contour levels that are not in that order.

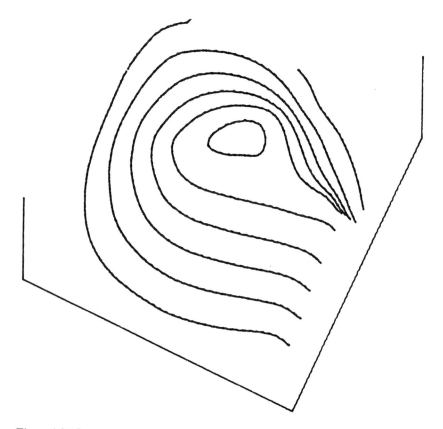

Figure 1.2.1 Isometric contour levels

SECTION 1.3

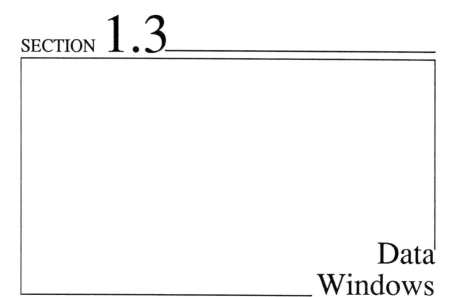

Data Windows

SECTION SUMMARY

Data windows are block subsets of the data set. They specify the scale and focus of the contour display.

WINDOWS

When a data set is too large to fit into available memory, or one wishes to study a subregion, a window can give a local view of the surface. The display window can be thought of as a frame around the generated view of the data. When it is made smaller, with the display being drawn at the same size, it effectively zooms in and magnifies that portion of the data set.

This requires that two concentric, temporary borders, or windows, be set up. Then a display of the representative surface is generated only within a square region which is the *display window*. However, the data used to compute the generated surface includes all the data within a larger, circular, *input data window*.

For example, Figure 1.3.1, on the next page, shows the data of TABLE 1, now with a smaller display window, and focused on the central region. The square border is scaled to 350 units on a side, and the lower left corner has coordinates (200,275). Some of the data lie outside the display window, but they are included in the interpolation unless they also lie outside the circular input data window.

Data Windows

Because the input data window is larger all around than the display window, contiguous display windows will agree at their edges, being based on the same data. In comparison, Figure 1.0.1, on p. 6, shows the same data set, but the square border is scaled to 750 units on a side and the lower left corner has coordinates (0,75). The display window that was chosen for that illustration totally enclosed the data, with room to spare.

SELECTING DATA WINDOWS

A display window is just that portion of your data set that you wish to "magnify" to fill your screen or plotter page. The programs allow you to select any size of square and position it anywhere over your data set. Then the region inside this square display window, if the input window contains more than three data, will be contoured and displayed.

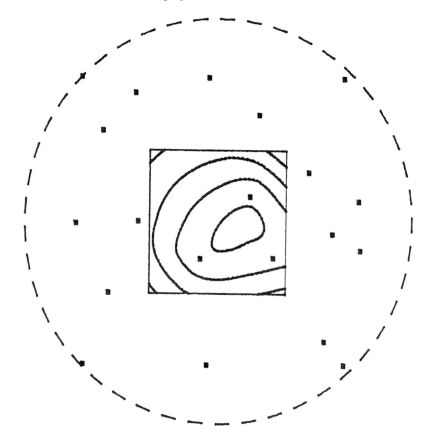

Figure 1.3.1 Input and display windows

Setting the size of the display window is done by changing the fourth line of the data set, shown in TABLE 1 on p. 8, from

	0	75	750	Lower left corner of window and width
to				
	200	275	350	Lower left corner of window and width

The radius of the input data window is the square root of **RD**, assigned on line 1100, times the width of the display window. This may need to be changed for irregularly spaced data.

Of course, the input window also can be made larger than the width of the data set, and then two types of output are available. CON2R.BAS, NEIGHBOR.BAS, TRIANGLE.BAS, and RECTANGL.BAS, cannot produce contours outside the convex perimeter of the data set, so the actual picture will be smaller than the display window. However, DISTANCE.BAS and FITFUNCT.BAS will fill the window even if it is larger than the data set.

The programs compute the average height of the representative surface, within the display window, each time a display is made. This is a precise estimate of volume under the displayed surface. When data sets are partitioned into blocks in a systematic manner, average surface height for each block provides a precise statistical evaluation of the surface. This is discussed fully on p. 27.

SECTION 1.4

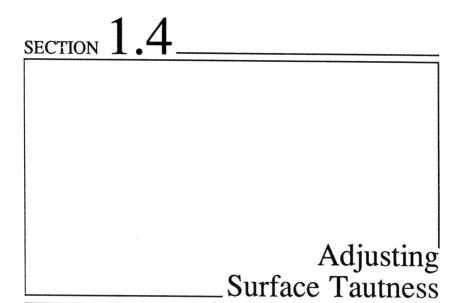

Adjusting Surface Tautness

SECTION SUMMARY

This section tells how to change the puffiness or tautness of the generated surface. Taut surfaces are conservative; rounded surfaces emphasize local trends in the data. The selection of appropriate tautness provides the greatest opportunity for an analyst to apply subjective influence to the interpretation of spatial data sets.

SURFACES

If a surface were formed by a rubber sheet that was stretched to meet all the data points, it would appear smooth everywhere between data points. However, it would form peaks, or pits, at the data points. Alternately, if it could be made of a single thin plate of spring steel, it would appear smooth everywhere including the data points. Although neither of these surfaces is realistic in itself, they serve as end members for the range of possible variation in generated surfaces.

Figure 1.4.1, on the next page, shows these two types of surface. Notice that, in the first example, most of the change of slope required to suit the variation in the data is incorporated in the surface at the data points and is expressed by the sharp peaks and pits; mathematically, the slope of the surface at the data points is undefined in the sense that no analytical derivative can be established. Of course, it is unlikely that this is the case for most real topographical functions.

However, for the alternative example, the surface is seen to have a continuous slope at the data points and to be more undulating than the elastic membrane type of surface. In this situation the change of slope required to suit the variation in the data is spread over the whole surface so that each location on the surface has close to the least possible curvature. Again, such a minimum curvature surface is unlikely for most real topographical functions.

Conceptually, there is a third type of surface which has all the change of slope along lines that form partitions between the areas surrounding the data points so that each datum effectively has a flat portion of surface in its immediate subregion. The slope of the surface at the data point is usually zero, although it may be tilted. This slope has a dominant local influence, but is sometimes adjusted to suit adjacent subregions with contrasting slopes. This surface also is unreal but it is useful for geometrically simplified generalizations of real topographical functions.

These three general types of surface serve to illustrate the range of possibilities for the shape of a surface that can be generated to represent a given data set. Of course, most real surfaces would be expected to be, in some sense, intermediate to these three examples. To allow for this, the programs have two parameters for adjustment of the type of surface generated. These parameters offer the possibility of effecting subtle changes and, as the diagram below shows, the apparent overall shape of the surface is not altered dramatically. Nevertheless, one should be aware that even such seemingly minor variations can lead to significant changes to the volume beneath the surface.

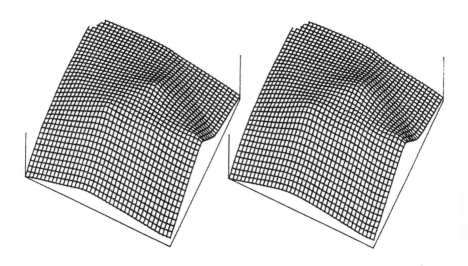

Figure 1.4.1 Surface tautness

SETTING TAUTNESS

The generated surface is adjusted by changing one, or both, of two blending parameters to obtain variations combining peaked, rounded, and flattened, terrain. The parameters **BI** and **BJ**, in line **1100**, are presently set to conservative values that give a fairly taut surface.

1100 . . . . . :BI=1.5:BJ=7: . . . .

Parameter **BI** performs in the approximate range of 1.0 to 2.5 while parameter **BJ** may vary in the approximate range of 3.0 to 10.0. For low values of the parameters, the surface is rounded and rolling; for higher values it tends to be taut with cusps at the data. Numerical overflow limits the effectiveness of upper values for these parameters, but a taut surface is obtained by setting **CL**=0, in line **1100**. The blending function is stable, in the sense that small changes in these blending parameters will cause only small changes in the generated surface.

The blending function is an **S**-shaped curve that controls the range of influence due to any particular data point for various adjacent positions. The parameters **BI** and **BJ** determine the actual shape of the S and this effects the tautness or puffiness of the resulting surface. A fuller discussion of blending functions begins on p. 163, and the parameters are described and illustrated on p. 170.

SECTION 1.5

Display Types

SECTION SUMMARY

This section discusses resolution of interpolation, gridding, planar contours, isometric contours, isometric orthogonal profiles, and stereograms.

OUTPUT RESOLUTION

Resolution is a term that refers to the amount of detail given by an observable difference between contiguous components. A representative surface is displayed by graphical expressions, usually isolines, that are abreviations of the surface, and the amount of detail they provide is the output resolution.

Curved lines, on a computer-generated display, must be approximated by a series of short straightline segments, or a series of pixels. For example, a plotter pen is set at a location that has been computed to be on a contour line; then it is instructed to draw a straightline to another nearby location that also has been computed to be on the same contour level. This is repeated until the whole length of the contour line is approximated by straightlines.

Of course, to make the contour look smooth, these segments must be short enough so that the change in direction, between any two consecutive segments, is small. The maximum length of a segment is limited by the step size of the interpolation algorithm. Most contouring programs are designed to allow adjustment of the step size and so change the smoothness of the curved lines.

Display Types
GRIDDING

Gridding is the general term applied to the procedure in contouring programs that provides interpolation at fixed intervals across the display window. These interpolation routines calculate the height of the surface at the nodes of a regular grid, and the interval between two adjacent nodes becomes the effective step size, mentioned on the previous page.

To achieve smooth curves, interpolation needs to be performed at closely spaced intervals. This requires computing time that increases as the square of the decrease in interval length. If the interval length is halved, interpolation will require four times as much computing. For many purposes a coarser treatment is more practical.

To alter the step size, reset parameter **GR**, in line **1100**, to any integer less than or equal to 40. For example, for a step size of 1/25, reset **GR** to

1100 . . . . . :GR=25: . . . .

The higher the value of **GR**, the more closely spaced will be the line segments in an isoline. The step interval is the inverse of **GR**; larger **GR** numbers produce smoother curves. However, if a value higher than 40 is desired, storage must be extended. See Section 2.7, on p. 178, for information on changing output array sizes.

Figure 1.5.1 illustrates the way in which the interpolation mesh (**GR**=20) is arranged over the display window (for a discussion of data windows, see p. 13). The interstitial triangles, defined by triplets of grid nodes, are as close to equilateral as is possible and still fill the square.

Figure 1.5.1 Interpolation grid nodes.

Gridding

The triangular grid, with **GR**=20, shown in Figure 1.5.1, requires 537 interpolations. This grid fills the unit square, and no location within the unit square is farther than 0.0278 from a grid node. However, if the unit square were gridded with a square grid, it would require 27x27=729 interpolations before no location would be farther than 0.0278 from a grid node. The triangular grid allows a saving of some 30% in the number of interpolations relative to a square grid.

The actual size of the plot, as drawn on the plotter or screen, is set with the parameter **SZ**, in plotter units or screen pixels, in line 1100. For example,

1100 . . . . . :SZ=340: . . . .

Usually, a trial run at a coarse interpolation interval can reveal more quickly the location of anomalies or regions of the data field that require additional data. Then, with sufficient data, the interpolation interval can be reduced for a second run to provide a smoother map.

PLANAR ISOLINES

The output can be obtained in several formats. The conventional contour map is a plan view of the level curves. Figure 1.0.1, on p. 6, shows such a planar set of isolines. This plan view format is implemented when parameter **DP**=0, in line **1100**.

ISOMETRIC ISOLINES

Planar isolines are not always easy to understand because both hills and hollows appear as concentric loops of isolines. A pseudoperspective view of the isolines reveals immediately which of two adjacent isolines is higher. This is obtained by rotating the data set to give an isometric view, for example see Figure 1.2.1 on p. 11. This format may be thought of as a "wireframe" isometric display. The isometric format is used when parameter **DP**=1, in line **1100**. The azimuth and tilt angles are set by **AZ** and **TL** in line **1100**.

ISOMETRIC ORTHOGONAL PROFILES

A "fishnet" display of the surface, shown in isometric view, is a third display format. This form of isometric crosshatching, formed by two perpendicular sets of profiles across the representative surface, is most useful because the change in slope, in any subregion, is indicated. For example, flat and level portions of the surface are shown by a group of squares; however, these squares become distorted as the slope changes, and the change from square indicates the direction and magnitude of the slope change.

Display Types

Figure 1.4.1, on p. 17, shows two examples of surface tautness as displayed by orthogonal profiles. In the programs, orthogonal profile format is obtained when parameter **DP=2**, in line **1100**.

STEREOGRAMS

Pairs of isometric views of isolines, or orthogonal profiles, may be displayed as a stereogram. Such a pair of views, with slightly different view points, allows a three-dimensional visualization of the generated surface. Figure 1.5.2 is a crossed stereogram that uses orthogonal profiles.

Stereograms such as Figure 1.5.2 are easy to use; stereoglasses are not required. Simply allow the lines of sight to cross slightly, and this will cause three images to appear. The two views have been selected so that if your lines of sight cross somewhere midway between you and the screen, or page, three views will appear. The middle image gives the illusion of three dimensions because it is formed by mental combination of the two slightly different views.

To obtain a cross stereogram in the programs, set parameter **SR=1**, in line **1100**. This parameter will have no effect if **DP=0** or **CR=1**, that is, for plan views or colored output.

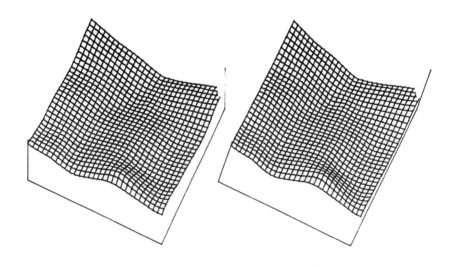

Figure 1.5.2 Stereogram

SECTION 1.6

Changing Color-fill Colors

SECTION SUMMARY

Setting alternate color combinations, for the color-filled displays, are discussed in this Section.

COLORS

Systems with color monitors can produce colored contour displays on the screen. Assigned colors are applied between the contour lines, which then are marked only by the contrast between the regions of color. Color is turned on by setting parameter **CR=1**, in line **1100**. An advantage of color is that it allows easy and effective hidden line elimination in the isometric view. However, color is not available for the isometric orthogonal profiles display or for stereograms.

Systems with only sixteen colors allow too few isoline intervals to be widely useful so the programs provide a method for blending two adjacent colors and this nearly doubles the possible number of isolines. Color blending is done by alternately using either color depending on the signs of the pixel coordinates. If both coordinates are odd, or even, the first color is used; if one coordinate is odd and the other is even, the second color is used.

A checker-board effect, which is consistent across the entire display, results from this approach. This blending effect is achieved in the programs in line **6110**, using the variables **YG** and **XG** from lines **6000** and **6020**.

25

Changing Color-fill Colors

COLOR SPECTRUM

The order of the colors is controlled by a list of integers in array **SP%(16)**, read from the data block at line **1040**. The present order is

1040 DATA 0,8,1,4,5,9,6,12,13,2,7,3,10,11,14,15

These numbers may be interchanged as desired to obtain satisfactory contrasts. The fifteen solid colors that are available plus the checker-board blending of each adjacent pair allow a maximum of twenty-nine distinct shades. This determines a limit of twenty-eight contour levels, when color is used.

Zero, the color number of black, should be kept in the first position so that it will be read into **SP%(1)**. This position in the spectrum is the default color for surface values above or below the contoured range.

The screen mode, and the text and background colors, are set in line **5080**.

   5080 SCREEN 9:COLOR 2,0: . . . . *suits the* EGA *graphics card*

The color of the border, presently yellow, may be reset at lines **5130** and **5150**. The color of the isolines, also yellow, may be reset at lines **6580** and **6600**.

```
5130 . . . . . :LINE(x,y)-(u,v),14: . . . .
5150            draws a line from screen coordinates (x,y) to
6580            (u,v) in color no. 14.
and
6600
```

The color of the plotted data points, presently red, is set in line **5580**.

```
5580 . . . . . :PSET(x,y),4: . . . .
                pixel (x,y) is given color no. 4.
```

In programs NEIGHBOR.BAS and TRIANGLE.BAS, the color of the triangles, presently blue, is set in lines **5710** and **5740**.

```
5710 . . . . . :LINE(x,y)-(u,v),9: . . . .
5740            draws lines, in color no. 9, to outline the riangles.
```

SECTION 1.7

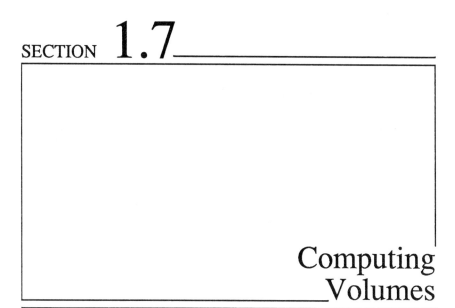

Computing Volumes

SECTION SUMMARY

This section discusses several basic geometrical concepts necessary for analysis of topographical data; area of a parallelogram, area of a triangle, volume of a prism, volume under a representative surface, circumcenter of a triangle, volume of a parallelepiped, volume of a tetrahedron, and circumcenter of a tetrahedron.

INTRODUCTION

Using a computer to analyze observations of a spatial function requires that the computer be programmed to perform geometrical operations. Computational geometry is a rapidly developing subdiscipline, and there is room to mention only a few elemental operations in this book. However, these basic concepts are useful in establishing a rigorous derivation and justification of the design and implementation of a computer program.

Of course, computational geometry is not geometry in the classical sense of straightedge and compasses. The geometrical ideas must be expressed in algebraic terms and the computer is programmed to execute these expressions. In this section, a number of geometrical constructions are described in algebraic form, and these expressions form the mathematical and logical skeletons of the algorithms discussed in other sections.

Computing Volumes

AREA OF A PARALLELOGRAM

Two pairs of parallel lines enclose a parallelogram and computing the area of such a shape is an important operation. The problems of computing the area can be appreciated by working through a calculation. The parallelogram in Figure 1.7.1, below, is a general example, and the area can be determined with two tools; (i) the area of a rectangle is the product of the lengths of two perpendicular sides; (ii) the area of a right angle triangle is one half the product of the lengths of the two perpendicular sides.

Figure 1.7.1 shows the area of the parallelogram is just the area of the large rectangle, $(a+c)*(b+d)$, less the area of the two small rectangles, $2(b*c)$, less four triangles, $2(a*b/2)$ and $2(c*d/2)$. This is

$$a*b + a*d + b*c + c*d - a*b - b*c - b*c - c*d = a*d - b*c$$

which is the area of a parallelogram.

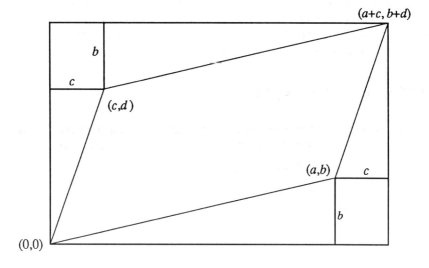

Figure 1.7.1 Area of a parallelogram

The four sides of a parallelogram are formed by two pairs of identical line segments. Notice, however, that the complete shape and position of the parallelogram can be determined by symmetry if only one side from each identical pair is known.

Suppose the Cartesian coordinates, (x,y), of the four vertices are known. Then knowledge of one vertex can be considered redundant because it can be recovered whenever it is required. This may seem obvious, but it is useful to show it to be true by calculating the area of the parallelogram while using only three vertices.

Area of a Parallelogram

Suppose the three vertices have coordinates (x_1,y_1), (x_2,y_2), and (x_3,y_3), then one side is the vector

$$(a,b) = (x_2,y_2) - (x_1,y_1) = (x_2-x_1, y_2-y_1)$$

and the other side is the vector

$$(c,d) = (x_3,y_3) - (x_1,y_1) = (x_3-x_1, y_3-y_1)$$

Then, although the fourth vertex is not represented here, the area is given by

$$\text{Area}_p = \text{ABS}((x_2-x_1)*(y_3-y_1) - (y_2-y_1)*(x_3-x_1)) = a*d - b*c$$

If the two edges defined by these three vertices are considered to be row vectors in a 2 by 2 matrix

$$\begin{matrix} x_2-x_1 & y_2-y_1 \\ x_3-x_1 & y_3-y_1 \end{matrix}$$

then the formula says that the area of the parallelogram is the absolute value of the determinant of this matrix. The determinant will be a negative value, whenever the numbered order of the three vertices is clockwise, but the area must be non-negative, so the ABSolute value function is required to make this expression for area correct for all situations.

AREA OF A TRIANGLE

The formula for the area of a parallelogram gives an immediate area for the triangle formed by the three vertices because it is one-half of the parallelogram.

$$\text{Area}_t = \text{ABS}((x_2-x_1)*(y_3-y_1) - (x_3-x_1)*(y_2-y_1))/2$$

is the area of a triangle with vertices (x_1,y_1), (x_2,y_2) and (x_3,y_3), in clockwise, or counterclockwise, order.

VOLUME OF A TRIANGULAR PRISM

Just as any polygon may be subdivided into triangles, so any polyhedral shape may be subdivided into triangular prisms. For this reason triangular prisms are often used for analysing solid bodies. Then the volume of any polyhedron may be determined by summing the volumes of its triangular prisms.

The volume of a triangular prism, with three parallel edges, is the area of its perpendicular triangular crossection times the height of the prism. However, if the triangular top, or base, is tilted, so that three vertical edges of the prism have different heights, then the height of the prism is the average height of the vertical edges. One of the parallel edges may have a zero length.

Computing Volumes

Suppose that the three top corners of the triangular prism have three-dimensional Cartesian coordinates (x_1,y_1,z_1), (x_2,y_2,z_2), and (x_3,y_3,z_3), and the lower three have coordinates (x_1,y_1,z_4), (x_2,y_2,z_5), and (x_3,y_3,z_6), as shown in Figure 1.7.2. Then the volume of the triangular prism is

$$\text{Vol}_{tp} = \text{ABS}((x_2-x_1)*(y_3-y_1) - (x_3-x_1)*(y_2-y_1))$$
$$*(z_1+z_2+z_3-z_4-z_5-z_6)/6$$

Notice that this formula depends upon the three vertical edges being parallel, and the crossection area is perpendicular to those three parallel edges.

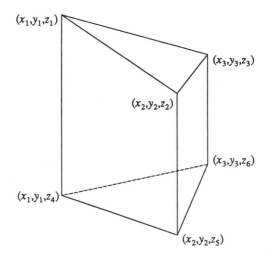

Figure 1.7.2 Triangular prism volume

VOLUME UNDER A SURFACE

There are several reasons to compute the volume under a representative surface; the most obvious is that one wishes to know the amount of material indicated by the topographical data set. Resource assessment, and excavation cost estimates, are examples, and such volume estimates ususally are the ultimate reason for gathering topographical data.

Aside from such practical and operational motivations, the volume under a computer-generated surface is a precise method of evaluating and comparing interpolation and gradient estimation methods. The volume under specific subregions in the data can differentiate readily between surfaces that look the same when they are displayed but are subtly different in the generosity or conservativeness of curvature. Badly behaved interpolation methods may become dramatically obvious when volume estimates are considered.

CONTOURING: A Guide to the Analysis and Display of Spatial Data
by David F Watson ISBN 0 08 0402860

Disclaimer

Neither Pergamon Press nor the author nor any employer of the author shall be liable for any special, indirect, consequential, incidental or other similar damages suffered by the user or any third party, including, without limitation, damages for loss of profits or business or damages resulting from use or performance of the software, the documentation, or any information supplied by the software or documentation, whether in contract or in tort, even if Pergamon Press or its authorized representative has been advised of the possibility of such damages; and Pergamon Press shall not be liable for any expenses, claims or suits arising out of or relating to any of the foregoing.

User Assistance and Information

The programs that accompany my book are intended to be workshop programs in the sense that they were prepared for other programmers to alter and use for experimentation. Although they can be used for direct data processing, the conventional safeguards and error trapping that one would expect to find in user-friendly commercial software could not be included because this would inhibit experimentation. You will find a reference to this on page 3.

I would like to point out that my book is intended to provide an understanding of the reasoning and logic underlying contouring techniques. Therefore, it is distinctly different from a commercial contouring package which is intended to deliver a contour map with minimal intervention from the user.

I will be pleased to help any readers in any way that I am able to, and I look forward to the suggestions and criticism that are sure to arrive.

Please contact me at P.O. Box 58, Mosman Park, WA 6012, Australia.

NUMERICAL INTEGRATION

The general approach to calculating volumes under the surface representing a topographical data set is simply to sum a large number of piecewise volume estimates. The generated surface, within the display window, is subdivided into triangles. Then the volume of each triangular prism is calculated by using a horizontal or tilted plane for its base, and a tilted plane, that approximates the generated surface, as a top to the prism. The estimated volume under a surface is given by the total of the volumes of these triangular prisms. The average height of the surface is this volume divided by the area covered by the prisms.

The actual volume computation is a trivial summation of prism volumes. When interpolation has been performed at the nodes of a triangular mesh (see Figure 1.5.1, on p. 22), the volume of a given triangular prism is the area of the triangle by the average of the three interpolated values at the triangle vertices, as Figure 1.7.2 shows. Of course, the triangular area is the same for all the prisms from a regular mesh, and is determined by the mesh size. The precision of the volume estimate is governed by the mesh size parameter, **GR**, in line **1100**.

If the triangular mesh is sufficiently closely spaced, the sum of all the triangular prisms is a reasonable estimate of the volume under the generated surface. Obviously, the smaller the triangular patches are, the more accurate the estimate, up to the point where accumulated roundoff errors become significant. This limit is unlikely to be reached if the calculations are done on 16 bit machines.

The estimated volume is output in line **5790**, and can be turned off by inserting a REMar k before the PRINT instruction. For example

5790 . . . . . :REM PRINT "AVERAGE etc . . .

The results of numerical integration can be summarized by a listing of subtotals; simply the volumes under a set of systematic partitions of the data region. So numerical integration provides a basis for rigorous comparison between surfaces generated by various schemes. For related discussions on numerical output, see p. 42 and p. 185.

CIRCUMCENTER OF A TRIANGLE

This is the center of the circle that goes through each of the vertices of a triangle, and computing the coordinates of this center also is an important operation.

Notice that all locations on each of the three lines that bisect perpendicularly the triangle edges are equal distances from the pair of vertices that they bisect. Because the circumcircle goes through each vertex, and its center is an equal distance from all three vertices, this center must lie on the intersection of all three perpendicular bisectors. However, any two of these bisectors will establish the location of the circumcenter.

Computing Volumes

The equation for a line on the plane is $Ax + By = C$, so the perpendicular bisectors of two edges of a triangle with vertices (x_1, y_1), (x_2, y_2) and (x_3, y_3) form a 2-row by 3-column matrix.

$$(x_2 - x_1)x + (y_2 - y_1)y = (x_2 - x_1) \cdot (x_2 + x_1)/2 + (y_2 - y_1) \cdot (y_2 + y_1)/2$$
$$(x_3 - x_1)x + (y_3 - y_1)y = (x_3 - x_1) \cdot (x_3 + x_1)/2 + (y_3 - y_1) \cdot (y_3 + y_1)/2$$

Then the intersection, (x, y), is established by solving this system of simultaneous linear equations. For this two-dimensional example, using Cramer's rule (page 88), the values of x and y are

$$x = (((x_2 - x_1) \cdot (x_2 + x_1)/2 + (y_2 - y_1) \cdot (y_2 + y_1)/2) \cdot (y_3 - y_1)$$
$$-(((x_3 - x_1) \cdot (x_3 + x_1)/2 + (y_3 - y_1) \cdot (y_3 + y_1)/2) \cdot (y_2 - y_1)))$$
$$/((x_2 - x_1) \cdot (y_3 - y_1) - (x_3 - x_1) \cdot (y_2 - y_1))$$

$$y = ((x_2 - x_1) \cdot ((x_3 - x_1) \cdot (x_3 + x_1)/2 + (y_3 - y_1) \cdot (y_3 + y_1)/2)$$
$$-((x_3 - x_1) \cdot ((x_2 - x_1) \cdot (x_2 + x_1)/2 + (y_2 - y_1) \cdot (y_2 + y_1)/2)))$$
$$/((x_2 - x_1) \cdot (y_3 - y_1) - (x_3 - x_1) \cdot (y_2 - y_1))$$

This is the approach used in the programs; at line **3310**, new circumcircles are determined for a datum and each contiguous pair of its natural neighbors.

The incenter of a triangle is the center of the largest circle that will fit into the triangle. Three lines, that each bisect one of the interior angles of the triangle, meet at the incenter. If (x_I, y_I) is the incenter of a triangle with vertices (x_1, y_1), (x_2, y_2), and (x_3, y_3), then

$$x_I = (d_{2,3} x_1 + d_{3,1} x_2 + d_{1,2} x_3)/p$$
$$y_I = (d_{2,3} y_1 + d_{3,1} y_2 + d_{1,2} y_3)/p$$

where $p = d_{1,2} + d_{2,3} + d_{3,1}$ and $d_{i,j}$ is the length of the edge from (x_i, y_i) to (x_j, y_j).

AREA OF A TRIANGLE IN 3D

Similar to triangles in the (x, y) plane, the area of a triangle in 3-dimensional space is one-half the area of the parallelogram defined by three 3-dimensional vertices. However, the calculation of such a parallelogram is more involved and is achieved by taking the cross product of two edges. The components of a cross product are determinants but now there are three instead of one, and the cross product is the root of the sum of these squared determinants. Because of the square, there is no problem of negative results. The area of a triangle, having Cartesian coordinates (x_1, y_1, z_1), (x_2, y_2, z_2) and (x_3, y_3, z_3), is given by

$$\text{Area}_t = ((((y_2 - y_1) \cdot (z_3 - z_1) - (y_3 - y_1) \cdot (z_2 - z_1))^2$$
$$+ ((z_2 - z_1) \cdot (x_3 - x_1) - (z_3 - z_1) \cdot (x_2 - x_1))^2$$
$$+ ((x_2 - x_1) \cdot (y_3 - y_1) - (x_3 - x_1) \cdot (y_2 - y_1))^2)^{1/2})/2$$

This expression gives the area as one-half the length of the vector that is the cross product of two vectors which are the sides of the triangle. The cross product vector, (x_n, y_n, z_n), is perpendicular to the triangle, and the components of this normal vector are

$$x_n = ((y_2-y_1)*(z_3-z_1) - (y_3-y_1)*(z_2-z_1))$$
$$y_n = ((z_2-z_1)*(x_3-x_1) - (z_3-z_1)*(x_2-x_1))$$
$$z_n = ((x_2-x_1)*(y_3-y_1) - (x_3-x_1)*(y_2-y_1))$$

Considered in terms of the number of arithmetic operations that the computer has to perform, it may be thought that the above expression for the area of a triangle is a less efficient computation than Heron's formula, which is

$$\text{Area}_t = (s*(s-a)*(s-b)*(s-c))^{1/2}$$

where

$$a = ((x_2-x_1)^2 + (y_2-y_1)^2 + (z_2-z_1)^2)^{1/2}$$
$$b = ((x_3-x_1)^2 + (y_3-y_1)^2 + (z_3-z_1)^2)^{1/2}$$
$$c = ((x_3-x_2)^2 + (y_3-y_2)^2 + (z_3-z_2)^2)^{1/2}$$
$$s = (a+b+c)/2$$

However, when the orientation of the triangle in 3-dimensional space also is required, as it is in the programs, the former expression is more suitable because it provides the normal vector.

CIRCUMCENTER OF A TETRAHEDRON

The circumcenter of a tetrahedron is just the center of the sphere whose surface coincides with all four vertices of that tetrahedron. To determine this center, consider first, a plane in 3-dimensional space that bisects perpendicularly the edge between two vertices. A tetrahedron has six edges, so six such planes are defined by any tetrahedron. Because every location on any of these planes is an equal distance from each of the pair of associated vertices, two planes that share a vertex must intersect at a point that is an equal distance from all three vertices.

By continuing with this approach, it can be seen that some point that lies on all six planes must be equidistant to all four vertices. This is the center of the circumsphere, whose surface coincides with all four vertices, and all six perpendicularly bisecting planes coincide at this circumcenter.

However, any three of the bisecting planes are sufficient to establish the location of the circumcenter by computation. Similar to the circumcenter of a triangle, on p. 31, the planes, expressed in the form $Ax + By + Cz = D$, are set up as a 3-row by 4-column matrix. The coordinates of the center are determined by solving this system of simultaneous linear equations; the solution is the intersection of the perpendicularly bisecting planes. Cramer's Rule (p. 88) is somewhat cumbersome to apply to a 3 by 4 matrix, and then such a system is solved by Gaussian elimination, beginning at line **6690**, in the programs.

CIRCUMCENTER OF A TRIANGLE IN 3D

Calculating this circumcenter is similar to the example of a tetrahedron, discussed on the previous page; the intersection of three planes must be established. Two of these planes are perpendicularly bisecting planes for pairs of data, just as for the center of a planar triangle, see p. 31. However, the third plane is defined by all three data, and it is the plane containing the triangle because the circumcenter must lie on this plane. The solution again is given by a system of three simultaneous linear equations representing the intersection of these three planes.

VOLUME OF A PARALLELEPIPED

Three sets of parallel lines define a parallelepiped. But just as the two pairs of lines for the parallelogram can be determined from three points (p. 28), so these three quadruplets of parallel lines are determined by four given points.

Similarly to the area of a parallelogram, the volume of a parallelepiped can be calculated as the absolute value of the determinant of any set of three row vectors formed by the four vertices. For example the three vectors determined by subtracting the first vertex from each of the other three vertices

$$(x_2-x_1, y_2-y_1, z_2-z_1)$$
$$(x_3-x_1, y_3-y_1, z_3-z_1)$$
$$(x_4-x_1, y_4-y_1, z_4-z_1)$$

allow the volume of the parallelepiped to be calculated as

$$\text{Vol}_p = \textbf{ABS}\ ((x_2-x_1) * ((y_3-y_1)*(z_4-z_1) - (y_4-y_1)*(z_3-z_1))$$
$$-(y_2-y_1) * ((x_3-x_1)*(z_4-z_1) - (x_4-x_1)*(z_3-z_1))$$
$$+(z_2-z_1) * ((x_3-x_1)*(y_4-y_1) - (x_4-x_1)*(y_3-y_1)))$$

The **AB**Solute value function forces the sum of determinants to be positive.

VOLUME OF A TETRAHEDRON

The formula given above for the volume of a parallelepiped also provides an immediate volume for the tetrahedron formed by the four vertices because it is one-sixth of the associated parallelepiped volume.

$$\text{Vol}_t = \text{Vol}_p/6$$

This may be compared to the area of a triangle which is one-half of the associated parallelogram area. The factor in each dimension is just the inverse of the dimensional factorial, that is, $1/n!$, where n is the spatial dimension. This factor also was seen in the volume of a triangular prism on p. 30.

GEOMETRICAL STATISTICS

The geometrical concepts mentioned in preceding paragraphs are some of the tools that are useful in the analysis of topographical data, and in the construction of a geometrical interpretation of a data set. Although some apparently statistical expressions are used in the analysis of topographical data, it is important to notice their geometrical interpretations because these provide some insight into spatial relationships. For example, the simple arithmetic average of three values can be used to give the volume of a triangular prism, when the area of the triangular cross section is known (see p. 29).

$$\text{Average} = \frac{1}{3} \left(\sum_{i=1}^{3} z_i \right) = \sum_{i=1}^{3} z_i / 3$$

This average is the height, at the cross-sectional centroid, of a tilted plane that forms the top of the triangular prism. The barycentric coordinates (see p. 76) of the centroid are ($1/3$, $1/3$, $1/3$), and these also are the proportions of the area of subregions of the cross section in a volume calculation.

Isted's Formula for average grade, G_A, given grade and thickness data, G_i and T_i, is an apparent statistical formula that has a geometrical derivation (see p. 186), and expresses a geometrical grade-volume product when both T_i and G_i vary linearly across the triangular prism. Isted's Formula for average grade is

$$G_A = \frac{1}{4} \left(\frac{\sum_{i=1}^{3} T_i G_i}{\sum_{i=1}^{3} T_i} + \sum_{i=1}^{3} G_i \right)$$

SECTION 1.8

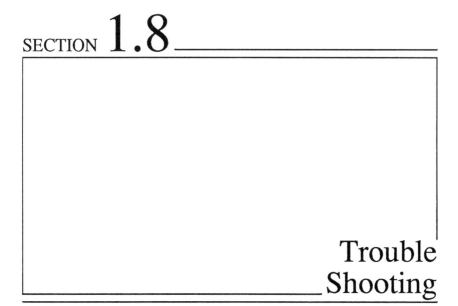

Trouble Shooting

SECTION SUMMARY

This section is a list of possible problems that may occur with the programs, and some suggested remedies.

DATA SET PROBLEMS

These probably are the most frequent cause of program failures. Insufficient data, meaning that there is less data in the input file than indicated by the first number (see p. 8), usually is made obvious by an appropriate message from the operating system. However, a miscount in the number of contour levels declared may not be so apparent. Echo printing of the contour levels, and the input data, will make these problems transparent.

WORKING STORAGE EXCEEDED

The programs have been set up with storage for a maximum of 60 data, solely to restrain the overall program size. When more than this declared amount of storage is used, it may not always be obvious that array sizes have been exceeded because other data may become overwritten; this can cause a range of error messages.

Trouble Shooting

If larger data sets are to be treated, then several arrays must be extended, and they are listed here. Array **P1#** contains the topographical data and estimated gradients (for gradient estimation, see p. 85).

1010 DIM P1#(63,6), . . . ,RO(60), ...

The first dimension of **P1#** must be 3 more than the number of data; the width is fixed at 6 and that will not need to change. Array **RO** contains error bounds or residuals for each datum, and must be extended for the desired amount of data. Working storage is in arrays **P2#**, **P3%**, and **P4%**, and the subscripts of the initialization loops also must be extended.

1010 DIM . . . . . :P2#(237,3), ...
1030 DIM P3%(237,3),P4%(237), ...

1600 FOR I1%=1 TO 237: . . . .
3120 FOR I0%=1 TO 237: . . . .

Array **P2#** contains the natural neighbor circumcenters and squared radii (for a discussion of natural neighbors, see p. 57), array **P3%** contains the triplets of data indices that define each natural neighbor circumcircle, and **P4%** holds a list of available rows of **P2#** and **P3%**; this is a form of garbage collection. The loops at lines **1600** and **3120** initialize array **P4%**. The length of these arrays and the range of the loops cannot always be predicted precisely because they vary with the configuration of the data. If the data are scattered uniformly, this length may be as low as twice the number of data plus one. When natural neighbor gradients are required for uneven arrangements of data, these arrays and loops may need one-half as much again.

In addition, arrays **P5#**, **S4**, and **P6%**, also require extension to suit the number of data.

1010 DIM . . . ,P5#(60), . . . ,S4(60), . . .
1030 DIM . . . ,P6%(60), . . .

These last mentioned arrays are working storage for local coordinate calculations. The number of such coordinates never exceeds the number of data, and for data which is scattered uniformly, it is usually much less. Other such arrays are **R3#**, **R9#**, and **B#**.

1020 DIM . . . . ,R3#(60,60), . . . ,R9#(60),B#(60)

These arrays are used to solve systems of simultaneous equations. The length is required to be equal to the amount of data for the FITFUNCT.BAS programs. In the other programs, the length need only equal the maximum number of natural neighbors about each datum, which is usually less than thirteen.

However, for anisotropic data, such as traverse data, the maximum number of natural neighbors can be high. Exceeding storage on these arrays usually causes a "division by zero", or "overflow", error message.

The contour levels are stored in array **P8**, before and after they are normalized, and **P8** also is used to hold the orthogonal profiles, if required. The number of profiles, **NC1%**, is determined in line **1430** and is relative to the display window output size **SZ**.

1010 DIM . . . ,P8(80),S1(2033), . . .

The interpolated grid values are stored in array **S1**. The size of this output database must be increased if **GR**, in line **1100**, is set to a value greater than 40. **GR** is the number of triangles along the horizontal edge of the data display window. If **GR** is required to be greater than 40, then the size of the output grid storage, **S1**, must be changed to be no less than

$$INT(GR/3^{\wedge}0.5+1)*(2*GR+3)+GR+1.$$

INPUT DATA WINDOW

If the data input window (see p. 13) is too small, it may have less than four data, and the programs will terminate due to the statement in line **1360**. The only surface attained by these programs, with less than four data, is a plane.

The radius of the data input window is specified by **RD**, on line **1100**, and set at 1.4 times the display window size (the set of three numbers that immediately precede the data, see p. 8).

SECTION 1.9

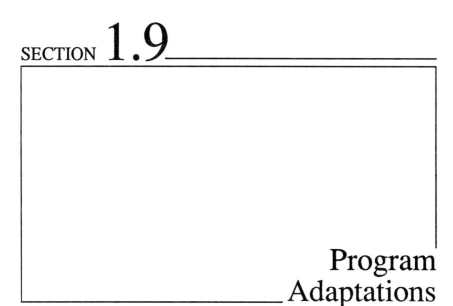

Program Adaptations

SECTION SUMMARY

The programs are intended to allow experimentation and comparison of computer contouring procedures. This section discusses a few of the ways in which the programs may be adapted for additional experimentation.

COMBINING PROGRAMS

The functions provided by these programs may be combined and interchanged to provide a system most suited to a particular source and type of data. Line numbers and variables have been assigned to avoid conflicts for any combination.

THREADING ISOLINES

When isolines are drawn by a plotter, it often is thought desirable to sort the line segments into continuous curves so that the pen can draw a complete contour with one pen-down and one pen-up instruction. Obviously this consideration is not relevant for screen displays, but for plotter output, contour segments can be stored in output order as they are generated, although considerable storage is required. This improves plotted output appearance for many plotters. Threading is discussed more fully on p. 180.

Program Adaptations

HISTOGRAMS

Histograms are convenient diagrams for summarizing the representative surface for a data set. They show the way in which the surface elevation is distributed over a range of height intervals. When histograms are used to describe a surface, the increments of surface are sorted into height classes. The total surface area, in each class, is expressed as a proportion of the display area, within that interval of height. For a contrasting approach, see "Block Averages" on the next page.

Obviously, the precision will depend upon the increment size, the inverse of the number of classes. If only a few classes are used, the histogram will be "box-like". The more classes are used, the smoother will be the description of height distribution.

Histograms may be made easily, in the programs, by forming subtotals of the average height and the area of triangles from the output grid, and then converting to proportions or percentages. The variables **SM** and **AR** are used to obtain average height over the display window, and so subtotals of the values **SM** and SA, in line **5860**, may be accumulated for each class interval.

OGIVES

Computing a cumulative distribution is done by taking sums of the percentages in the histogram. The ogive is given the same set of intervals, and the percentage for each ogive class is the sum of all the classes in the histogram to the left of, and including, the given interval. Figure 1.9.1 shows a histogram and the corresponding ogive. They summarize a distribution of surface area over seven intervals of height.

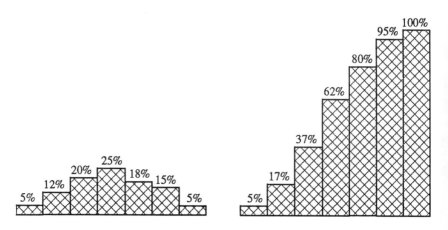

Figure 1.9.1 Histogram and cumulative distribution

BLOCK AVERAGES

Another approach for summarizing the behavior of a surface is computing average heights for subregions. Just as each interval of the histogram shows the proportions of area within an interval of height, so each block in a set of block averages shows the average height within that block.

The entire display window is divided into blocks. The sum of triangular prism heights in each block then may be expressed as the average height within each interval of area; for comparison with histograms, see p. 42. If only a few blocks are used, again the results will be "box-like", but now as a surface. The more blocks that are used, the smoother will be the description of the spatial distribution. Figure 1.9.2 shows four-by-four block averages of **A:\HILL.DAT**

8.55	11.85	10.09	5.10
11.42	19.14	21.74	13.27
10.89	21.27	24.08	21.32
6.31	11.70	12.32	9.32

Figure 1.9.2 Array of block averages

COMBINING SURFACES

The set of interpolated grid nodes, being the output database, in array S1, is a convenient method of storing a computed surface. The regularity of the output grid allows that the x and y values need not be stored and, being a triangular grid, it is 30% more efficient than a square grid (see p. 22 and p. 179). Several such surfaces with the same grid size may be added, subtracted, multiplied and divided by applying the operation to corresponding grid nodes, and contouring the result.

PART 2
Principles
Of Contouring

OVERVIEW OF PART 2

This part presents a description and derivation of the rationale and concepts used in computer generation of contour displays for topographical data sets. The most fundamental of these concepts is the assumption of continuity because it allows us to theorize the construction of a surface that conforms to discrete data. On this assumption rests the expectation that the data will be more comprehensible if, firstly, the generated surface represents the data well and, secondly, the surface can be seen and measured well. In this part, assuming continuity in the data source, various techniques of surface construction and display are considered in terms of their plausibility, efficacy, algorithmic genealogy, and characteristic peculiarities.

SECTION 2.0

General Concepts Of Contouring

SECTION SUMMARY

This section discusses the nature of contour displays. First, the terms "contour" and "topographical data" are specified, and the history and genealogy of contour displays are mentioned. Then the surface forms underlying isoline and isochor maps, and the development of isolines from isochors, are discussed. The general approach to developing contour lines is described, and some major difficulties of contouring are mentioned. Data patterns considered are regular gridded, scattered, and profile or traverse, data. Interpolation procedures are based on fitted functions or weighted averages, and depend upon the assumption of continuity in the data source. A distinction is made between the precision of the data set, and the precision of interpolation.

INTRODUCTION

The word "contour" is both a noun, in the general sense of surface shape or particular sense of local shape descriptor, and a verb, in the general sense of forming a surface or in the particular sense of generating a local shape descriptor. In this last sense, it also is a generic term that refers to a wide group of manual and automatic procedures for developing, measuring, and displaying, surface shape inferred from the variability of topographical data. By general usage, contouring is the inferrence of surface shape that is assumed to be representative of a data set.

Gradual variation in a bivariate function has a natural expression as the shape of a surface in 3-dimensional space. When a spatial function is known only at a few places, its estimated behavior within a region usually is portrayed by a contour display of a representative surface. Graphical depictions of such surfaces, expressing the topographical variation, are contour displays. A well known example is the isoline map in which smoothly curving lines trace level paths over the topography. Although contouring is a small, and possibly overlooked, aspect of cartography and geographical information systems, it is important because it interprets the spatial variation, and so has the greatest single influence on the accuracy of the resulting map or database.

TOPOGRAPHICAL DATA

Of the many forms of multivariate data that occur, one type is made distinct by being a compound of two independent locational variables, and one dependent or functional variable. These data are descriptions, at particular locations, of a spatial function that exhibits gradual variation, and has a natural representation as the shape of a surface. For example, the point where a drill hole intersects a particular stratum has both geographic coordinates of location, and a depth below the surface. The depth value is understood to be dependent locationally on the geographical coordinates, in the sense that depth to the stratum is variable.

Topographical data, as far as algorithms are concerned, are simply sets of ordered triples of numbers in which one number is dependent functionally on the other two numbers. Although some authors restrict the term "topographical data" to land surface elevation data, in this book its use simply implies that one component of each datum has a functional dependence on the others.

Analysis of topographical data involves detailed, local, examination and portrayal of the characteristics of spatial variability implied by the data. The general compatibility of topographical data sets with particular probabilistic models depends upon subjective interpretation of several nonlocal aspects, such as a regional mean value, and so the application of parametric statistical models to topographical data (see the survey by Rock, 1988) is not pursued in this book.

HISTORY OF CONTOURING

For hundreds of years before computers became available, the manual preparation of maps that expressed the shape and content of the landscape, and the distribution of its resources, was a highly specialized form of draftsmanship. Such maps had many uses, from archival, through economic, military, navigational, political, to zoological, and others. The message that these maps carried was an expression of "how much of what is where", and this point distinguishes contour maps from maps that tell merely "how to get from where to where".

History of Contouring

The origin of the art and practice of making and reading manual contour maps lies in nautical history: two fundamental problems of early seafarers led to the development of modern displays of spatial variation. The vital questions "how deep is the water?" and "which way is north?" required a more dependable medium than human memory, so a pictorial record was developed.

Robinson (1971) credits the first contour map to Bruinss in 1584; it charted the depth of water, with isolines termed isobaths, and so portrayed the variation of the sea bottom. The variation in compass north relative to true north was a directional and so, apparently, a distinct problem. Robinson (1971) points to Halley's magnetic chart in 1701, using isolines termed isogones, as an early example. Sager (1983) gives several historical references to early maps, and attributes the first published contour map of landscape topography to P. Buache in 1752. Since then many people, from generals to geologists, have had need for displays of representative surface contours. With the arrival of computers in the 1940's, John von Neumann (see Panofsky, 1949) suggested automatic interpolation of meteorological data. Computer contouring had begun.

Reading contour maps required some skill, but it is considerably easier to learn to interpret the shape of a surface from a map of the contours than to learn the manual art of drawing a representative map from a limited amount of data. The actual thought processes involved in manual contouring were not well-defined. Contour generation required the development of an ability to visualize a surface that curved smoothly through nearby data, but how this visualization was obtained was not specified as an algorithm.

Writing before computers were used for contouring, Wright (1942) considered, correctly, that no map could be wholly objective, and uninfluenced by human shortcomings. Many other authors have pondered about making maps objectively. For more recent examples, Wren (1975) recognized that the subjective nature of manual contouring was inimical to precision in maps, and Morrison (1974) noted difficulties encountered in attempts to evaluate algorithms for automatic production of contour maps.

ISOLINE MAPS

Topographical data, obtained by measuring values of a spatial function at particular locations, are punctulate data. The information in each datum concerns only the point in space addressed by the locational coordinates. The variability of a set of measurements may be represented and displayed by constructing a representative surface over the region.

On that hypsometric surface, contour lines, or isolines (literally lines of equalness, meaning that all points on a given isoline have equal height), may be traced to indicate the relief in the surface and how it varies. Isolines, at regular height intervals, illustrate the surface shape, to the practiced eye, by their more closely spaced appearance wherever the surface slopes more steeply.

A set of isolines presents the illusion of a smoothly varying surface that is continuous, and the density of isolines shades the map proportionally to steepness of the surface. The map expresses the shape and contours of a surface which, in turn, describes the spatial variability in the topographical data.

A particular property of isolines is that, similar to a shoreline, they always are perpendicular to the direction of steepest slope of the representative surface. This implies that all isolines are closed loops around hill or basin forms, although such closure is not always complete on a given map. It also implies that isolines never bifurcate or form a three-way junction; however, four-way junctions do occur at cols and passes. Whenever the spatial function is single-valued, that is, has no overhanging cliffs or caves, isolines never cross one another.

DOT DENSITY MAPS

Another type of display that gives the illusion of a smoothly varying surface may be produced by placing myriad dots on the map. The local density of dots is made proportional to height so that the density of dots may be used to infer the relative local height of the surface. This gives a qualitative impression of the surface that may be appreciated quickly. However, such a map is difficult to interpret precisely.

Because of the many density calculations required, computers were highly appropriate and, indeed, necessary, for the production of dot density maps. An early description of computer construction of dot density maps was given by Simpson (1954).

ISOCHOR MAPS

The location of a datum provides it with a punctulate property, and this book is concerned predominantly with maps produced from punctulate topographical data. However, another form of topographical data is obtained by summation or aggregation of punctulate, or event, data. Subtotals are accumulated for particular regions, and are expressed as regional mean values or densities relative to specific areas.

Such data are termed choropleth data (literally place-quantity data, indicating that both the area and the sum are involved in each datum). Robinson (1971) emphasized this distinction between choropleth and punctulate data in his historical review of contour maps.

Each choropleth datum may be visualized as a polygonal prism whose base is a polygon that represents the size and shape of the region. The upper surface is a flat and level plane at a height that represents a volume proportional to the summation over the region, or equivalently, the average quantity in the region. Hilbert (1981) refers to this concept as a polygonal-histogram or prism map.

Isochor Maps

On a map, such regions are isochors (literally places of equalness, meaning that all locations within an isochor have equal value). It should be noticed that this equalness over the region is an artifact of the information gathering process, and not a property of the original spatial function.

The spatial behavior of choropleth data may be represented and displayed by a map that is subdivided into regions representing particular height intervals, and characteristically color-coded or hatched to mark the boundaries by contrast. In comparison with an isoline map, the surface expressed by an isochor map is piece-wise level and the boundaries formed by choropleth data are not isolines because the map surface is multivalued along these lines. Therefore, isochor maps show only a broad approximation of the surface shape appropriate to the spatial function because the surface has, effectively, been terraced in the process of illustrating the chorography of the selected intervals.

An isochor map displays density in the space of the independent variables, whereas an isoline map displays variability in the space of the dependent variable. To emphasize this distinction, it may be put another way. On the one hand the surface represented by an isoline map portrays the spatial variation between adjacent punctulate data with no information about the boundaries of regional classes, while on the other hand the surface represented by an isochor map portrays the boundaries between adjacent regional classes with no information about spatial variation within an isochor.

Additionally, MacKay (1951) has pointed out that isochor maps, because choropleth data are aggregated by region, are inherently less reliable than isoline maps produced from absolute quantities because the error of the ratio (forming the choropleth datum) is a product of both the errors of the quotient and divisor.

The distinction between punctulate and choropleth data is important to computer contouring because it expresses itself in the contouring algorithms required. Producing isochor maps from choropleth data requires a classifying and categorizing approach that is distinct procedurally to the synthesizing and melding required for isoline maps of punctulate data.

Of course, isochor maps can be produced readily from punctulate data by all the methods that produce isoline maps. The isochor map is abstracted from the interpolating surface, as generated for the punctulate data, by summing surface values over the area of each region, or alternatively, determining regional boundaries by categorizing surface elements for given intervals (see "Block Averages", p. 43). However, when information about some spatial function has been aggregated to form choropleth data, reconstructing the surface that would have represented the information before it was aggregated is more problematical. As well as the increased unreliability, local trends in the spatial function have been obscured by the collection process.

The general approach is to assign a center to each isochor, and then an interpolating surface is generated (Section 2.5, p. 101) for the centered choropleths, treated as punctulate data. The results, however, depend upon the appropriateness of the assigned centers, and the choice may be highly subjective.

A more objective approach is possible. Starting from an isochor map, which may be thought of as a set of polygonal prisms with specified volumes, a smoothly changing surface is sought which bounds or covers an equivalent volume for each polygon, and agrees, in height and slope along the boundaries, with its neighbors.

Such volume preserving algorithms, termed "pycnophylactic" (Tobler, 1979b), are considered necessary because they maintain the broad contours implied by the choropleth information (Lam, 1983). Recovering a more detailed representative surface for the punctulate data used to make an isochor map, is considerably more difficult.

Wahba (1981), and Dyn and Wahba (1982), require the representative surface to satisfy certain minimum curvature criteria, while matching, or approximately matching, the volumes of the isochors. They term their surface a LaPlacian histospline. Unfortunately, this surface can be negative, for a part of an isochor, even though the original spatial function is strictly nonnegative.

Fisher (1982, p.42) discusses the development of a surface that equals the value of the choropleth at a "base center" of the isochor, presumably assigned. However, this surface would not go negative because its lowest point would be equal to the least choropleth, which is nonnegative. However, a computer algorithm for this approach is not obvious.

Goodchild and Lam (1980) extend the problem to calculating the choropleth of a region that overlaps some, or all, of a set of smaller regions with established choropleths. In effect, they seek the volume under a certain region of an interpolating surface reconstructed from several isochors. This might be done by distance weighting of assigned centers, but these authors use an area-based weighted average (see Section 2.3, p. 75) and so do not need to assign centers to the isochors. For this approach to be satisfactory, the choropleths should be formed from a spatial function that is constant across each subregion and multivalued at the boundaries. In a similar approach, Kennedy and Tobler (1983) use a weighting based on the length of adjacent edges.

OTHER MAPS

The variability of a spatial function, expressed by topographical data, is most conveniently visualized as a surface of some, usually restricted, form. The shape and contours of that surface can be displayed in other ways besides the planar views given by isoline, dot density and isochor maps. By changing the perspective, the 3-dimensional nature of the surface is made more obvious.

For example, isometric views of isolines or isochors, hill shading methods that give an illusion of hollows and hills in the surface, stereograms that allow a 3-dimensional image to be synthesized by the observer, closely spaced orthogonal profiles that appear as a net lying on the surface, as well as crossections or profiles, may be constructed once a representative surface has been generated.

COMPUTING CONTOURS

Starting from the understanding that isolines, because they are level lines, intersect the lines of steepest assent at right angles, a contour segment is determined by estimating first the directions of steepest slope. For a small neighborhood of data points, a draftsman may do this "by eye" and so estimate the direction of level curves that intersect the slopelines perpendicularly.

With more objectivity, the directions of maximum slope may be estimated by comparing a series of 3-dimensional slope calculations between pairs of adjacent data. The slope is the height difference multiplied by the inverse of the distance between the data. The contour line must intersect this maximum slope line at right angles.

Although, generally, more than three data are immediate neighbors of any map location, the draftsman's approach to isoline construction may be demonstrated within a triangle formed by any three adjacent data. Because a direction on the plane may be designated by its 2-dimensional slope, which is the y component divided by the x component, the direction of steepest 3-dimensional slope is easily calculated by using the cross product of any two triangle edges. Call the data $\{x_i, y_i, z_i\}$, $i=1-3$; the direction of steepest slope, s, is obtained from the projection of the perpendicular (p. 33) to the triangle.

$$s = ((z_2-z_1)*(x_3-x_1) - (z_3-z_1)*(x_2-x_1)) / ((y_2-y_1)*(z_3-z_1) - (z_2-z_1)*(y_3-y_1))$$

If this triplet of height data is represented by a planar surface then a straight segment of constant-height contour line is defined by the intersection of the triangle with a level line, such as a shore line, and that line will be perpendicular to the direction of steepest ascent, s. Of course, a curved surface over the triangle is the general situation, and then the contour line must be curved accordingly to maintain perpendicularity with each maximum slope line.

Although manual construction of isoline maps was subjective, and difficult to define, two clear-cut maxims can be recognized. First, isolines are required to be perpendicular to the direction of steepest slope, and second, the estimation was done on a changing subset of adjacent data. Even so, hand-drawn contour maps are individualistic, so that inconsistency and irrepeatability in construction, and hence in interpretation, is a source of considerable ambiguity.

Mainframe computers became generally available in the early 1960's, and were welcomed warmly by those who saw, for the first time, the possibility of consistent and reproducible map construction, with both speed and a saving in labor. Expressing the draftsman's skills, however, in algorithmic terms suitable for execution by a computer, was surprisingly difficult and complex. Even duplicating the appearance of manually drawn maps, ambiguous as they were, may have seemed beyond the capabilities of the programmers and their computers. The attractions of computer contouring remained, however, despite the avalanche of less than satisfactory maps produced by nascent algorithms.

MAJOR DIFFICULTIES OF CONTOURING

To give an overall impression of the dimensions of computer contouring and the evolution of contouring procedures, a broad outline of general problems concerning data types, interpolation paradigms, plausibility, and reliability of contour displays, follows.

Topographical data occur typically in one of three distribution patterns. Not unusually the data are observations at the nodes of a regular grid. This has the advantages that the data are distributed evenly over the region and computer storage is compact because the location of each datum is implicit in the order of storage. An important advantage is that there is a fixed number of adjacent data in the immediate neighborhood of each datum, and this allows a uniform approach to interpretation. Wallis (1976) dates the use of a topographical grid, or graticule, to China in the first century. Grids usually are rectangular but may be triangular (see p. 22 and p. 179).

Not infrequently, however, the data are scattered, either because of difficulties in the data collection process, or to meet a statistical criterion intended to prevent bias. Three times as much storage is required for such data because the locational coordinates of each observation must also be stored. Additionally, an irregular number of adjacent data, at irregular distances, surround each datum and this makes some interpolation methods inapplicable.

The third distribution pattern concerns topographical observations that are closely spaced along widely spaced traverses or profiles. Data storage requirements are similar to scattered data, with the possibility that regular spacing along the traverse allows some saving in storage of locational coordinates. This is the most difficult data type to contour automatically because local information is available only in the direction of the traverse.

Many methods that work well for gridded data either are inapplicable or are unsatisfactory for scattered and traverse data. This impediment motivated the early development of intermediate procedures intended to convert scattered observations to gridded data by estimating the topographic height at the nodes of a regular grid superimposed on the region. This is interpolation, and there are two interpolative approaches, *fitted functions* and *weighted averages.* These are explored in detail in Section 2.5, p. 101.

The single objective of contour displays of topographical data is the expression of the shape of a representative surface that portrays plausibly the spatial variation of the data. To say that a surface is plausible, in this book, is to say only that its behavior is not recognizably inconsistent with the data.

In this sense, plausibility depends, first of all, on the assumption of continuity in the data source. Specifically, this implies that if enough data could be gathered, the data themselves would define completely a surface that is continuous, in general, although not necessarily so at any particular location. If one does not feel that this criterion applies for a given data set, then computer contouring techniques would likely give misleading results and should not be employed.

This particular specification of continuity is used here because it shows that problems of interpolation between data occur only when the data are sparse with respect to height variation. Many authors have noted that with sufficient data any interpolation method gives satisfactory results.

As a corollary to this definition of continuity, the ability to make an acceptable surface for sparse data provides a useful standard for evaluating contouring methods. More complex definitions of continuity, such as that given by Sabin (1985), are used for surface design (p. 103), but exploratory surfaces require continuity in the data source only.

The sparseness of the data may be thought of as the precision of the data set, as a whole. This form of precision is distinct to the precision of individual measurements. For example, if one were estimating the volume of a large round hill, fewer data would be required to give a result of specified accuracy than if the same volume was expected to be contained in a series of smaller irregular hills. Although measurements are made with the same precision in each example, more measurements must be made in order to define the more intricate shape of the group of small hills to the same degree of accuracy in the volume estimate.

Therefore, estimates of the precision of a topographical data set, which may be thought of as its resolution, determine the level of precision available to conclusions based on the data. Of course, one never is justified in extracting more significant digits than are used in the initial measurements.

Notice that data set precision should not be confused with the computational precision with which the calculations are made, or with the display precision with which the representative surface is expressed. Specifically, this implies that closely spaced isolines provide a precise display of a representative surface, but say nothing whatsoever about the precision of the topographical data set, or the computational precision.

In this respect, computer contouring differs strongly with manual contouring. Manual contouring is a subjective process, and degrades the information available from the data by the inadvertent inclusion of human error. This motivated authors of books on manual contouring to recommend a minimum interval between isolines to deter map users from inferring conclusions that were excessively precise, and unjustifiable in terms of data set precision, from the display precision.

Possibly the idea continues to have adherents. However, computer contouring algorithms may be assembled with sufficient care that significant construction errors are not incorporated. The computer drawn contour map simply is a physical expression of the original data in terms of the selected algorithm. This indicates that the display precision, should be as high as is practical; there is no need to consider a minimum isoline interval.

More importantly, such questions of display precision underline the crucial choice of interpolation algorithm, often made by default. The numerical precision of the algorithm is the single most relevant factor that is within the control of the spatial analyst., although nothing can improve poor data.

General Concepts Of Contouring

Lee (1985) has discussed several methods for measuring the accuracy of computer drawn maps. Review and discussion of historical and philosophical aspects of contouring may be seen in Bugry (1981), Cayley (1859), MacGillivray, Hawkins, and Berjak (1969), Maxwell (1870), Sager (1983), and Schmid and MacCannell (1955).

SECTION 2.1

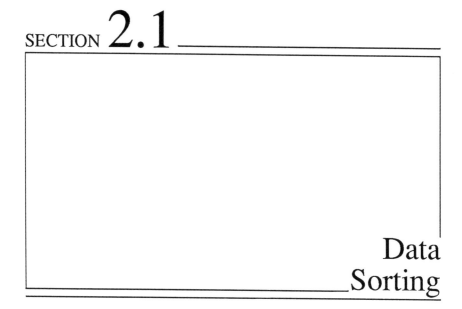

Data Sorting

SECTION SUMMARY

Sorting establishes the proximal order among a set of topographical data. This is required in order to select particular triplets, quadruplets or neighborhood subsets of adjacent data. These subsets are used to estimate local trends or gradients, compute local coordinates, and to interpolate the data set.

INTRODUCTION

In order to contour a topographical data set, by any method based on subsets, information on the adjacency or proximal relationships among the data points is required. Of course, when the data are located at fixed intervals, such as the nodes of a regular grid, these relationships are visually obvious, and computationally implicit in the order of indices. However, when the data comes from either scattered locations or traverses, the data set must be sorted spatially to extract this proximity information. Lack of this information, as Tobler (1979a) has said, makes interpolation difficult.

Unlike 1-dimensional sorting, spatial sorting is not a well-known procedure. Multiple application of univariate methods is not satisfactory because data components cannot be given equal treatment; one component must be given presidence over the others. To find the proximal order of the data, it necessary to use a method that detects adjacency relationships without directional bias.

PROXIMAL ORDER

The proximal order of a datum, with respect to the other data, is the set of all pairwise proximity relationships involving data that are nearby, in a sense to be made precise in the next few paragraphs. The proximal order of nongridded data generally is conceptualized and expressed as a triangulation or tessellation, and there are four kinds of proximal order present in the literature.

The types of proximal order are distinguished by arbitrarily or systematically determined pairwise proximity relationships within the data. Of the four kinds of proximal order, the most important is natural neighbor proximal order. In the following subsections, after a brief description of proximal order types, three general classes of algorithm for extracting natural neighbor order, and three forms of natural neighbor lists, are discussed.

(1) Arbitrary Proximal Order.

This is the order developed by a cartographic draftsman when he triangulates manually a set of topographic data. He declares pairwise proximity relationships by connecting pairs of data with straightlines until all possible triangles have been drawn. However, these lines do not necessarily connect a datum to all of its natural, or even near, neighbors.

Arbitrary triangulations can be obtained automatically, but tend to produce triangles that are not as equidimensional as they might be. Some long, thin, triangles can be removed by altering iteratively the choice of edges according to a rule such as taking the shorter of two intersecting quadrilateral diagonals.

(2) Optimal Proximal Order.

This is a triangulation with a minimum sum of edge lengths, and it is referred to as a minimum weight triangulation. Pairwise proximity relationships are determined solely by the relative distance between the pair, that is, triangulation edge length. Special configurations of data may have two or more distinct orderings with the same sum of edge lengths. A published algorithm (Klincsek, 1980) for this triangulation has efficiency of $O(N^3)$.

(3) Greedy Proximal Order.

This triangulation also is formed by the shorter edges but in the sense of a local minimum. Specifically, for topographical data, no edge is used if there is a shorter edge that would properly intersect it, that is, other than at its end points. Again, special configurations may have two or more distinct orderings.

A Greedy sort is implemented easily by ordering all possible edges by length and, starting with the shortest edge, including each edge that does not properly intersect any edge included already. This approach is $O(N^2 \log N)$ because the $(N^2 - N)/2$ edges must be sorted by length before the Greedy triangulation can be obtained.

(4) Natural Neighbor Proximal Order.

This triangulation is obtained by the empty circumcircle criterion. Specifically, natural neighbor proximal order is established by determining all circles that pass through three or more of the data and such that no datum lies inside any circle. In other words, this is the set of all circles defined by the data, but that have no data within them. This results in a configuration of overlapping circles with data at their intersections, as shown in Figure 2.1.2 on p. 62.

Each circle through a particular datum is defined by two other natural neighbors of that datum, and no other data are closer to the circumcenter than these three. All the natural neighbor pairwise proximity relationships involving that datum are established by all such circles through it. These circles have an average radius that is less than it would be for any other selection of triplets and the triangles defined by these triplets are the most equiangular possible.

The concept of natural neighbors arises from the notion that a pair of neighbors share a mutual interface, or region, that is equally close to each of the pair and all other neighbors are no closer. This criterion defines the natural neighbor pairwise adjacency relationship.

Polygonal tessellations of natural neighbors always are convex, that is, natural neighbor (Voronoi) polygons never have reentrant angles at their vertices. The uniqueness and usefulness of natural neighbor order was recognized by workers in the natural sciences early in the century. They used it to estimate graphically quantities of rainfall, and volumes of ore deposits.

Computing the natural neighbor order of a multivariate data set is an exercise in computational geometry. Geometric algorithms that determine (i) the nearest neighbor of a datum, (ii) all natural neighbors of a datum, and (iii) the convex hull (p. 102) of a data set, all develop and exploit the natural neighbor order of the data.

Aside from such theoretical interests, the exercise is important because a complete knowledge of the natural neighbor order of a data set is a prerequisite for precise evaluation of local trends in locationally dependent functions. Not only applications such as contouring and computing regional averages depend upon schemes to organize, or sort, the data in a manner that reveals local order; many engineering problems may be treated by finite element analysis (see, for example, Botkin and Bennett, 1985) over a set of nodes connected by a triangular mesh. Also, questions of adjacency amongst atoms and molecules arise in several areas of chemistry and physics.

Data Sorting

This sorting problem occurs in many fields of applied science and published attempts to find solutions and apply them occur in association with such terms as area-of-influence polygons, Delaunay triangulations, Dirichlet, Thiessen, and Voronoi, tessellations, and Wigner-Seitz cells.

The forms of proximal order are not always distinct but Figure 2.1.1, below, illustrates a situation where each of five possible triangulations of the points **A**, **B**, **C**, **D**, and **E**, is different. The Optimal triangulation selects edges **AC** and **CE** because this gives a minimum sum of lengths. The Greedy triangulation selects edge **BD** because it is the shortest and then chooses edge **AD** because it is the shorter of the two remaining nonintersecting edges. The Delaunay triangulation selects edges **AC** and **AD** because only the triangles **ABC**, **ACD** and **ADE** have empty circumcircles. The edges on the convex hull occur in all three triangulations, in the absence of additional data points, of course.

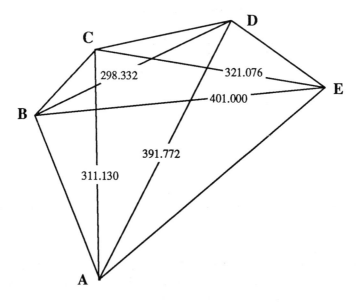

Figure 2.1.1 Systematic triangulations

Figure 2.1.1 also illustrates the ambiguity in iterative application of the "shortest quadrilateral diagonal" rule. It is known that this rule does not always produce a Delaunay triangulation. In fact, however, it does not even produce a unique triangulation in all situations, because it is order dependent. Suppose that the initial triangulation selects edges **BE** and **CE**. If an iterative pass begins with edge **BE**, the result will be the choices **AC** and **CE**. On the other hand, if the iterative pass begins with edge **CE**, the result will be the choices **BD** and **AD**. No further exchanges will be made. Note that neither of these end results is the Delaunay triangulation.

TABLE 2   Interior Angles from Figure 2.1.1

Data	x	y	
A	172	451	
B	81	671	
C	163	762	
D	350	800	
E	480	711	
ACD	28.7	99.5	51.7
ABC	20.5	116.7	42.7
ADE	23.4	81.9	74.7
Delaunay triangulation, rmsd = 32.9			
ABC	20.5	116.7	42.7
ACE	52.2	78.4	49.4
CDE	21.1	133.6	25.3
Optimal triangulation, rmsd = 39.0			
BCD	23.1	142.3	14.6
ABD	49.3	93.6	37.1
ADE	23.4	81.9	74.7
Greedy triangulation, rmsd = 39.4			
ABE	72.7	73.0	34.3
BCE	43.7	121.1	15.1
CDE	21.1	133.6	25.3
Unnamed triangulation I, rmsd = 41.0			
ABE	72.7	73.0	34.3
BDE	20.6	119.0	40.4
BCD	23.1	142.3	14.6
Unnamed triangulation II, rmsd = 43.0			

This particular example demonstrates fundamental differences in the various forms of proximal order, and also serves to illustrate the "most equiangular" property of the Delaunay triangulation. The interior vertex angles of the Delaunay triangulation have the least root mean squared deviation. The two remaining unnamed triangulations are included for completeness.

The distinction between "thin" and "good" triangles also may be seen here. The empty circumcircle criterion selects against large circles so the Delaunay triangulation has the smallest mean circumcircle diameter and, in this sense, is most compact. It is these qualities of compactness and equiangularity that make natural neighbors so useful for local coordinates (p. 81), for neighborhood-based interpolation (p. 155), and for local measures of variability (p. 171).

Any three or more data are natural neighbors if no other datum lies within the circle that goes through them. In Figure 2.1.2, using the first thirteen data from Table 1 (p. 8), the natural neighbors of point A are B,C,J,K, and L, whereas the natural neighbors of L are A,C,D,E,F,G,K, and M. The fourth list in Table 3 (p. 64), shows these natural neighbors in groups of triplets.

Data Sorting

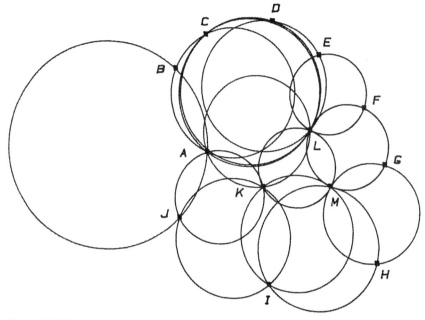

Figure 2.1.2 Natural neighbor circumcircles

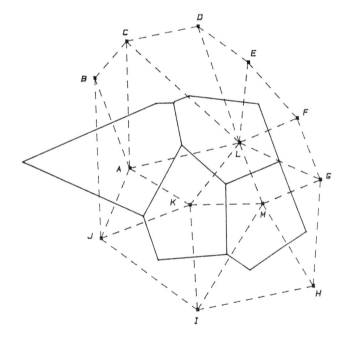

Figure 2.1.3 Delaunay and Voronoi tessellations

On average, a datum in a topographical data set is expected to have six natural neighbors. It may have as few as three in the interior of the data set, and as few as one if it is on the boundary of the set. There is no upper limit for special patterns of data, but more than thirteen natural neighbors are seldom seen in evenly scattered data.

Natural neighbor subsets can be illustrated by the set of circumcircles in Figure 2.1.2, as well as by the Delaunay and Voronoi tessellations of Figure 2.1.3, on the opposite page. The Voronoi tessellation (solid lines) illustrates that each datum has a unique natural neighbor region, defined by its Voronoi polygon, and bounded by the halfway interfaces of that datum with each of its natural neighbors.

The Delaunay triangulation (dashed lines) is the geometric dual of the Voronoi tessellation. Each triangle is defined by three natural neighbors and the center of their circumcircle is the shared vertex of the Voronoi polygons associated with each of the three.

EXPRESSING PROXIMAL ORDER

Each datum in a multivariate data set has a number of these proximity relationships with nearby data, and any systematic listing of these relationships expresses the spatial order of the data set. There are several ways that this proximal order can be expressed by a list.

(1) The spatial order of a topographical data set is expressed most simply by a list of the pairwise proximity relationships among the data; such a list can be made for any form of proximal order. This is the least complex way of tabulating proximal order, but it is seldom used, probably because practical applications require more complex local subsets of these relationships.

Natural neighbor proximal order, however, has four other types of lists of proximity relationships. They are classified here by increasing information content, and their complexity is described in terms of natural neighbor circles, with interpretations using the Delaunay and Voronoi tessellations.

(2) After the list of the pairwise proximity relationships, the second least complex proximity list is a list of the triplets of data that define each circumcircle. These three neighboring data share a Voronoi vertex and define a Delaunay triangle.

(3) The third type is a list of the triplets of circles that share two data with a given circle. These circles define the Voronoi vertices that are contiguous with a given Voronoi vertex, and the Delaunay triangles that share an edge with a given Delaunay triangle.

(4) Fourth is a list of the variable number of circles that pass through each datum. These circles involve all the natural neighbors of a datum, and define the Voronoi polygon associated with the datum, or all the Delaunay triangles that share a given datum.

Data Sorting

(5) A list of the sets of circumcircles that share at least one datum with each circumcircle. These determine the set of Voronoi vertices on the triplet of polygons that share a given Voronoi vertex, or all the Delaunay triangles that share a datum with a given triangle. TABLE 3 shows only the second, third and fourth lists for the first thirteen data given in TABLE 1 on p. 8 and shown in Figures 2.1.2 and 2.1.3 on p. 62.

TABLE 3 Proximal Order Lists

Triplet index	Second list	Third list	Fourth list
1	ABC	2,3	1,2,3,4,5
2	AB J	1,4	1,2
3	ACL	1,5,6	1,3,6
4	AJK	2,5,13	6,7
5	AKL	3,4,15	7,8
6	CDL	3,7	8,9
7	DEL	6,8	9,10,11
8	EFL	7,9	11,12
9	FGL	8,10	12,13,14
10	GLM	9,11,15	2,4,13
11	GHM	10,12	4,5,13,14,15
12	HIM	11,14	3,5,6,7,8,9,10,15
13	IJK	4,14	10,11,12,14,15
14	IKM	12,13,15	
15	KLM	5,10,14	

The numbers in the first column are indices to the second list which is the natural neighbor triplets in the data. The entries in the third list are the sets of indices of triplets in the second list that share a pair of data with that triplet. The entries in the fourth list are the indices of all the triplets that share a datum.

These lists may have portions that are nonunique or redundant, that is, triplets **FGL** and **GLM** refer to the same natural neighbor circle (p. 62) as would the alternate triplets **FGM** and **FLM**. Any of these lists may be derived from any other by searching, and one or more are generated by each of the natural neighbor sorting algorithms mentioned next.

NATURAL NEIGHBOR SORTING ALGORITHMS

In general, establishing natural neighbor order can be treated as a task of inserting each successive datum into an existing natural neighbor proximity list by deleting, and adding, entries. However, by first identifying the subset of natural neighbors for a given datum, solutions to the inserting task always deal with a constant, or expected, size subset of data and therefore have an efficiency that is linear in N, the number of data; spatial sorting is performed in two steps.

(1) Identifying Subsets.

The method of locating subsets determines the execution time efficiency. Such subsets may be obtained by three general approaches as follows.

(a) Checking all $N^2 - N$ possible pairs of data; of course this method is $O(N^2)$. See Watson and Smith ($n=3$, 1975), Finney ($n=3$, 1979), Tanemura, Ogawa, and Ogita ($n=3$, 1983), and Langridge ($n=2$, 1984), among others, for examples.

(b) Performing a directed walk across the fourth natural neighbor proximity list. Starting from an arbitrary datum, move to whichever of its neighbors is closest to the new datum, recursively, to eventually reach the nearest neighbor of the new datum. A given walk is expected to involve $O(N^{1/n})$ neighbors and therefore N walks (to insert N data) have an expected time of $O(N^{(n+1)/n})$. Green and Sibson ($n=2$, 1978) and Bowyer ($n=2,3,4$, 1981) use this technique.

(c) Most directly, partitioning the data into a regular array of bins with time $O(N)$. Lee and Schachter ($n=2$, 1980), Bentley, Weide, and Yao ($n=2$, 1980) and Maus ($n=3$, 1984) have used this method. The subset of bins, whose contents are used as a subset for inserting a given datum, necessarily contains more data than just the set of natural neighbors of the datum, to ensure that all such neighbors are considered. Thus, although this approach is linear in N, the constants in the efficiency expression for this algorithm would be large.

(2) Inserting Into A Subset.

Inserting a datum into a set of natural neighbors requires (i) locating and deleting previously computed entries in the natural neighbor proximity list(s) whenever the new datum lies inside an existing circumcircle, and (ii) determining all new circumcircles involving the new datum among its natural neighbors.

Inserting generally is approached by checking all circumcircles to select those that contain the new datum and so must be deleted, then checking all possible circumcircles that could be defined by the new datum and an irregular configuration of a variable number of neighbors; a complicated procedure, especially in higher dimensions. Cline and Renka ($n=2$, 1984), for example, divide the inserting task into four subprocedures and list 27 steps. Solutions to the inserting task are -

(a) Arbitrarily triangulate the new datum with a subset of possible natural neighbors and, iteratively, exchange quadrilateral diagonals for shorter diagonals. This is intended to improve the triangulation by reducing the number of long, thin, triangles. Such an approach was used by Lawson ($n=2$, 1977), Moore ($n=2$, 1977), Mirante and Weingarten ($n=2$, 1982), and Magnus, Joyce, and Scott ($n=2$, 1983), but it does not always obtain natural neighbors (p. 60).

Note that when some authors write of "optimizing" an arbitrary triangulation, their criteria indicate that they mean Natural Neighbor Proximal Order rather than Optimal Proximal Order (p. 58).

Data Sorting

(b) Sort the subset of neighbors by distance from the datum to be inserted; then starting with the two nearest neighbors and the datum, consider their circumcircle. If no other datum in the subset is closer than the circumcircle radius to the circumcenter, then that group of three data are natural neighbors. Otherwise, other pairs of neighbors must be considered until all possibilities are exhausted. Examples of this approach include McLain (n =2, 1976), Lee and Schachter (n =2, 1980), Maus (n =3, 1984), and Cline and Renka (n =2, 1984). Fortune (1987) introduces a sweepline algorithm that determines the transformed Voronoi diagram in $O(N \log N)$ time.

(c) Perform a directed walk around the new datum on the third and fourth natural neighbor proximity lists. Green and Sibson (n =2, 1978) and Bowyer (n =2,3,4, 1981) begin an inserting procedure from the nearest neighbor of the new datum. Then the third list provides a cross reference to the three entries in the fourth list which give all the natural neighbors that must be checked to update the natural neighbor proximity lists. These neighbors are accessed in a cycle around the new datum.

(d) Another approach involves recursive subdivision of the data set until some empty circumcircles are revealed, then merging subsets of data with application of the empty circumcircle test; this again is $O(N \log N)$. Shamos and Hoey (n =2, 1975), Lewis and Robinson (n =2, 1978), and Lee and Schachter (n =2, 1980), have described such a technique. Some details of the merging procedure are given by Oxley (n =2, 1985). Dwyer (n =2, 1987) provides a divide-and-conquer algorithm.

(e) The second proximity list may be generated by treating the insertion task as a search for natural neighbor circumcircles that contain the new datum, rather than for individual natural neighbors. To insure that each datum is contained in some circumcircle, the second list is initialized with a circumcircle through three arbitrary control points whose convex hull (p. 102) encloses a region that contains the data set.

The three pairs of indices, that can be selected from the triplet associated with each circumcircle containing the new datum, are stacked temporarily, and any pairs that occur twice are deleted. The pairs that remain, after all current circumcircles have been checked, make an unordered list of the pairs that form new triplets with the new datum. Adding these triplets to the second list completes the inserting task. This algorithm is outlined in pseudo-algorithmic language on the next page.

This approach also can be used to delete a datum from the second natural neighbor proximity list simply by natural neighbor sorting the neighbors of the deleted datum. This establishes the circumcircles that would exist among the natural neighbors of the datum in its absence.

Programs CON2R.BAS and NEIGHBOR.BAS use this last mentioned algorithm to sort the data, establish gradient subsets and interpolation subsets; TRIANGLE.BAS uses this algorithm to sort the data. The triangles are displayed if **TT**=1, in line **1100**.

Natural Neighbor Sorting Algorithm

procedure natural.neighbor.insert (triplist, datum)
comment insert a new datum by updating a list of triplets of data
array triplist
var datum
begin
   array templist
   var jpair
   for i = 1 to rows(triplist)
   begin
      if new datum is inside circumcircle triplist(i)
         begin
            for j = 1 to 3
         begin
            jpair = jth pair of triplet in triplist(i)
            if jpair is already in templist
               then delete jpair from templist
               else add jpair to templist
         end
      delete triplist(i)
      end
   end
   for each pair in templist
   begin
      concatenate new datum to form new triplet
      add triplet to triplist
   end
end procedure

This algorithm also determines all the vertices on the boundary of the convex hull (p. 102) by deleting all new triplets that do not include a control point. Because fewer triplets need to be checked, the ordered extreme points of the convex hull are obtained in less time than triangulating the entire data set.

Sorting Efficiency

The efficiency of this algorithm is established by noting that it is initialized with an arbitrary circumcircle and, each time a datum is added to the network, exactly two additional circumcircles are generated. Searching through these circumcircles for an intersection with the next datum is the only step that is not linear in N because the size of the temporary list will vary about the expected size of six. When N data have been inserted, the number of circumcircles examined is the sum of the first N odd integers, that is, N^2, so the algorithm is $O(N^2)$.

Data Sorting

However, these estimates of efficiency are average-case expressions for unordered data, and efficiency will improve if the data are preprocessed. Therefore, ordering the data by the first component would make this algorithm $O(N \log N)$, and bin sorting the data makes it $O(N)$.

Degeneracies

The Delaunay triangulation, described by the second proximity list, is not always unique because more than three data may lie on a circle, and then, locally, the triangulation becomes nonunique. This so-called degenerate condition is illustrated on the plane by four co-circular points, F, G, L and M in Figure 2.1.2 on p. 62. These four points have two possible triangulations that satisfy the empty circle criterion (although the natural neighbor order and the Voronoi tessellation, represented by the fourth proximity list, remain unambiguous).

From the viewpoint of computer contouring, this problem is more apparent than real. In theory a number is considered to have an infinite precision, but in the computer it is represented by a limited string of bits. Observations and measurements usually have even fewer digits. When an measurement is placed into the computer, it is the default convention to fill the least significant bits with zeros. Although the user is unable or unwilling to measure with sufficient precision to know the value of these bits, he proceeds under the assumption that the value of these bits does not have great importance, and their probable value is zero anyway.

So in practice, these degeneracies are prevented easily by giving each datum a small random displacement. In effect, zeros are not assigned to the least significant bits but they are filled with random numbers whose probable value is zero. Then more than three data will never have the same circumcircle and the triangulation will always be unique.

This is done in the programs, in line **1460**. A random number with zeros in the first four positions to the right of the decimal point is randomly added, or subtracted, from the locational coordinates of each datum. This makes the probability of more than three data sharing any circumcircle to be practically zero, but does not degrade the data significantly, or distort the map, because the perturbation is so small.

P1#(I1%,1) = P1#(I1%,1) + 0.0001∗(RND - 0.5)
P1#(I1%,2) = P1#(I1%,2) + 0.0001∗(RND - 0.5)

Some reviews of data sorting have been provided by Boots and Murdock (1983), Green and Sibson (1978), Hayes and Koch (1984), Nagy (1980), Sawkar Shevare, and Koruthu (1987), and Watson (1985).

SECTION 2.2

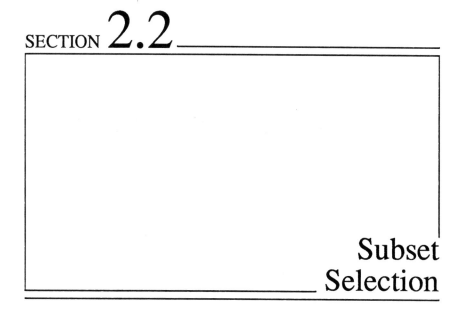

Subset Selection

SECTION SUMMARY

Data sets are partitioned into running or disjoint subsets for estimating gradients, computing local coordinates, local interpolation, and for economy of computation. Running subsets may be selected on the basis of fixed distance, fixed area, fixed number of nearby data, or natural neighbors. Disjoint subsets are used for some types of piecewise interpolation.

INTRODUCTION

A subset of data is obtained whenever some of the data are extracted preferentially from the set, and used as if the remainder did not exist. Subset selection often is the first procedural step in the sequence of operations involved in contouring. There are three main uses for subsets of data in contouring algorithms. Two of these occur in weighted average interpolation methods, and the third use is with fitted functions.

(1) Some weighted average gradient interpolation methods require subset selection for gradient estimation. If the slope of the representative surface at each datum is estimated from the surrounding data, it can be treated as an estimate of local trend at that datum. Then a weighted average of these trends can provide a surface that interpolates both the height, and the estimated gradient, at each datum.

(2) Weighted average interpolation methods also require subset selection for computing local coordinates (p. 75). In triangle-based interpolation methods, three data that lie at the vertices of a triangle must be selected for each interpolation point, to obtain barycentric coordinates. A four-datum subset is used by interpolation methods that are applied to rectangularly gridded data, to obtain rectangular coordinates. In the more general situation, the natural neighboring data (six, on average) of an interpolation point are used as a subset for natural neighbor local coordinates (p. 75).

(3) Piecewise patches formed by fitted function methods require running subsets of data, or disjoint subsets of three or four data. Also, the amount of data that can be fitted usually is limited, so when a data set is too large for available memory, or for a interpolation technique such as solving a set of simulataneous equations, subsets of data are selected by region.

An obvious source of difficulty in selecting a subset, from scattered or traverse data, is determining whether or not a given datum should be included in the subset. Although a cluster of adjacent data are easy to identify when they are arranged in a fixed pattern, assessing the eligibility of an individual datum with respect to the interpolation point is more complicated and ambiguous for irregularly spaced data.

The usual dissatisfaction with subset selection is either too much, or too few, data to obtain both a detailed and reliable surface. For uniformly gridded or evenly scattered data, an appropriate fixed subset size usually can be determined by trial and error. However, the question of subset size is more difficult when the data are located irregularly because a size that suits one portion of the data set does not suit another portion.

SUBSET REQUIREMENTS

The general requirements of a subset selection method are straightforward, and cover three aspects. These concern the size of the subset, the azimuthal distribution, and the collinearity of any two or more data relative to the interpolation point.

(1) The selection method should not allow excessively large or small subsets for the following reasons. If the selected subset is large, the combined effect will tend to submerge the local detail and, therefore, defeat one of the purposes of using subsets. On the other hand, if a subset is too sparse, there will not be enough information to obtain a robust and stable surface.

(2) The subset should be distributed uniformly about the interpolation point as nearly as possible. If the data in an interpolation subset are predominantly in fewer than four quadrants about the interpolation point, the interpolating procedure is supported inadequately from one side and the resulting surface will vary unsatisfactorily, even erratically, as the running subset is changed for the next interpolation.

(3) Individual data points should not mask other data in the subset. This occurs, for example, whenever the mildly positive influence of a nearby datum is more or less canceled by the strongly negative influence of a more distant datum behind it. The situation occurs when two or more data are in line with the interpolation point. Subsets need to be selected so that the maximum of local information can be extracted without being submerged in the effects of somewhat less local, and so less pertinent, information.

SELECTION CRITERIA

No method of subset selection seems to be best in all situations, and a particular criterion may be selected only because it is least inappropriate. Three general approaches, fixed number, fixed area, and natural neighbors, can be considered in relation to their most appropriate use.

(1) Fixed Number Subsets.

The problem of excessively large, or insufficiently small, subsets may be approached by selecting the nearest fixed number of data because this causes the running subset to always be the same size. However, the unsatisfactory effect is that the nearest fixed number of scattered data often are not arranged evenly around the interpolation point in regions where the density of data is changing.

Therefore, effects due to uneven distribution of data about the interpolation point remain, and this method is particularly susceptible to problems of masking of nearby data by more distant data. Selecting a subset with a fixed number of data is suitable only when the data are distributed evenly over the region, so partitioning by fixed number suits systematic piecewise surface patch methods. For example, rectangular patch methods may use four, or sixteen, gridded data.

(2) Fixed Distance Subsets.

By this criterion, subsets are selected on the basis of fixed distance by including every datum that is closer than some fixed distance from the interpolation location. All the data within a circular region centered on the interpolation location belong to the subset.

This approach works well for data that are uniformly and regularly spaced so that subsets are nearly always the same or similar size. However, whenever the data are more dense in one region of the data set than in another, this approach tends to select too many in the densest region, and too few in the sparse area. Goodin, McRae, and Seinfeld (1979) discuss the choice of an optimum radius of influence based on average distance between pairs of adjacent data.

Subset Selection

Preferably, the fixed distance may be varied, according to local density of data, to prevent excessively large or small subsets. This, however, requires monitoring of either density or number of data obtained in adjacent subsets. If the data are more densely spaced in one direction than another, fixed distance subsets may include too many data and still not provide an adequate local subset with respect to the sparse direction.

Fixed area subsets are a conceptual variation of fixed distance that allows an anisotropic subset to be selected. The use of fixed distance subsets implies that all the data within a circular window are included in the subset. To allow for data that are unequally dense in some directions, a variation on fixed distance allows the number of data in the running subset to be controlled by the size and shape of the window that includes the subset data.

The window containing the data to be included in the subset may be rectangular, or elliptical (see Smith, 1968). The orientation of the rectangle, or ellipse, is chosen so that the maximum and minimum widths correspond to the directions of least and greatest density of data. The ratio of width to length for these windows should be similar to the ratio of directional densities so this approach remains unsuitable if the data density is strongly anisotropic or this ratio varies.

Selecting all data from a fixed area is required when the data set is too big to be handled as a whole by the chosen algorithm, and must be treated in two or more blocks. Fixed area partitions also are necessary when the data come from regions that are distinctly different in character and therefore require separate treatment. If smaller subsets are required, then fixed area is most suitable for data that are more dense in one direction than the other. The shape and orientation of the subset area is selected to suit the anisotropy of sampling.

(3) Natural Neighbor Subsets

Selection by fixed criteria does not work well for data sets whose locational density does not vary consistently with the variability in height values. This data condition is termed heteroscedasticity, and it is commonplace in real data sets. For example, if a grid of data were obtained from a region containing both smooth, moderately level ground, and rough, mountainous terrain, then the level area can be interpolated more confidently than the hilly area.

Similarly, if two similar regions were sampled by different schemes, one more dense than the other, the uneven amount of information in the combined data set implies that a subset size suitable for one subregion is not suitable for the other.

This is a simplification of a general problem concerning the uneven distribution of information in real data sets. The size and shape of the region covered by the subset needs to be adjusted to suit the local configuration of data locations, and this is done by natural neighbor subsets. A subset size that adjusts automatically is necessary for these irregular data distributions.

The smallest satisfactory gradient, or interpolation, subset contains all the natural neighbors of a given interpolation point. When, in addition, all natural neighbors of this minimum subset are involved through the gradient estimates, all local trends in the data about the interpolation point become manifest and available.

Natural neighbor subsets can be obtained automatically and are useful in three distinct ways.

(a) They provide a subset of data around a given datum for gradient estimation (see p. 85).

(b) They provide a subset of data around the interpolation point for interpolation (see p. 101).

(c) They provide the natural neighbor local coordinates (p. 75) required for both interpolation and gradient estimation.

Subsets are selected in the programs by fixed distance for the data input window (at line **1220**, by comparing each datum to squared distance **RD**), and by fixed area for the display window, at line **1190**, using variables **XS**, **YS**, and **DT**, as noted on p. 13 . Natural neighbor subsets are determined in all the programs at lines **1580** and **3880**, used for calculating local coordinates at line **2180**, and used for gradient estimation, at line **2360**.

SECTION 2.3

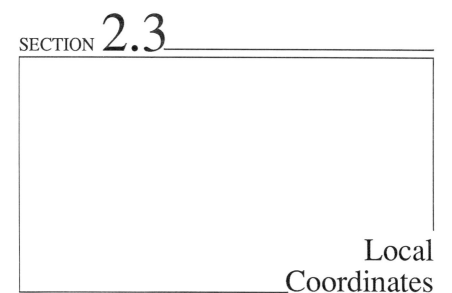

Local Coordinates

SECTION SUMMARY

Local coordinates are a means of defining a given position, or location, relative to nearby reference points without regard for the regional, or geographical, coordinates of the location. Local coordinates may be based on two, three, four, or more, reference points.

INTRODUCTION

Local coordinates are a method of identifying numerically particular locations relative to a small number of nearby reference points, in contrast to Cartesian coordinates which are relative to the Cartesian reference frame. In other words, local coordinates label a location, with respect to a given set of control locations, in terms that carry no explicit information about their global, geographic, or absolute position.

Local coordinates are indispensible in computer contouring because they allow general procedures that adjust automatically to variable local conditions. They are particularly useful for types of contouring where the interpolated value depends only on a local subset of data. This is due to a special property of local coordinates; their numerical values range between zero and one, and always sum to one. This allows each local situation is normalized to a standard geometrical condition.

Local Coordinates

BINARY LOCAL COORDINATES

The simplest form of local coordinates involves two control points; binary coordinates relate a point on a line to two other points, and so are limited to one-dimensional situations. For example, as Figure 2.3.1 shows, the position of **X**, on the line between **I** and **J**, subdivides that line segment into two portions whose lengths are fractions of the length of the original segment. Those two fractions, of course, add to one. The position marked by **X** has local coordinates (0.6,0.4). The local coordinate of the point at **X** with respect to reference point **I** is $w_I=0.6$ and the local coordinate of the point at **X** with respect to reference point **J** is $w_J=0.4$.

Notice that each of these coordinates is proportional both to the length of the line segment on the other side of **X**, and to the inverse of the distance from **X** to the associated control point. An alternative interpretation is that **X** is the balance point between a weight of 0.6 at **I** and a weight of 0.4 at **J**.

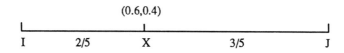

Figure 2.3.1 Binary local coordinates

One advantage of these coordinates is that when data points **I** and **J** have associated heights, then the height at **X** is a weighted average of **I** and **J**, that is, the sum of $w_I=0.6$ times the height at **I** and $w_J=0.4$ times the height at **J**. Because the sum of the binary coordinates, w_I and w_J, is one, the height at **X** is termed a linear combination of the heights at **I** and **J**. If **X** and **I** coincide, then the binary coordinates of **X** are $w_I=1$ and $w_J=0$, and the height at **X** is equal to the height at **I**. If **X** is moved along the line between **I** and **J**, then the height at **X** traces a straightline between the height at **I** and the height at **J**.

Binary local coordinates are used in manual interpolation (p. 112), and in distance-based, and rectangle-based, computer methods. They are conceptually important because they are a simple example that illustrates the general structure of local coordinates and their influence on computer contouring methods.

BARYCENTRIC COORDINATES

A general and widely known form of local coordinates relate a given location to three data points. The three data, or reference points, may be conceptualized as lying at the vertices of a triangle, as Figure 2.3.2, on the next page, shows.

Barycentric Coordinates

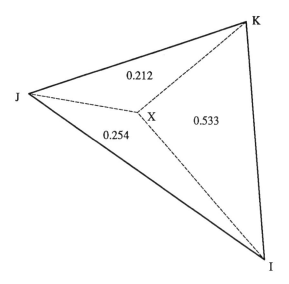

Figure 2.3.2 Barycentric local coordinates

Calculating these coordinates is straightforward. Any interpolation point, say X, within the triangle IJK, subdivides it as shown in Figure 2.3.2. The areas of these subtriangles, XIJ, XJK, and XKI, each calculated as fractions of the area of the triangle IJK, are the barycentric coordinates of the interpolation point for each opposite vertex, respectively.

Each coordinate refers to the data point that is opposite that subtriangle so that w_K, the coordinate with respect to data point K, is proportional to the area of XIJ in Figure 2.3.2. Specifically, the barycentric coordinate of X with respect to I is w_I=0.212, the barycentric coordinate of X with respect to J is w_J=0.533, and the barycentric coordinate of X with respect to K is w_K=0.254. Of course the sum of these barycentric coordinates is one because the sum of the areas of the subtriangles is equal to the area of the original triangle.

Barycentric local coordinates provide the primary method for relating an interpolating function to a triangular patch. They can be thought of as mutually representing the normalized position of the given point relative to each of the three sides of the triangle and their opposite vertices. Their actual values, though, are proportional to area. In this sense they are truly bivariate, compared to binary coordinates which are proportional to distance and so are univariate.

The term *barycentric* acknowledges the fact that the point labeled X is the centroid, or center of gravity, of a three-body system, if the local coordinates are proportional to the magnitudes of point masses located at the respective vertices of the triangle. A specific calculation of these coordinates for the point, X, with respect to the triplet, IJK, shown in Figure 2.3.2, is provided on the next page.

Local Coordinates

$$w_I = ((J_x-X_x)*(K_y-X_y) - (K_x-X_x)*(J_y-X_y))/\text{Det}_{IJK}$$
$$w_J = ((K_x-X_x)*(I_y-X_y) - (I_x-X_x)*(K_y-X_y))/\text{Det}_{IJK}$$
$$w_K = ((I_x-X_x)*(J_y-X_y) - (J_x-X_x)*(I_y-X_y))/\text{Det}_{IJK}$$

where subscripts x and y are the Cartesian coordinates and

$$\text{Det}_{IJK} = (J_x - I_x)*(K_y - I_y) - (K_x - I_x)*(J_y - I_y)$$

Then by substituting **I** = (350,100), **J** = (97,276), and **K** = (330,360) from TABLE 1 (p. 8), and with **X** = (210,260)

$$w_I = .212, \quad w_J = .533 \text{ and } \quad w_K = .254$$

The cyclic order, (clockwise or counter-clockwise), in which the three data points, **I, J,** and **K,** are labeled, though often important, is not a problem in this calculation.

Barycentric coordinates may be determined for an interpolation point with respect to any three control points in any configuration, except all in a line. This calculation also may be performed for interpolation points outside the triangle, and then one or two of the barycentric coordinates will be negative. However, the sum of the three coordinates will remain unity. This implies that the one or two positive barycentric coordinates will be greater than unity and, moreover, will increase as the interpolation point is moved away from the data. Therefore, although barycentric coordinates are geometrically valid anywhere on the plane, they are useful for computer contouring only inside a given triangle because they need to increase as the interpolation point is moved toward the data.

RECTANGULAR LOCAL COORDINATES

Another popular form of local coordinates relates the interpolation point to four data points at the corners of rectangle, as shown in Figure 2.3.3, on the next page. Because the grid of data usually is chosen to be square with the geographical coordinate system, the rectangular local coordinates of an interpolation point can be obtained easily by first calculating r and s. Using the unit square as an example, although the geographic coordinates of any rectangle could be substituted for the values at the corners of the unit square, the rectangular coordinates for a location **X**=(x,y)=(0.333,0.6), within the unit square MGFL are

$$w_M = (1-r)*(1-s) = 0.267$$
$$w_G = r*(1-s) \quad\quad = 0.133$$
$$w_F = r*s \quad\quad\quad\quad = 0.200$$
$$w_L = (1-r)*s \quad\quad = 0.400$$

where

$$r = (X_x - M_x)/(G_x - M_x) = (0.333 - 0.0)/(1.0 - 0.0) = 0.333$$
$$s = (X_y - M_y)/(L_y - M_y) = (0.6 - 0.0)/(1.0 - 0.0) = 0.6$$

The corners may be labeled in either cyclic order. Each of the four coordinates is proportional to the area of the subrectangle on the diagonally opposite side of the point, **X**. The coordinate, w_M, for the data point, **M**, is the ratio of the area of the subrectangle diagonally bracketed by the interpolation point, **X**, and the data point, **F**, to the area of the rectangle.

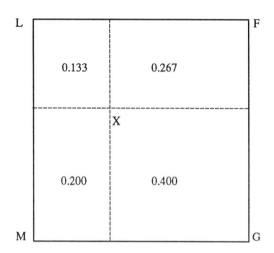

Figure 2.3.3 Rectangular local coordinates

Bayer (1985), Rogers and Adams (1976), Stelzer and Welzel (1987), Tobler (1979a), and Tobler and Kennedy (1985), are a few of the authors that have discussed the use of rectangular local coordinates.

PARALLELOGRAM COORDINATES

Although barycentric coordinates may be applied to irregular triangles as well as regular triangular grids, rectangular coordinates can be used only with rectangles and orthogonal grids. When the rectangular grid of data points is not set square with the geographical coordinate system or the two sets of grid lines are not orthogonal, the simplified calculation of r and s for rectangular coordinates must be expanded in order to determine their values.

Parallelogram coordinates are a generalization of rectangular coordinates. Their calculation is more complex, and they are required infrequently. However, they are conceptually important because their construction emphasizes the area-based nature of rectangular coordinates, which is not obvious and is often overlooked. For example, consider four data points at the corners of a parallelogram, and labeled in cyclic order in either direction, as **SRQP** in Figure 2.3.4 on the next page.

Local Coordinates

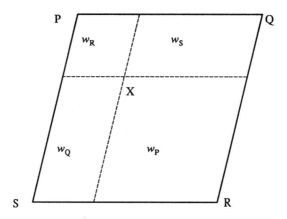

Figure 2.3.4 Parallelogram local coordinates

The interpolation point, $X=(x,y)$, may be seen to be the vector sum of any control point or parallelogram corner and two, as yet unknown, signed segments of the parallelogram edges so it must satisfy two equations in r and s such as

$$S_x + r(R_x - S_x) + s(P_x - S_x) = X_x$$
$$S_y + r(R_y - S_y) + s(P_y - S_y) = X_y$$

where subscripts x and y are the Cartesian coordinates. Rearranging and solving with Cramer's Rule (p. 88) gives

$$r = ((X_x - S_x) * (P_y - S_y) - (X_y - S_y) * (P_x - S_x))/d$$
$$s = ((R_x - S_x) * (X_y - S_y) - (R_y - S_y) * (X_x - S_x))/d$$

where

$$d = (R_x - S_x) * (P_y - S_y) - (R_y - S_y) * (P_x - S_x)$$

Then again, as for rectangular coordinates (compare p. 78),

$$w_S = (1-r)*(1-s)$$
$$w_P = r*(1-s)$$
$$w_Q = r*s$$
$$w_R = (1-r)*s$$

are the parallelogram coordinates of **X** with respect to parallelogram **SRQP**.

This derivation of parallelogram local coordinates shows the values of r and s to be the ratios of areas of parallelograms (p. 28). The value for d is the area of the parallelogram **SRQP**, the value for r is the area of the subparallelogram to the left of the broken line through X divided by d, and the value for s is the area of the subparallelogram below the broken line through X divided by d. Since a rectangle is a parallelogram, rectangular local coordinates also are area-based variables although their definition on p. 78 appears to be univariate.

NATURAL NEIGHBOR COORDINATES

The most general example of local coordinates concerns a variable number of three or more data points. The data points must be natural neighbors of the interpolation point. For an explanation of natural neighbors, see p. 62. Typically, six natural neighbors surround a particular location on a topographical surface. For the sake of graphical clarity, Figure 2.3.5 shows only five neighbors, but the behavior is typical for any number of data.

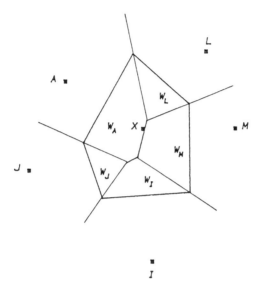

Figure 2.3.5 Natural neighbor local coordinates

This system of local coordinates, due to Sibson (1980b), relates the position of the interpolation point to each datum in the neighborhood subset, as shown in Figure 2.3.5.

The coordinates, ranging between zero and one, are proportional to the areas defined on natural neighbor regions for each of the data, roughly comparable to the way that the barycentric coordinates used in triangle-based interpolation are proportional to the areas of subtriangles.

Computationally, a neighborhood coordinate is formed from the overlap of the Voronoi proximal polygon surrounding the interpolation point with the Voronoi proximal polygon that would surround a datum in the absence of the interpolation point. For example, the area labeled w_A in Figure 2.3.5 shows the overlap of the polygon associated with the interpolation point X and the Voronoi polygon associated with datum A. The area of this region becomes the coordinate of X with respect to datum A after normalization (division by the sum of all overlaps with natural neighbors of X).

Local Coordinates

Such area-based coordinates are superior to distance-based coordinates for multivariate data because, when the data are unevenly dense, the coordinates vary anisotropically between two adjacent interpolation points. Distance-based coordinates make no allowance for the distances to the other data; that is, distance-based interpolation is not sensitive to a changing configuration of data.

COMPUTING NATURAL NEIGHBOR COORDINATES

Figure 2.3.5 on the previous page shows that natural neighbor coordinates are proportional to areas of overlap of Voronoi polygons. However, they require the computation of Voronoi polygon tessellations with and without the interpolation point. Further, computing these coordinates from the Voronoi polygons is a complex procedure because of the variable shape, and number of edges, for each overlapping Voronoi portion.

A more systematic approach generates the coordinates as a sum of triangular areas obtained from each natural neighbor circumcircle. For example, a subset of data known to be natural neighbors of X, such as points A, I, J, L, and M, in Figure 2.3.6, has the point X inside each of the circumcircles formed by the triplets of data {A,I,J}, {A,I,M}, and {A,L,M}.

Taking each of these data triplets in turn, the Voronoi vertex or circumcenter of each pair of data and X, is computed. For the triangle whose circumcenter is AIJ, these circumcenters are XAI, XAJ, and XIJ, as shown in Figure 2.3.6.

Computed areas of triangles are either positive or negative, depending on the clockwise or counterclockwise order of the vertex labels (p. 29). The signed areas of the triangles formed by each pair of these circumcenters with the circumcenter of the original triplet is added to a subtotal accumulated for each of the natural neighbors. The result is the area of overlap for each datum. The procedural steps are

(1) For each data triplet whose circumcircle contains X, compute the circumcenter (p. 34) of X and each pair of the three data; these are Voronoi vertices. Then compute the signed area of the triangle (p. 29) formed by the data triplet circumcenter and each pair of the three Voronoi vertices from (i), taken in cyclic order, and add to the subtotal for the respective datum. Although some components of each subtotal may be negative, the final sums will always be positive.

(2) For each neighbor of X, divide the accumulated signed areas for that neighbor by the total of accumulated signed areas. This is the neighborhood coordinate of X with respect to that neighbor. Natural neighbor coordinates are calculated in the programs at line **2170** for gradient estimation, and at line 3880 for interpolation.

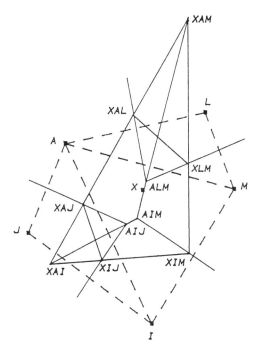

Figure 2.3.6 Computing natural neighbor coordinates

SECTION 2.4

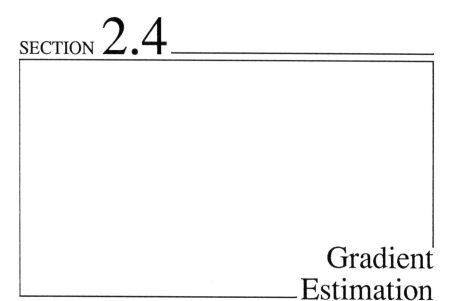

Gradient Estimation

SECTION SUMMARY

This section describes methods of gradient estimation, using least squares planes, quadratic surfaces, minimum curvature splines, vector cross products, and linear neighborhood-based interpolation. Gradient estimates are developed from subsets selected by fixed distance, fixed number, or natural neighbors.

INTRODUCTION

Topographical data are considered to be descriptions, at discrete points, of a spatial function that exhibits gradual variation. Such spatial variation has a natural representation as the shape of a surface, and the difference in descriptions of two adjacent data is an estimate of that variation. If the data are so dense that their differences provide satisfactory descriptions of the local variation, then the slope of the surface between two such data may be described by a straightline.

Unwin (1974) defines the gradient of a surface as the maximum slope, and so a gradient implies a direction of steepest slope. A gradient estimate at a data point is an estimate of the maximum slope of a plane that is tangent to the surface at that datum. Because the surface usually is not known, other than by the data, an estimated gradient at a given datum necessarily is an estimate of the slope of some plausible surface that suits nearby data points. Unwin (1981) and Shafer, Kanade, and Kender (1983) discuss gradients and their relationship to surfaces.

Gradient Estimation

Without gradient estimates, a surface may fit the data exactly but be overly contorted between data. However, a surface that conforms to both the topographical data and estimated gradients, incorporates local trends as expressed by the gradients, and is thereby restrained in its behavior between data.

Wolfe (1982) has remarked on the improvements to accuracy and robustness when gradients are used. Because the appropriateness of the trends made manifest by a representative surface, and the volume under that surface, are influenced by the gradients, the choice of gradient estimation methods is an important aspect of computer contouring.

SURFACE TYPES

First, to appreciate the comparative effects of different gradient estimates on the interpolation, consider the broad range of possible surface types that might be generated to suit a data set, relative to three extreme examples. These example surfaces usually are unacceptable for real data, but are useful as reference shapes to categorize generated surfaces.

(1) The surface may be taut between data points, like a rubber sheet, with sharp peaks or pits at the data points; this is the most conservative interpretation of the data when measurement error is known to be small. It makes no use of gradient information. Although the generated surface is smooth between data points, sharp cusps are caused by the change in slope at each datum. This surface represents the situation where almost all the change in slope required by the data is accommodated at the data points.

(2) When the observed surface is known or expected to have a gradient at each datum, the representative surface may be shaped to conform to the gradient in the region surrounding each datum. Of course, this requires large changes in slope to occur along the intermediate region between the gradients of any two adjacent data. This surface represents the situation where almost all the change in slope is accommodated between the data points.

(3) A third surface type represents the situation where the change in slope is distributed as evenly as possible over the whole region. Such minimum curvature surfaces, although apparently never occurring in nature, produce the smoothest and most attractive displays. For spatial functions that are not restricted to minimum curvature, however, this third type of representative surface is less conservative than those that may incorporate strong curvatures where implied by the data.

By controlling the range of influence of the gradient estimates, a representative surface can be made to vary from extremely taut, through smoothly curved, to rounded, and even puffy. This may be done for any set of topographical data, no matter how numerically erratic or noisy it may be. The surface, at any particular location, is a combination formed from nearby data and estimated gradients using weights determined by a blending function (see Section 2.6, p. 163).

Surface Types

An additional use for gradient estimates is in calculating the volume under the gradient plane, bounded by the Voronoi polygon about the datum (for Voronoi polygons, see p. 59). When this is done for all the Voronoi polygons in the data set, the sum of the volumes is a better estimate than that given by flat-top or constant value proximal polygon (p. 107) volume estimates, because local trend information is implied by the gradient estimates.

GRADIENT SUBSETS

Gradients are estimated at individual data points by using a subset of data that are selected because they are, in some sense, nearby. The ways of selecting such subsets are covered in Section 2.2, p. 69. A selected type of surface is obtained that suits this subset. Its gradient, at the datum, is used as the estimated gradient. This is the orientation of a surface patch that is considered to be tangent to the spatial function, and so expresses the local variation around the datum.

TANGENT PLANES

To illustrate a plane that is tangent to a spatial function, the set of six data points in Figure 2.4.1 could be the locations around the brow of a hill where measurements of elevation have been made. A smooth continuous surface through these six elevations has a tangent plane at the data point **K**, for example. Of course, this plane would be easy to determine if the surface were known, but an estimate of the tangent plane is needed before the surface can be generated.

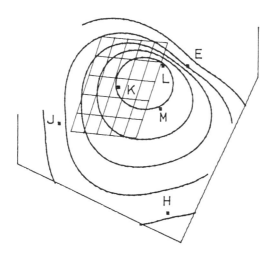

Figure 2.4.1 Tangent to a surface

Gradient Estimation

Gradients are described conveniently by a vector that is perpendicular to the tangent plane. For bivariate data, this requires three numbers, and it is the relative proportions among these three, rather than their magnitudes, that determine the direction of the vector and fix the orientation of a perpendicular plane.

Gradient estimates are inferred from the surrounding data, and so their reliability depends upon the local roughness in the data. Hobson (1972) has outlined some definitions for a general concept of the roughness of a surface and described three quantitative approaches to assessment of roughness: "(1) comparison of estimated actual surface area with the corresponding planar area; (2) estimate of 'bump' or elevation frequency distribution; and (3) comparison of the distribution and orientation of approximated planar surfaces within sampling domains."

The set of estimated tangent planes, identified by the methods discussed here, describe the general behavior of the spatial function for any subset of data and so are within the scope of Hobson's third approach. Examples of the other two approaches also are given (see p. 95 and p. 98).

Least Squares Gradients

This method requires that the subset of data to be used for gradient estimation at a datum be fitted with a plane whose orientation satisfies the "least sum of squared differences" criterion. This implies that the squared differences between the height at each datum and the height of the fitted plane at that datum's location, for all data in the subset, are the controlling influences on the orientation of the plane. The sum of the squares of these differences is made as small as possible for a gradient plane chosen by this criterion.

Least squares gradients are computed by linear regression. Data component sums and products are accumulated in a matrix that is set up as a system of simultaneous equations and solved by Cramer's rule (p. 88). In the example below the summation sign, sigma, indicates the sum of k numbers, each subscripted by an index i.

To estimate a gradient for a subset of data, (x_i, y_i, z_i), $i=1 - k$, the 3x3 matrix on the left and the column on the right show the sums, and sums of products, that must be accumulated. Matrix multiplication of this 3x3 matrix with the intermediate column vector (a_1, a_2, a_3), results in the column vector on the right.

$$\begin{pmatrix} k & \Sigma x_i & \Sigma y_i \\ \Sigma x_i & \Sigma x_i^2 & \Sigma x_i y_i \\ \Sigma y_i & \Sigma x_i y_i & \Sigma y_i^2 \end{pmatrix} \begin{pmatrix} a_1 \\ a_2 \\ a_3 \end{pmatrix} = \begin{pmatrix} \Sigma z_i \\ \Sigma x_i z_i \\ \Sigma y_i z_i \end{pmatrix}$$

Then the coefficients of the gradient plane are the components of the solution (a_1, a_2, a_3). These may be computed from the accumulated sums by the following expressions.

$$a_1 = (\sum z_i * (\sum x_i^2 * \sum y_i^2 - \sum x_i y_i * \sum x_i y_i)$$
$$- \sum x_i * (\sum x_i z_i * \sum y_i^2 - \sum y_i z_i * \sum x_i y_i)$$
$$+ \sum y_i * (\sum x_i z_i * \sum x_i y_i - \sum y_i z_i * \sum x_i^2)) / d$$

$$a_2 = (\ k\ * (\sum x_i z_i * \sum y_i^2 - \sum y_i z_i * \sum x_i y_i)$$
$$- \sum z_i * (\sum x_i * \sum y_i^2 - \sum y_i * \sum x_i y_i)$$
$$+ \sum y_i * (\sum x_i * \sum y_i z_i - \sum y_i * \sum x_i z_i)) / d$$

$$a_3 = (\ k\ * (\sum x_i^2 * \sum y_i z_i - \sum x_i y_i * \sum x_i z_i)$$
$$- \sum x_i * (\sum x_i * \sum y_i z_i - \sum y_i * \sum x_i z_i)$$
$$+ \sum z_i * (\sum x_i * \sum x_i y_i - \sum y_i * \sum x_i^2)) / d$$

where

$$d = \ k\ * (\sum x_i^2 * \sum y_i^2 - \sum x_i y_i * \sum x_i y_i)$$
$$- \sum x_i * (\sum x_i * \sum y_i^2 - \sum y_i * \sum x_i y_i)$$
$$+ \sum y_i * (\sum x_i * \sum x_i y_i - \sum y_i * \sum x_i^2)$$

These expressions for the coefficients of the linear regression matrix use a ratio of determinants. They are tedious to calculate by hand but very convenient when computer code has been written. The gradient estimate is the plane $z = a_1 + a_2 x + a_3 y$, which, as far as gradients are concerned, is described conveniently by its pole $(-a_2, -a_3, 1)$, a vector that is perpendicular to the gradient plane.

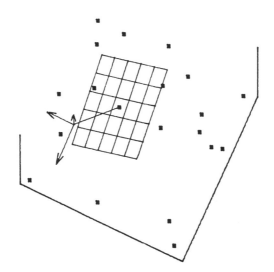

Figure 2.4.2 Least squares plane gradients

Least squares estimates of gradient planes are an option in the programs (except CON2R.BAS) for comparison with other gradient estimation methods. Set the parameters **CL**=1 and **SL**=2, on line **1100**, to turn on this estimator.

Gradient Estimation

This gradient estimation method may be applied to any subset of data obtained by the subset selection methods of Section 2.2, p. 69. However, it works best for natural neighbor subsets (p. 72), because data that are more distant than the natural neighbors of a datum are given excessive influence on the orientation of the gradient plane by the squared difference mechanism.

A related approach was described by Mansfield (1980) using an inverse distance weighted least squares estimate of partial derivatives from directional divided differences. For each neighbor of the datum (x_0, y_0, z_0), in the gradient estimation subset (x_i, y_i, z_i), $i=1 - k$, where d_i is the distance from (x_i, y_i) to (x_0, y_0) and sigma indicates summation over the k data, accumulate the matrix

$$\begin{matrix} \sum (x_i-x_0)*(x_i-x_0)/d_i^2 & \sum (x_i-x_0)*(y_i-y_0)/d_i^2 \\ \sum (x_i-x_0)*(y_i-y_0)/d_i^2 & \sum (y_i-y_0)*(y_i-y_0)/d_i^2 \end{matrix} \begin{matrix} a_1 \\ a_2 \end{matrix} = \begin{matrix} \sum (x_i-x_0)*(z_i-z_0)/d_i^2 \\ \sum (y_i-y_0)*(z_i-z_0)/d_i^2 \end{matrix}$$

The solution to this system is the column vector (a_1, a_2) and, by Cramer's rule (p. 88),

$$a_1 = ((\sum (x_i-x_0)*(z_i-z_0)/d_i^2 \cdot \sum (y_i-y_0)*(y_i-y_0)/d_i^2) - (\sum (y_i-y_0)*(z_i-z_0)/d_i^2 \cdot \sum (x_i-x_0)*(y_i-y_0)/d_i^2)) / d$$

$$a_2 = ((\sum (x_i-x_0)*(x_i-x_0)/d_i^2 \cdot \sum (y_i-y_0)*(z_i-z_0)/d_i^2) - (\sum (x_i-x_0)*(y_i-y_0)/d_i^2 \cdot \sum (x_i-x_0)*(z_i-z_0)/d_i^2)) / d$$

where

$$d = (\sum (x_i-x_0)*(x_i-x_0)/d_i^2 \cdot \sum (y_i-y_0)*(y_i-y_0)/d_i^2) - (\sum (x_i-x_0)*(y_i-y_0)/d_i^2 \cdot \sum (x_i-x_0)*(y_i-y_0)/d_i^2)$$

Notice the distinction between d_i which is the distance from (x_i, y_i) to (x_0, y_0) and d which is the determinant in the denominator. Again, this gradient estimate is described conveniently by its pole $(-a_1, -a_2, 1)$, a vector that is perpendicular to the gradient plane. This estimate is fast, and is most reliable for gridded data.

To mention a few other applications of least squares planes, Batcha and Reese (1964) used a least squares fitted plane on a running subset to interpolate at grid nodes. Pelto, Elkins, and Boyd (1968) used a distance weighted least squares plane on a running subset of trend analysis residuals. Davis (1985) combines least squares fitting with knowledge of the expected error. Lowden (1985), estimates partial derivatives by distance weighted least squares quadratics (see next).

Least Squares Quadratics

The "least sum of squared differences" criterion may be used with polynomial surfaces of higher degree in a similar fashion to that of least squares planes, and least squares quadratic surfaces are used often.

This approach fits the gradient estimation subset with a rectangular hyperboloid (see p. 111 and p. 139), which may be expressed as

$$F(x,y) = a_1 + a_2 x + a_3 y + a_4 xy$$

For a gradient estimation subset, $\{x_i, y_i, z_i\}$, $i=1-k$, and where sigma indicates summation over the k data, set up the following matrix as a set of simultaneous equations and solve it for the parameters of the hyperboloid. The matrix is

$$\begin{pmatrix} k & \sum x_i & \sum y_i & \sum x_i y_i \\ \sum x_i & \sum x_i^2 & \sum x_i y_i & \sum x_i^2 y_i \\ \sum y_i & \sum x_i y_i & \sum y_i^2 & \sum x_i y_i^2 \\ \sum x_i y_i & \sum x_i^2 y_i & \sum x_i y_i^2 & \sum x_i^2 y_i^2 \end{pmatrix} \begin{pmatrix} a_1 \\ a_2 \\ a_3 \\ a_4 \end{pmatrix} = \begin{pmatrix} \sum z_i \\ \sum x_i z_i \\ \sum y_i z_i \\ \sum x_i y_i z_i \end{pmatrix}$$

The solution of this system of simultaneous equations is the column vector, (a_1, a_2, a_3, a_4), obtained by Gaussian elimination with pivoting, at line **6690**, in the programs. The gradient estimate is the slope of the tangent plane whose pole, $(-a_2-a_4 y, -a_3-a_4 x, 1)$, is a vector that is perpendicular to the rectangular hyperboloid at the location (x, y).

Similar to least squares gradient planes, this method may be applied to any subset of data obtained by the subset selection methods of Section 2.2, p. 69. Natural neighbor subsets (p. 72) are preferable because other data, more distant than the natural neighbors, exert a degree of influence at the point of tangency that is disproportionate to their spatial relationship.

For comparison with other methods, least squares estimates of gradient quadratics are an option in the programs (except CON2R.BAS). Set the parameters **CL**=1 and **SL**=4, in line **1100**.

Squared difference methods are fast, once the estimation subset has been selected. However, three disadvantages of these methods are relevant to their use in gradient estimation.

(1) The slope of a smooth surface at a datum is similar to the slopes at nearby locations than to those at more distant points. However, the difference between the least squares fit and the data is weighted by squaring, for both nearby and distant data. Therefore, a disproportionate influence on the orientation of the fitted function is exercised by the distant data, which should be less significant.

(2) Suppose the data in a gradient estimation subset lie close to some tilted plane, except for one datum which has a considerable offset from the plane. The square of the large offset will dominate the sum of squared differences, and the outlier will exert an unduly large influence on the orientation of the estimated gradient.

(3) A general problem with nonlinear polynomial surfaces is that unexpected shapes may be developed to accommodate the configuration of data. Although the surface is acceptably close to the data, between data points implausible extremes or extraneous inflection points may be created, and in so doing, influence the slope at a datum in a manner unjustified by the data.

In other words, by fitting higher than first degree polynomials, the least squares method selects a gradient surface that gives a minimum squared difference but confers implausible humps or hollows to the constructed surface. It cannot be considered to be a reliable estimate of gradients for sparse data.

Surkan, Denny, and Batcha (1964) apparently introduced the concept of fitting a least squares plane to a local subset to obtain a representative surface for topographical data. However, least squares gradient planes are examples of bivariate linear regression, and so the origins of this concept are in the last century (see Riggs, Guarneiri, and Addelman 1978). Mclean (1974) extended the concept to polynomials of higher degree. Schut (1976) describes a proposal by Bauhuber, Erlacher, and Gunther, in 1975, to extend this approach to gradient estimation. Lawson (1977) described a six-parameter quadratic polynomial estimate, fitted by distance-weighted least squares. Sibson (1981) used spherical quadratics fitted to each datum and its natural neighbors. Harada and Nakamae (1982) use a weighted average of the tangent planes to spherical quadratics through nearby data.

Spline Gradients

Instead of minimizing the sum of squared differences between the data and a fitted surface, one may fit the data exactly and minimize the curvature of the fitted surface. This approach establishes a surface that fits the gradient subset with a surface that has the least amount of curvature possible.

Computing a spline gradient at a datum, which has $k-1$ other data in its subset, requires using a matrix that is $k+3$ square. For each datum in turn, term it $\mathbf{P}_1$ at location (x_1, y_1), and all its natural neighbors, $\{\mathbf{P}_2, \mathbf{P}_3, \ldots, \mathbf{P}_k\}$, set up the matrix of simultaneous equations shown below. This matrix is the same as that given for minimum curvature interpolation on p. 122. In the programs, this system of simultaneous equations is solved by Gaussian elimination with pivoting, at line **6690**.

Each $C(\mathbf{P}_i-\mathbf{P}_j)$ is a function of the distance on the x, y plane between $\mathbf{P}_i$ and $\mathbf{P}_j$. This basis function is defined on line **1060** in the programs. Two functions are provided, $d^2 \log d$ and $d^2 (\log d - 1)$; they may be set or canceled by REMarks. There are several possibilities for this function, discussed more fully on p. 122.

$$\begin{pmatrix} 1 & P_{1x} & P_{1y} & 0 & C(P_2-P_1) & \cdots & \cdot & C(P_k-P_1) \\ 1 & P_{2x} & P_{2y} & C(P_1-P_2) & 0 & \cdots & \cdot & C(P_k-P_2) \\ 1 & P_{3x} & P_{3y} & C(P_1-P_3) & C(P_2-P_3) & \cdots & \cdot & C(P_k-P_3) \\ 1 & P_{4x} & P_{4y} & C(P_1-P_4) & C(P_2-P_4) & \cdots & \cdot & C(P_k-P_4) \\ 1 & P_{5x} & P_{5y} & C(P_1-P_5) & C(P_2-P_5) & \cdots & \cdot & C(P_k-P_5) \\ \cdot & \cdot & \cdot & \cdot & \cdot & & & \cdot \\ \cdot & \cdot & \cdot & \cdot & \cdot & & & \cdot \\ 1 & P_{kx} & P_{ky} & C(P_1-P_k) & C(P_2-P_k) & \cdots & \cdot & 0 \\ 0 & 0 & 0 & 1 & 1 & \cdots & \cdot & 1 \\ 0 & 0 & 0 & P_{1x} & P_{2x} & \cdots & \cdot & P_{kx} \\ 0 & 0 & 0 & P_{1y} & P_{2y} & \cdots & \cdot & P_{ky} \end{pmatrix} \begin{pmatrix} b_1 \\ b_2 \\ b_3 \\ a_1 \\ a_2 \\ \cdot \\ \cdot \\ a_{k-3} \\ a_{k-2} \\ a_{k-1} \\ a_k \end{pmatrix} = \begin{pmatrix} P_{1z} \\ P_{2z} \\ P_{3z} \\ P_{4z} \\ P_{5z} \\ \cdot \\ \cdot \\ P_{kz} \\ 0 \\ 0 \\ 0 \end{pmatrix}$$

Tangent Planes

As can be seen in the matrix, the sum of the a_i is equal to zero, as is the sum of the products of the a_i and P_{ix}, or the a_i and P_{iy}. The unknown values a_i and b_j are obtained by solving the system of linear equations and these become the coefficients of the surface for an arbitrary location. Then the minimum curvature surface, say $F(x,y)$, for k data is

$$F(x,y) = b_1 + b_2 x + b_3 y + \sum_{i=1}^{k} a_i C(d_i)$$

where

$$d_i = ((P_{ix} - x)^2 + (P_{iy} - y)^2)^{1/2}$$

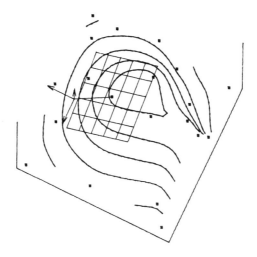

Figure 2.4.3 Minimum curvature spline gradients

Minimum curvature splines have considerable intuitive appeal because they avoid the extreme curvatures that occur with other methods. Their use as gradient estimates, though not widely practised, would seem to be appropriate since a minimum curvature surface distributes the slope changes as widely and evenly as possible.

Figure 2.4.3 attempts to show such a surface but convincing proof of minimum curvature can only be seen by considering mathematical expressions of the surface (Meinguet, 1979); these show that the integral of the second derivative is minimized.

The gradient of a minimum curvature surface is extracted by applying the solution parameters to the datum location and two nearby points. If the datum has locational coordinates (x, y), then the two other locations are taken conveniently as $(x+.0001, y)$ and $(x, y+.0001)$. The slope of the triangular planar patch between these three points is taken as an approximation of the gradient at (x, y).

Gradient Estimation

The gradient estimate is the slope of the tangent plane whose pole, $(g_1, g_2, 1)$, is a vector that is approximately perpendicular to the minimum curvature surface at the location (x, y). The values for g_1 and g_2 are

$$g_1 = -(F(x+.0001, y)-F(x, y))/.0001$$
$$g_2 = -(F(x, y+.0001)-F(x, y))/.0001$$

Minimum curvature gradient estimates are an option in the programs to allow comparison with other methods. Set the parameters **CL**=1 and **SL**=3, on line 1100, to turn on this gradient estimator.

In the programs, minimum curvature gradient estimates are obtained from natural neighbor subsets. However, this method also may be applied to any subset of data obtained by fixed distance, fixed area or fixed number (p. 69). Unlike least squares gradient estimation methods, spline gradients are not affected strongly by extra data in the gradient subset. The surface at a given datum is determined locally, and individual distant data have only moderate effect.

The most appropriate subset would be all the natural neighbors of a datum and all their natural neighbors (p. 72). Nevertheless, minimum curvature surfaces, as estimates of gradient, may develop a surface that gives the minimal curvature necessary to fit the data but allow an implausible hump, or hollow, between the data. This may distort the slope at a datum unless large subsets are used. However, larger subsets use more execution time because, again unlike least squares methods, a large matrix must be manipulated. For these reasons minimum curvature cannot be considered to be a reliable estimate of gradient for sparse data.

Barnhill (1983) describes a multiquadric (p. 121) gradient estimator that behaves in a similar manner to the minimum curvature estimate used in the programs, and Stead (1984) also discusses this estimate.

Cross Product Gradients

When the gradient estimation subset contains only the datum and its natural neighbors, the datum can be thought of as surrounded by triangles, each formed by the datum and a pair of its natural neighbors. These triangles have various orientations in three-dimensional space, depending upon the relative positions and heights of each pair of neighbors. Any form of average of these triangles can provide an estimate of gradient at the datum when all the natural neighbors are known.

To develop an average that reflects both the size and orientation of the triangles in a triangulation, each triangle, including the measured heights, is treated as vectorial data in three-dimensional space. Then the cross product (see p. 33) of any pair of sides from each triangle is computed. This three-dimensional vector is perpendicular to the triangle, and its length is twice the area of the triangle.

For example, each triangle $\{P_1, P_2, P_3\}$, with Cartesian coordinates of the vertices are $P_1=(x_1, y_1, z_1)$, $P_2=(x_2, y_2, z_2)$, and $P_3=(x_3, y_3, z_3)$, the cross product vector, (x_p, y_p, z_p), is given by

$$x_p = ((y_2 - y_1)(z_3 - z_1) - (y_3 - y_1)(z_2 - z_1))$$
$$y_p = ((z_2 - z_1)(x_3 - x_1) - (z_3 - z_1)(x_2 - x_1))$$
$$z_p = ((x_2 - x_1)(y_3 - y_1) - (x_3 - x_1)(y_2 - y_1))$$

This vector will have a negative z-value if the triangle vertices are ordered clockwise. Because this is the height component, the cross product must be converted into the opposite, positive sense. If z_p is less than zero, then multiply all three components, x_p, y_p and z_p, by negative one.

To obtain the gradient estimate at a datum, the vector sum of the cross products (each taken in the positive direction with respect to the height), is accumulated for all the triangles that involve that datum. This is the pole

$$(\sum x_{pi}, \sum y_{pi}, \sum z_{pi})$$

with the summation over all triangles in the natural neighbor gradient subset. The estimated gradient plane is perpendicular to this vector sum, which can be thought of as a spatial average of slopes and sizes of the surrounding triangles.

Because the area of a triangle is one-half the length of its cross product, the length of the vector sum of cross products over the sum of the lengths of cross products is a ratio of areas in the sense of Hobson's (1972) first roughness parameter (see p. 88). This is a comparison based on the ratio of the sum of the area of the triangles as projected onto the tangent plane over the sum of the areas of the triangles.

So this procedure also allows an estimate of data sufficiency in the sense that it measures the roughness of the data set from the local perspective of each datum. This roughness index (p. 171) then is used in the construction of the surface to influence the peakedness or roundedness in the neighborhood of each datum, as described in Section 2.6, p. 163.

The disadvantage of cross product gradients is their dependence on the triangulation. They are ambiguous because another triangulation may be equally justifiable but will give a different gradient. This effect is illustrated in Figure 2.4.4 and Figure 2.4.5 on the next page. The data points are the same but a different triangulation shows a small but possibly significant difference in the cross product gradients.

Cross product gradients are included in the programs. To turn on these gradient estimates, set the parameters **CL**=1 and **SL**=1, on line **1100**. The set of triangles surrounding a datum is obtained by the code starting on line **1580**; all data that do not share a triangle with a given datum are ignored and so a small set of triangles (typically six) is determined quickly. Cross product gradients are produced easily once the natural neighbor subset has been extracted, and this code starts at line **2480**.

Gradient Estimation

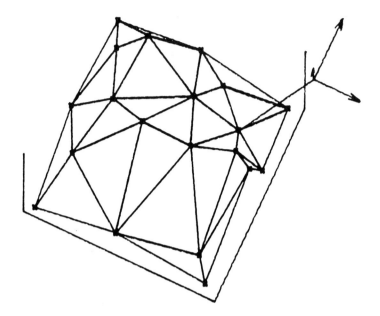

Figure 2.4.4 Cross product gradients #1

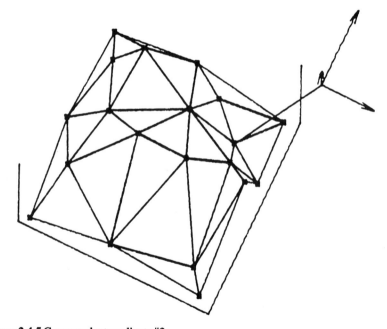

Figure 2.4.5 Cross product gradients #2

Tangent Planes

Just to mention a few of the efforts in the development of cross product, and similar, gradient estimates, Klucewicz (1978) calculates the average of the partial derivatives obtained from each of the triangles having the datum for a vertex. Franke (1982b) reviews several gradient estimates, and recommended an estimate of partial derivatives by a weighted average of the slopes of planes through neighboring points in the triangulation. This gives the same result as a cross product gradient if all the triangles are formed by natural (Voronoi, see p. 59) neighbors.

Nielson (1983) uses an iterative solution to establish an average of the partial derivatives obtained from each of the triangles having the datum for a vertex. Akima (1978, 1984) estimates cross products from n nearest neighbors, but, again, this may not include all the natural neighbors for scattered data.

Turner and Miles (1967) used a parametric spherical mean direction for unit vectors that are normal to triplets of gridded data. Philip and Watson (1982b) computed a parametric spherical mean direction for the cross product vectors of a natural neighbor gradient subset, and used it as a pole for the tangent plane. This approach is not justified because the small number of cross products (typically six) is an insufficient statistical sample.

Stead (1984) discusses a gradient estimate obtained by using the surface generated by inverse distance weighting (p. 112) of nearby data but without the datum whose gradient is being estimated. Barnhill (1983a), Little (1983) and Stead (1984) refer to the "triangular Shepard's method" which is an inverse distance weighted average of the slopes of the triangular facets that surround each datum.

Neighborhood-Based Gradients

To compute a neighborhood-based gradient, linear neighborhood-based interpolation (p. 155) is performed at the datum and two nearby locations, using natural neighbor local coordinates (p. 81) as weights on the gradient subset. Because the datum itself is not used, this approach provides a surface whose slope depends only on the concomitant spatial relationships of its natural neighbors.

This surface is a linearly weighted average of heights of the natural neighbors and, in this sense, gives the average height at the datum whose gradient is being estimated. The slope of this surface, at the location of the given datum, is estimated by computing this surface for two other points close to the datum, and so partial derivatives are estimated.

The approach is similar to that of minimum curvature spline gradients (p. 92). If the datum has locational coordinates (x, y), then two other locations are conveniently taken as $(x+.0001, y)$ and $(x, y+.0001)$, as in line **2180**. Then the gradient estimate is the slope of the tangent plane. The pole, $(g_1, g_2, 1)$, of this plane is a vector that is perpendicular to the linear neighborhood surface at the location (x, y).

Gradient Estimation

The values for g_1 and g_2 are

$$g_1 = -(\mathbf{F}(x+.0001, y) - \mathbf{F}(x, y))/.0001$$
$$g_2 = -(\mathbf{F}(x, y+.0001) - \mathbf{F}(x, y))/.0001$$

where $\mathbf{F}(x, y)$ is the linear neighborhood-based interpolation at the point (x, y).

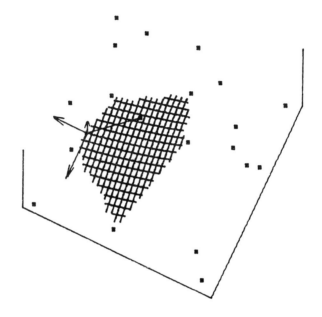

Figure 2.4.6 Natural neighbor gradients

Natural neighbor gradients are restricted to use with natural neighbor subsets. However, unlike cross product gradients, natural neighbor gradients are unique because they depend only on the number and configuration of natural neighbors and not on how these neighbors are triangulated. If a datum is not bounded fully by natural neighbors, because it is on the perimeter of the data set, a natural neighbor gradient cannot be computed, and a cross product gradient is used in the programs.

The difference between the height interpolated linearly without the datum, and the known height at the datum indicates the extent to which each datum is an outlier with respect to its natural neighbors. In this sense it is a signed estimate of the 'bump' roughness at the datum in the sense of Hobson's (1972) second example (p. 88).

Natural neighbor gradients are implemented in the programs, and are recommended because these gradients seem to be the most stable and reliable estimates available. Set the parameters **CL**=1 and **SL**=0, on line **1100**, for natural neighbor gradient estimates.

Natural neighbor gradient estimates require a triangulation of the neighbors of a datum *in the absence of that datum*. Then the circumcircles of these triangles are used to determine the natural neighbor local coordinates (p. 81) of the datum whose gradient is being estimated. The set of triangles surrounding a datum is obtained by the code starting on line **1580**; however, if a datum is on the perimeter of the data set, it will not have a natural neighbor gradient, and a cross product gradient must be determined. This requires a triangulation of the natural neighbors *and the datum*.

It is advantageous to generate both triangulations in pseudo-parallel, although only one will be used. Any data not involved in the triangulation that includes the datum (typically six data) may be ignored for the other triangulation; this keeps both triangulations small and, therefore, generated quickly. Cross product gradients, at line **2480**, are generated if the datum does not have a complete set of natural neighbors.

Some other published proposals for gradient estimates are mentioned next. Harada and Nakamae (1982b) illustrate a gradient estimate, for a planar curve, as an average of tangents to circles. Liszka (1984) suggests approximation of the surface by Taylor series, where the required derivatives are determined by least squares. Wolfe (1982) describes a technique for testing the accuracy of a gradient estimation scheme.

SECTION 2.5 Interpolation

SECTION SUMMARY

Most published interpolation methods are described in this section. These surface generation techniques are discussed in terms of their construction, the plausibility of the resulting surface, and its local fidelity to the data. Interpolation methods involve one of two contrasting general approaches, fitted functions or weighted averages, and are applied to two distinct interpolation problems, exploratory surfaces or created surfaces. There are three broad subsections. First, the concept of optimal interpolation is considered relative to difficult aspects that commonly require compromise. Second, manual methods are discussed to show the derivation of the automatic methods, which are cataloged in the third part.

INTRODUCTION

Interpolation, using a computer, is the performance of a numerical procedure that generates an estimate of functional dependence at a particular location, based upon knowledge of the functional dependence at some surrounding locations. It is only an informed estimate of the unknown.

There are many interpolation methods, and the efficacy of each method is related to its interpretation of the variability of the functional dependence in the data. For some methods, the interpolated values, although mutually consistent as a set, may have a tenuous, and even contentious, relationship to the data.

Interpolation

Notionally, when interpolation is performed at many closely spaced locations, a surface may be formed by filling the interstices between interpolation points with simple planar patches. This fabricated surface, representing the spatial variation in the data, is displayed and measured. This implies that the correct selection of interpolation technique is crucial to successful analysis of data.

Interpolation is an operation on multivariate data, so it has a natural spatial domain, the convex hull of the data set. The convex hull is the region enclosed by a connected set of straightline segments around the perimeter; none of these lines can be extended between any of the data because each datum lies either on one of these bounding line segments, or to only one side of each such line. The result is that the boundary of the convex hull has no indentations.

Some interpolation procedures may be applied to locations outside the convex hull of the data set, and the procedure is then termed extrapolation. However, the absence of data on some sides of the estimation location causes extrapolation to be a less reliable estimate; it is more of a guess than a deduction, but would have some use as an approximation at short distances from the convex hull.

Some representative surfaces for topographical data are so smooth that they differ distinctly from most, or all, of the data; these surfaces are thought of as approximations to the data set. For example, if the data are known to include a component of random or experimental error, such a surface may be judged to be a good representation of the spatial function, although contrasting with a representative surface that coincides with all the data; it is the spatial function that is being interpolated, rather than the data which only are approximated.

Of course, strictly exact interpolation of the data usually is not possible for numerical reasons. This suggests that the conceptual distinction between interpolation and approximation in computer-generated representative surfaces is only a matter of degree of approximation. From the viewpoint of contouring, both interpolation and approximation schemes develop surfaces that may be contoured, integrated numerically, and displayed by the same, or similar, computer procedures.

A critical understanding of automatic interpolation requires an appreciation of two major thrusts in the development and application of interpolation schemes because there is significant overlap in the theory behind both approaches.

An analyst, on the one hand, collects multicomponent data and constructs a representative surface by interpolation. On the other hand, a designer specifies the data and then creates a representative surface by interpolation. Put another way, in the first situation, the interpolation scheme generates an exploratory representative surface for a set of measurements; this surface models the data. In the second situation, the interpolation scheme generates a created representative surface for a set of specified values; this surface exemplifies the data. In both situations there is a fixed order for the components of a datum because the data, as a set, are used to infer a continuous functional relationship over the region. Therefore, interpolation schemes for either approach are similar because they must infer a functional shape from both the values and order of the components.

In contrast, the interpolation of choropleth data requires a considerably different approach than that required for the interpolation of punctulate data. Choropleth (place-quantity) data (p. 50) is obtained by summarizing samples from a spatial function over subregions, and is expressed as so-much-per-subregion. For such data the underlying spatial function may be discontinuous, or even discrete at a fine scale. However, only the large scale variation in the original spatial function can be recovered, with any interpolation method, as a smooth representative surface.

IDEAL INTERPOLATION

Some authors express the optimality of an interpolation method in terms of the spatial function from which the data were drawn. This may be done by specifying (i) a least maximum difference, (ii) a least sum of squared differences, or (iii) agreement up to the k th derivative at each datum.

Such criteria are relevant when the data source is a known analytical function. For example, a particular interpolation scheme may be said to have quintic precision, meaning that the scheme will reproduce any quintic, or lower order, polynomial exactly by using data from such a function.

Because a spatial function which is a source of topographical data usually is nonanalytical, and seldom well-known, the optimality of the generated surface usually must be inferred from intuitive geometrical considerations, and from experience with similar surfaces. However, the essential aspects of ideal interpolation can be summed up by the following generalizations.

(1) The interpolated surface fits the data to a nominated level of precision, that is, it agrees with the individual data points to within arbitrary, user prescribed, limits.

(2) The interpolated surface is single-valued, continuous, and smooth at all places, that is, it has a continuous and finite slope everywhere that interpolation is required.

(3) Each interpolated value depends only on a local subset of data, and the members of this subset are determined solely by the configuration of the data that are, in some sense, near the interpolation point.

Such a restriction is necessary to prevent the masking or submerging of low amplitude surface forms by more dominating features, and prevent widespread propagation of an inadvertent error in some datum. This insures that the generated surface is stable in the sense that a small change in any datum cannot cause a large change in the surface.

(4) The tautness of the surface between data points is adjustable relative to data configuration. For most interpolation methods, some control of tautness is necessary to prevent the surface oscillating between data points. Also, it is characteristic of some surfaces to be more rolling and rounded than others. Tautness control gives the flexibility to suit various surface types.

(5) The interpolation method can be applied to all configurations and density patterns of data, whether gridded, scattered, or traverse. It is necessary that the method be general with respect to sampling patterns before there can be consistency of interpretation, for example, between two data sets of the same region but with different sampling patterns.

DIFFICULT ASPECTS

All the major difficulties with computer interpolation are caused by insufficient data and observational error. The distance between the observations must be less than the width of the features that the map is expected to display. With sufficient precise data, any interpolation procedure will give good results because the sampled surface is known so well. On the other hand, domes, basins, ridges, saddle points, and so on, may be specified so poorly by sparse data that most, if not all, interpolation methods are unable to infer their presence. Even then, such inferences can be obtained only by using local estimates of trends; these indicate the short range behavior of the surface between adjacent data. Sparse data requires methods that incorporate the influence of local slopes.

GENERAL APPROACHES

Computer interpolation of topographical data can be obtained by two methodologically distinct approaches, conventionally referred to as fitted functions and weighted averages. Both streams of development, however, arise from the same philosophical spring which gave rise to the manual methods.

Interpolation begins with the notion that a topographical measurement, a datum, is a unit of information that describes a particular location and, with somewhat less certainty, a limited range of the surrounding region. This proximal region has been referred to as a kernel of influence. The surface representing a set of topographical data is a collective expression of these kernels.

Computer interpolation methods are techniques to determine the sum of particular forms of these influences, at arbitrary locations. The distinction between one method and another refers to the manner in which the influence of a datum is assumed to decline for more distant interpolation points, and the computational maneuvers necessary to process the elected influence function.

These methods can be classified broadly as those that, on the one hand, first determine the parameters of an analytic bivariate function that represents a regional, or local, aggregate of data influences. Then, using these parameters, the function is evaluated at a given location to obtain the height of the representative surface; these are *fitted function* methods. The alternative approach, to obtain a representative surface height at a given location, is by directly summing the data influences that are within range; these are *weighted average* methods.

Crain (1970) refered to the resulting surface types as mathematical surfaces and numerical surfaces, Alfeld and Barnhill (1984) termed them patch methods and point methods, and Cuyt (1987) termed them coefficient problems and value problems. They also may be thought of as analytical or discrete expressions.

FITTED FUNCTION INTERPOLATION

This class of interpolation techniques utilize surfaces that can be described by a set of coefficients in a polynomial function of geographical coordinates. These surface parameters usually are extracted by solving a system of linear equations expressing the combined influences of the data and the criteria controlling the fit of the polynomial function. Once the parameters, which in a sense are a summary of the data, have been determined, they may be applied easily to a series of locations for explicit evaluations of the surface.

There is an advantage in fitting a function as an initial step because it can impose a prescribed general behavior on the surface to override aberrant, anomalous, or noisy, data. Therefore, fitting a function is a smoothing approach, and some degree of local detail may be submerged in the overall surface. The complexity of the representative surface depends on the number of parameters, which is equal to the number of linear equations to be solved.

WEIGHTED AVERAGE INTERPOLATION

The alternative approach uses a direct summation of data influences at each interpolation point, without using an intermediate parametric surface. The weight applied to each datum is the evaluated influence, relative to the interpolation point, for that datum. A set of weights, or influence assessments, must be computed for each interpolation point.

A principal advantage in computing a weighted average surface is that local detail, implied by the small-scale trends in the data, can be developed to a degree of surface complexity not possible for a parametric surface with a reasonable number of parameters. This approach also is applicable to data sets of unlimited size because each computation involves only moderately sized subsets of data. Most importantly, a representative surface is produced that is dominated by local trends in the data.

Conceptually, the difference between the two approaches is that weighted average methods emphasize local detail whereas fitted functions methods summarize global behavior. Computationally, weighted average representative surfaces require considerably more time per interpolation point than do fitted function surfaces. The other side of the story is that fitted functions tend to overshoot in situations where a tighter, sharper curve established by weighted averages would give a more conservative surface.

Interpolation

MANUAL METHODS

Until the arrival of computers, interpolation for contouring was limited to methods that could be implemented by hand. Aside from the apparently spontaneous drawing of contours by an experienced draftsman who has developed an unspecified mental technique through practice, there are several manual methods of interpolating a data set in a manner that can be explained and taught, and that allow an estimate of volume.

Not surprisingly, the origins of computer contouring methods may be perceived easily in the traditional manual techniques for assessing topographical data. The surfaces developed by some of these manual methods were not as plausible as the better computer methods. However, this implausibility was less obvious because, although representative surfaces were used implicitly by these methods, those surfaces were seldom, if ever, portrayed fully, or even considered. Manual methods are discussed under the next seven numbered headings.

(1) Statistical Method.

The simplest of the manual methods, statistical interpolation was a straightforward average of the elevations obtained by weighting each of the N data by $1/N$. This single weight implies a level plane, at the average height, as a representative surface over the region. If the height measurements are $f(x_i, y_i)$, a level plane, **L**, over the region has an altitude

$$L = L(x,y) = \sum_{i=1}^{N} (1/N) f(x_i, y_i)$$

Although this level plane has poor local agreement with most of the data, as a global estimate of regional height it offers an estimate of volume which is near enough to be usable, as Schellmoser (1962) pointed out. Crain (1970) used this plane to initialize a set of grid nodes before solving iteratively for the representative surface.

Of all the interpolation techniques described in this section, this method is the simplest, the least informative about the contours implied by the data, and the most general in applicability. For example, the statistical method can be used on any pattern of data locations, whether gridded, scattered, or traverse, but few if any data will coincide with the level surface.

Such a surface cannot be contoured, of course, because there is no variation in a level plane, and so the spatial variability implied by the data cannot be developed or displayed. The statistical method deserves a place in this book only because it is the conceptual forerunner of all the other methods, both manual and computer, that are listed here.

Considered in terms that are relative to conventional contouring methods, the statistical method is both a fitted function and a weighted average. As a fitted function (a zero degree polynomial) it has only one parameter which summarizes the data. The fitting criterion is a zero sum of differences between the fitted surface and the data. As a weighted average it has only one weight, constant for all interpolation points; the interpolation criterion is constant weight.

If the data are spaced irregularly, such an arithmetical average can be severely misleading. The confusing results of arithmetic averages prompted the 31st President of the U.S.A., mining geologist Herbert Hoover (1909, p. 9), to expound upon the fundamental distinction between an arithmetical and a geometrical mean from the point-of-view of ore deposit estimation.

Aside from the statistical method, all other interpolation methods discussed in this section can be thought of as forms of geometrical averaging of topographical data. Several of these are somewhat analogous to statistical procedures however, and the Isted formula (p. 187), for example, expresses a geometrical average in an apparent statistical form.

(2) Proximal Polygons

In order to develop a surface giving better agreement with individual data points, the region may be subdivided into polygonal blocks, approximately centered around each datum. The subdivision boundaries are determined loosely by the locations of the data, and enclose a portion of the region in the proximity of each datum.

These proximal polygons may be treated as polygonal prisms by assigning the measured elevation for that datum as a level planar upper surface; all the blocks are given a standard basal surface. The inferred representative surface is a series of steps, such as a three-dimensional histogram, and local agreement with the data is, of course, perfect. These polygonal prisms, as a group, form an unclassed isochor map, because the data are not sorted into class intervals (see Gale and Halperin, 1982). A display of such a surface requires a continuous spectrum of color, or hachure, so that variation in the display feature appears to be directly proportional to the altitude of each plane. Simpson (1954) produced such a map of gravity anomalies, with the density of dots around each datum logarithmically proportional to the observational residuals.

When the subdivision boundaries are determined by other considerations, such as thematic units, the value attached to each polygon is a subregional aggregate, rather than a single observation at a location. Such choropleth data, and the associated boundary information, may be displayed as a perspective or isometric view of a three-dimensional histogram, termed a statistical map by some authors. Drury (1987) discusses developing and displaying spatial patterns in data when the region has been mapped subjectively by boundaries which separate different categories of surface.

Such constant value interpolation by proximal polygons comes closer to a prescription for a specific surface shape, able to portray the variability of the data, but, because the top surfaces of the polygonal prisms are flat, isolines follow the edges of the polygons and the surface as a whole is multivalued along these edges. However, the sum of volumes of the polygonal prisms does offer a better estimate of total volume than that given by the statistical method, and it can be seen to reflect the local variation in the data.

(3) Natural Neighbor Proximal Polygons

The major dissatisfaction with proximal polygons was made explicit when an alternate choice of polygon arrangement gave a significant difference in the estimated total volume. A unique set of polygons was required to give consistency to the interpolated estimates. This was made possible when Thiessen (1911) proposed using natural neighbor polygons. These polygons are unique and defined, for each datum, by the natural neighbors of that datum. See the illustration of natural neighbor (Voronoi) polygons on p. 62.

The interpolated surface was formed, as before, by the flat tops of a group of polygonal prisms, but the uniqueness of the polygons meant that the natural neighbor method provided reproducible sets of polygons. Of course, isolines on this surface also follow the edges of the polygons, and so natural neighbor proximal polygons are unable to display contours of local variability.

On the plus side, polygonal areas were now inversely proportional to the density of sampling, so yielding an estimate of volume that was closer to reality, especially for unevenly scattered data. However, the flat tops of proximal polygons tended to overestimate the broad peaks, and underestimate the regional troughs, that the experienced eye could infer from the data. So, although the global result generally was unbiased, locally, natural neighbor proximal polygon interpolation apparently was in error.

Of course, the tops of the proximal polygons do not have to be level. If an estimate of the gradient at each datum is available, a tilted plane can be fitted as a top to each polygonal prism. Such a surface (Watson and Philip, 1984b) is not attractive but slant-top proximal polygons are useful as a conceptual bound for surfaces, and provide an intermediate step in surface generation (p. 116).

(4) Triangulations.

The three manual methods, discussed already, are characterized by the level planar surfaces used to interpolate the data, and consequently, although the volume could be estimated easily, contoured displays of the data variability could not be developed. It also could be seen that the lack of gradual variation led to the local estimation bias exhibited by proximal polygons.

A much older method used triangular polygonal prisms. As Schmid and MacCannell (1955) later suggested, a variable surface should tend toward an average of adjacent subregions along the block boundaries. Clearly a method that allowed for gradual variation across a data set was required before that variability could be shown. By selecting three adjacent data, an interstitial triangular polygon was defined, and by using the heights at each datum, a tilted plane was assigned to the upper surface to form a slant-top triangular prism. Now the data set was subdivided into triangular prisms such that each triangle had a datum at each vertex, and each datum was involved in several triangles (p. 62).

The volume under the representative surface is the sum of the volumes of the triangular prisms. More importantly, this was the first entirely objective method that could allow isolines. These were given by the intersection of the triangular facets with specific horizontal planes (see Figure 2.5.1 on p. 110).

Interpolating the planar surface of a triangle only required applying barycentric coordinates (p. 77) to the data at the vertices of the triangle. This is a weighted average method and the elevation of the interpolated surface, say L(x,y), at some interpolation point (x,y) within the triangle is

$$L(x,y) = \sum_{i=1}^{3} w_i f(x_i, y_i)$$

where the weight w_i is the ith barycentric coordinate of the interpolation point with respect to the triangle, and $f(x_i, y_i)$ is the observed dependent variable at the data point (x_i, y_i). The sum of the barycentric coordinates always is one for any interpolation point. Therefore, each interpolated value lies on a plane that usually is tilted and coincides with the three data which lie at the corners of the triangle. Figure 2.5.1 on the next page shows such a surface built from a patchwork of these triangular facets.

Now a network of triangles allowed the construction of an isoline map of the region because the interpolated surface has a gradual variation. The relative positions of the isolines could be estimated from an assumption of uniform change across each triangle.

Working mechanically from these triangles, contour segments were formed from straightlines across each triangle. Because it is piecewise linear, the surface formed by this method has a discontinuous slope at the edges of each triangular facet. The interpolated surface differs linearly within any triangle and so changes of slope must be incorporated at the triangle edges. This is the most noticeable aspect of linear triangle-based interpolation when it is used for displaying contours, because this method produces contours with an angular appearance.

More subjectively, an experienced and skillful draftsman, working from a triangulation of the data, could thread the isolines sinuously through the network of triangles and produce an isoline map as smooth in appearance as any computer drawn curve, although it was tedious and time-consuming work.

Interpolation

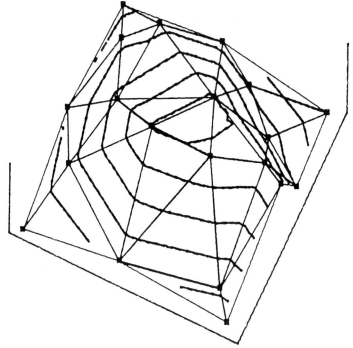

Figure 2.5.1 Triangle-based interpolation

Triangulations are related closely to proximal polygons because each triangle is a dual of a vertex in a tessellation of proximal polygons; this may be done in three distinct ways. If the edges of the triangles are bisected perpendicularly, the bisecting lines will meet at the circumcenter (p. 31) of the triangle. If the interior angles of a triangle are bisected, the bisecting lines will meet at the incenter (p. 32) of the triangle, the center of the incircle. If midpoints of each edge are connected by lines to the opposite vertices, these lines intersect at the centroid (see Kurtze, 1986). Lines connecting the circumcenters, incenters, or centroids, of contiguous triangles are the edges of proximal polygons.

Similar to arbitrary proximal polygons, arbitrary triangulations are not unique, and so allow different volume estimates. To obtain a unique result, Whitney (1929) used the Delaunay triangulation (p. 62). This triangulation is the geometrical dual of the natural neighbor proximal polygons. If the edges of a triangle in a Delaunay triangulation are bisected perpendicularly, the bisecting lines will meet at the Voronoi vertex, the center of the circumcircle. Lines connecting these vertices are the edges of natural neighbor proximal polygons.

The interpolated surface formed by linear triangular facets tends to underestimate the highs and overestimate the lows, conversely to the tendency of the proximal polygon methods; however, similarly to them, linear triangulations tend to be unbiased overall. Watson (1982) gives a FORTRAN program to triangulate and linearly contour scattered data.

(5) Data on Rectangular Grids.

Triangulation methods were required for manual contouring whenever the data were scattered. Obviously, the most convenient and orderly arrangement for a data set is when the data are arranged at the nodes of a rectangular grid. Moreover, data on a rectangular grid can be grouped conveniently in subsets of four data. Then each set of four, with their heights, forms a rectangular interstitial prism, although the top surface is not determined so easily as it is for triangular prisms.

This rectangular interstitial polygon generally does not have a single plane that fits all four heights, but the rectangle can be subdivided into two triangles. However, although the Delaunay triangulation is unique for scattered data, triangulations of data on a rectangular grid are usually ambiguous, because the triangular subdivision of four points on a circumcircle can be done in two ways, so offering either of two distinct pairs of triangular surfaces for each interstitial four-sided polygon.

Nevertheless, a unique triangular subdivision can be established by setting a pseudo data point at the centroid of the rectangle. The height of this added datum is the average of the four corner data. Then interpolation within each of the four triangular facets forms the upper surface of the interstitial polygon.

Additionally, a curved surface can be formed on this rectangle, and it is bounded by straightlines at the edges of the rectangular prism. This surface is curved smoothly across the interstitial polygon, although its slope contrasts, along the edges, with adjacent rectangular prisms.

Such a curved surface is formed by using the rectangular coordinates (p. 79) of the interpolation point. The w_i, where $i=1 - 4$, are rectangular coordinates for any location (x, y), within the rectangle. This is a weighted average method and the elevation of the interpolated surface, term it $F(x ,y)$, at some interpolation point (x, y) within the rectangle is

$$F(x,y) = \sum_{i=1}^{4} w_i f(x_i, y_i)$$

where the weight w_i is the ith rectangular coordinate of the interpolation point, and $f(x_i, y_i)$ is the observed height at the data point (x_i, y_i). The sum of the rectangular coordinates is always one. This causes the interpolated value to lie on a perpendicular pair of straightlines that, in turn, lie on straightlines along the bounding edges of the rectangular prism. This surface, termed bilinear, is a rectangular hyperboloid (see Fig. 2.5.3 on p. 139).

Probably the least attractive aspect of linear and bilinear interpolation of data on a rectangular grid, when used for displaying contours, is that the slope of the interpolated surface is discontinuous at the data points and along the edges of each rectangle, so causing jagged-looking contour lines unless the grid is dense. However, it is a fast and easy method.

Interpolation

(6) Inverse Distance Weighting

Linear triangular interpolation, and bilinear rectangular interpolation, are weighted average methods, producing an explicit interpolation surface that can exhibit contours because it changes gradually. Calculating a weighted average interpolation of more than three or four data was an obvious extension, even though it increased the amount of manual calculation. A variable number of weights was required.

The use of barycentric coordinates to interpolate triplets of gridded or scattered data, and the use of rectangular coordinates to interpolate gridded data, suggested that, for a more general approach, the weight applied to a more distant datum should be less than the weight for a nearby datum. So, making the weight inversely proportional to the distance was a natural choice. Horton (1923) gave the formula (in different notation) for the ith inverse distance weight, w_i, relative to three data.

$$w_i = (1/d_i) / (1/d_1 + 1/d_2 + 1/d_3) \quad \text{for } i = 1,2,3$$

d_i is the distance to the ith datum from the interpolation point. Of course, any number of data could be used. Using N data, where $f(x_i, y_i)$ is the observed dependent variable at the data point (x_i, y_i), the elevation of the interpolated surface, term it $F(x, y)$, at the interpolation point (x, y) is

$$F(x,y) = \sum_{i=1}^{N} w_i f(x_i, y_i) = \sum_{i=1}^{N} \frac{f(x_i, y_i)/d_i}{\sum_{j=1}^{N} 1/d_j}$$

The concepts behind this approach can be seen to derive both from the simple average obtained by the statistical method, which uses all data, and from the linear triangular and bilinear rectangular methods, where the weights sum to one. However, instead of using a weight such as $1/N$, or barycentric coordinates, the weights are made inversely proportional to the distance from each datum to the interpolation point.

This is a variation on the idea of binary local coordinates (p. 76), applied to any number of data, and then normalized to sum to one. Normalization implies that the weights tend to $1/N$ at large distances from the data so IDW surfaces tend to the mean level plane (p. 106).

The surface generated by using weights that are inverses of distance has the appearance of a taut rubber sheet, constrained to meet the data points, and forming cone-like pits and peaks at the data points. Although the slope changes gradually between data points, it is discontinuous at each datum. Similar to other manual methods, inverse distance weighting cannot infer a surface that is higher, or lower, than the maximum and minimum values of the data.

(7) Linear Regression

Manually implemented inverse distance weighted averaging of more than a few data is a tedious calculating chore. A fitted function interpolation method, fitting a tilted plane to more than three data by a least sum of squared differences criterion, was easier to apply. By accumulating sums, and sums of products, the coefficients of the fitted plane are obtained algebraically. The computational details are discussed on p. 88, where linear regression is used for gradient estimation.

COMPUTER METHODS

Manual interpolation methods are interesting because they contribute to an understanding of the origin, development, and characteristics, of computer methods. The roots of all the dominant concepts used in computer algorithms for bivariate data can be seen in the seven manual methods discussed above, and the development and evolution of these ideas through time as seen in the literature provides a facinating insight to scientific thought.

Automatic procedures may be classified into five broad categories deriving from these manual methods, and differing primarily in the type of data subset and the basis for interpolation. A data set can be treated either as a whole in a global sense or piecemeal by local subsets, and the data may be fitted by some form of function or averaged by some form of weighting. Further, the weighting may be distance-based or area-based, and the size of subset is a basis for distinction among area-based methods.

Distance-based weighted averages are the most venerable group in the literature, and it seems likely that the manual origins of this group predate the earliest publication of the method. Various computer adaptations of distance weighting are discussed under the first numbered heading on the next page.

The second category covers globally-based techniques for fitting functions to a set, or running subset, of data. These methods are conceptually transparent so they have become well-established and widely known.

The next three categories cover area-based methods. Many variations and adaptations cause these techniques to be the most numerous. The third numbered heading concerns weighted average techniques of interpolation using triplets of data. Triangles provide obvious and fundamental subdivisions of space.

Rectangularly gridded data, taken four at a time, allow many weighted average and fitted function methods that are not applicable to other data arrangements; these are mentioned in the fourth group.

The fifth category is neighborhood-based interpolation. This is the most general approach and it offers the most flexibility, although it is the least well-known. All forms of data, with any degree of difficulty, and requiring any surface type, can be contoured by neighborhood-based methods.

Interpolation

(1) Distance-Based Methods.

Methods that use distance-based weighting of the data follow the assumption that each datum has a local influence that diminishes with distance and becomes negligible beyond a limiting radius. So the physical shape of the influence attributable to a given datum is a radially symmetric cone or bell, centered at the datum. The representative surface at any location is the sum of these influences at that point.

(a) Inverse Distance Weighting Observations (IDWO)

Weighting the observations inversely to their distance from the interpolation point became the first computer interpolation method, and is widely used, being embedded in most commercial contouring packages as a first stage gridding procedure for scattered data. This method is based on the sound and intuitively appealing notion that values at nearby data points are more significant than more distant observations when an estimate is to be made at an arbitrary location. The "influence" of a datum at an interpolation point is assumed to be inversely proportional to the its distance from the interpolation point.

When the influence of a datum decreases with increasing distance, the assigned weight behaves like a decay function, having a maximum of one at zero distance and being positive for all distances less than some set value. Interpolation by inverse distance weighting involves calculating the sum, at a given location, of the decaying influence function for each datum. Other decay functions, besides inverse distance, are important to interpolation and are discussed on p. 120.

Two major problems were evident with this approach: (i) the discontinuous slope of the surface at each data point was unappealing because continuity of slope was attained on all other parts of the surface, and (ii) the effect of high or low distant values tended to submerge local detail because the averaging effect was global.

These difficulties seemed to be remedied neatly by using the inverse of the distance squared, instead of the inverse distance. Now the interpolated surface has a continuous slope everywhere and, because more distant data points are assigned a weight closer to zero, local detail is not so muffled so strongly. Using the cube of the distance is even more effective.

Taking the square of the distance, before inverting, and in general, taking the pth power of the distance, firstly provides a method of varying the resulting surface to suit the data. Secondly, changing the value of p helps to show the nature of inverse distance weighted average surfaces. The regional emphasis may be shifted from global with p small to local with p large. The ith datum weight is

$$w_i = d_i^{-p} / \sum_{j=1}^{N} d_j^{-p} \qquad i = 1, \ldots, N$$

When $p=1$, Figure 2.5.2 shows a profile of the cone-like surface that is formed at a data point. Around a data point, the interpolated result approaches the shape of a cone, whose apical angle is determined by the relative elevations of the surrounding data points.

When $p>1$, the surface is flat and level at each data point. The size of this flat spot increases as p increases until the surface becomes an arbitrarily close approximation to the surface obtained by flat topped natural neighbor proximal polygons (p. 108). Steep scarps are formed at the edges of these polygons and the surface is a series of flat terraces. This is caused by a dominating weight (close to unity) being applied to the nearest data point, and negligibly small weights applied to all other data, for high values of p.

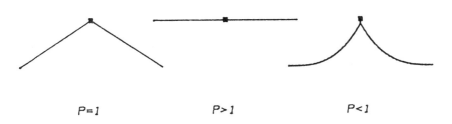

Figure 2.5.2 Inverse distance parameters

On the other hand, when $p<1$ all weights tend to be more nearly equal; of course, when the interpolation point is close to a data point, the weight for that point is close to one, and the other weights tend to zero. This causes the interpolated surface to cusp sharply at each data point, and otherwise approximate the global mean level plane (p. 106) for the data set. The spike-like pits and peaks at the data points cause a series of concentric isolines, around each datum, to be developed by a contouring program. The effect of decreasing p, from $p>1$ to $p<1$ is to retract the scarps at the polygonal edges to form spikes at the data points. Figure 2.5.2 shows profiles through a data point for large ($p>1$), and small ($p<1$), values of p.

Using values of p that are orders of magnitude greater than or less than 1, shows that IDWO interpolation is bounded by the surface formed by natural neighbor proximal polygons (p. 108), and by the surface formed by the statistical method (p. 106). Also, similarly to those surfaces, an IDWO surface is confined to the range of the data; interpolated values cannot lie above or below the maximum and minimum measured values. For this reason this method cannot be fully responsive to local trends, and unsampled peaks cannot be inferred although they may be implicit in the data. Flat but tilted regions are interpolated poorly by IDWO.

Moreover, because this is a distance-based interpolation procedure, each datum has a radially symmetric influence and linear features such as ridges and valleys are obscured by this enforced isotropy and the terracing effect.

Use program DISTANCE.BAS to obtain inverse distance weighted surfaces. On line **1100**, set **CL**=0 and **PW**=3. This provides a surface which is a weighted average of the data. With **PW**=3, the weighting is inverse distance cubed. For other powers of inverse distance less than nine, set **PW** to that value.

Inverse distance weighted interpolation is handicapped by its essentially one-dimensional nature; it cannot form ridges or domes from sparse data because the isotropic influence given to each datum produces radially symmetric peaks or pits about each datum. However, this effect may be moderated by incorporating local trends expressed by the data. To do this, gradients are required; gradient estimation is discussed on p. 85.

(b) Inverse Distance Weighted Gradients (IDWG)

Interpolation methods mentioned to this point, produce values that are linear combinations of the data points. Further, although the surface is continuous, the slope at each datum is either zero or nonexistent, and the IDWO surface ($p>1$) has a stepped, terraced appearance that is incongruous for some spatial functions.

A second form of inverse distance weighting applies inverse distance weights to estimated tangent planes through the data, instead of the observations, termed a projected inverse distance weighted average. The altitude of the representative surface at any point is a weighted average of the altitudes of the tangent planes at that location.

For high values of the inverse weighting parameter, p, the surface now approximates slant topped polygonal interpolation (p. 108). For lesser values, the surface is smooth but with tilted flatspots at the data. Although the sum of weighted gradient planes provides a continuous surface, unsampled peaks are clearly overestimated, in comparison with inverse distance weighting of the data where unsampled peaks are underestimated.

(c) Blended Inverse Distance WeightedSurfaces

This over- and underestimation suggests that a combination of these surfaces may be appropriate. Simple inverse distance weighting of the data provides the foundation for a representative surface; then it is amended by adding proportions (see the blending function, p. 163) of the differences between it and the nearby gradient planes. When either of the two surfaces, inverse distance weighted observations (p. 114) or inverse distance weighted gradients (above) is used by itself, then all the change in variation implied by the data necessarily is accommodated in a band along the polygon boundaries, with flat, level or tilted, regions around each datum. However, the blended combination of these two methods provides a representative surface with variation distributed more evenly.

The IDWO interpolation, gradient planes, and roughness indices (p. 171) are combined to provide a curvilinear interpolation of the data by using the roughness index as a parameter to regulate a spatial blending between the gradient planes and IDWO interpolation; that is,

$$F(x,y) = L(x,y) + \sum H(w_i, r_i)*(S_i(x,y) - L(x,y))$$

where $F(x, y)$ is the combined surface at some point (x, y), $L(x, y)$ is the IDWO interpolation at (x, y), w_i is the weight (p. 112) for the ith datum, and r_i is the roughness index of the data point at that vertex. $S_i(x, y)$ is the altitude, at (x, y), of the ith gradient plane, and $H(w_i, r_i)$ is the blending function (p. 163) evaluated at coordinate w_i, with a roughness index r_i. The summation is over N data.

This results in a smoothly varying surface that fits all data points exactly, and displays any existing indication of unsampled highs or lows between the data points. The slope is continuous because it is the sum of continuous functions. In particular it is responsive to local trends in the data.

More importantly, IDWO and IDWG estimates provide interpolation limits for the resulting representative surface. The estimated tangent plane at a datum and the IDWO surface establish bounds for the combination surface around that datum; the surface is assuredly not less and not more complex than these respective limits. Because the difference between the gradient plane and an IDWO surface is zero at the data points, and because the blending function (p. 163) ranges between zero and one, with its first derivative being zero at these limits, the resulting surface is tangent to the gradient plane; the higher derivatives of the interpolated surface go to zero at each data point. This representative surface is local in the sense that the result at any point depends only on the measured values at the three nearest data and the neighbors of those data. Typically, this would be twelve for scattered data, and sixteen for gridded data.

IDWO is known as Shepard's Method after Shepard (1968), but there are earlier reports including Horton (1923), Bergthorsson and Doos (1955), Krige (1964), Ojakangas and Basham (1964), Surkan, Denny, and Batcha (1964), Weaver (1964) Carlson and others (1966), Crain and Bhattacharyya (1967).

IDWO has been reinvented many times, and studied by many others such as Arthur (1973), Barnhill, Dube, and Little (1983), Barnhill and Stead (1984), Berry (1987), Braile (1978), Cortey (1982), Devereux (1985), Foley (1980), Farwig (1986a, 1986b), Foley and Nielson (1980), Foley (1984, 1987), Gordon and Wixom (1978), Jancaitis and Junkins (1973), Junkins, Miller, and Jancaitis (1973), Kane and others (1982), Kok and Begin (1981), Lancaster and Salkauskas (1981), Lodwick and Little (1970), MacGillivray, Hawkins, and Berjak (1969), McCullagh (1983), McLain (1974, 1976), Morrison (1974), Palmer (1969), Piazza, Menozzi, and Cavalli-Sforza (1981), Ripley (1981), Rom and Bergman (1986), Smith (1968), Sutcliffe (1976), Swain (1976), Tobler (1979a, 1979b), Tobler and Kennedy (1985), Van Kuilenburg, and others (1982), Wahba (1981), Walters (1969), Watson and Philip (1985), and Yeo (1984).

Interpolation

Some variations on the IDW method are mentioned here. Johnston and Harrison (1984) discuss discount weighted moving averages, and Goodin, McRae, and Seinfeld (1979) compare modified forms. Little (1983) points out that IDW is a convex combination of Taylor interpolants; the gradient estimate serves as the first derivative. He also introduces an inverse distance weighted sum of the planes determined by triangular facets. Liszka (1984) uses a Taylor expansion to derive an expression for the derivatives of a cubic polynomial, fitted by least squares, to running subsets of data weighted by inverse distance cubed.

Blended inverse distance weighted surfaces are available in program DISTANCE.BAS. Set CL=1 and PW=3 (or any value between 1 and 8), in line 1100. Roughness indices are computed at line 2590, after gradients are estimated.

(2) Fitted Function Methods

Fitted function is a generic term that refers to a class of computer interpolation methods which use a polynomial expression for an analytic surface that fits the data, either globally or in local patches, and based on either disjoint or running subsets of data. These methods are applied in two stages; first the parameters of the function are determined and then interpolation is done at particular locations by using these parameters. More general than the set of methods that may be applied to gridded data (described on p. 111 and p. 137), fitted functions may be applied to scattered and traverse data as well, although the results are often judged to be unsatifactory because the surface behaves unexpectedly between data.

The polynomial surface is a linear combination of elemental surfaces termed basis functions (p. 164), expressed in coordinates of location. A well-known example of basis functions are the power functions

$$Z=1, \ Z=x, \ Z=y, \ Z=xy, \ Z=x^2, \ Z=y^2, \text{ and so on,}$$

where x and y are Cartesian coordinates. Each power function can be thought of as a surface in its own right. For example, $Z = x$ is a tilted plane that increases in height with increasing x, but it is not tilted in the y direction. The linear combination of such surfaces, for a given data set, often is obtained by solving a set of simultaneous equations.

Various polynomials can be used for the elemental surfaces, and fitted function interpolation methods are classified according to the type of basis function involved. The general approach is to select a family of surfaces whose members are specified by a set of parameters. Then the parameter values are selected so that the resulting surface interpolates the data.

Power basis functions and trigometric basis functions are the conventional elemental surfaces. Some basis functions may have nonzero altitude at any point over the (x, y) plane, and so are termed nonfinite to distinguish them from functions that are zero everywhere outside a specified region. The fitting techniques are discussed under the next six alphabetized headings.

(a) Lagrange Interpolation

This is the classical method for fitting a polynomial function through univariate data. Although it usually is not used for interpolation of topographical data, the interpolation principles involved occur in several other methods so it is instructive to consider this approach. Lagrange interpolation also is interesting in contrast to other methods of surface fitting because it exaggerates problems that often arise, although to a lesser extent, with other techniques.

For an example of Lagrange interpolation, consider four altitudes measured at locations along a traverse, say $\{x_i, z_i\}$, $i=1 - 4$. An interpolated value, $F(x)$, at some point x, along the traverse is

$$\begin{aligned}F(x) &= (x-x_2)/(x_1-x_2) * (x-x_3)/(x_1-x_3) * (x-x_4)/(x_1-x_4) * z_1 \\ &+ (x-x_1)/(x_2-x_1) * (x-x_3)/(x_2-x_3) * (x-x_4)/(x_2-x_4) * z_2 \\ &+ (x-x_1)/(x_3-x_1) * (x-x_2)/(x_3-x_2) * (x-x_4)/(x_3-x_4) * z_3 \\ &+ (x-x_1)/(x_4-x_1) * (x-x_2)/(x_4-x_2) * (x-x_3)/(x_4-x_3) * z_4\end{aligned}$$

This example is a cubic interpolant. For N ordered, univariate data, the Lagrangian interpolant is degree $N-1$. It always is possible to select such a polynomial to fit any data set exactly at the data points; however, the behavior of Lagrange interpolants between data points often is not what might be expected. Higher degree polynomials may peak or trough excessively between any two data and this implies that features revealed by the interpolation may be more an artifice of the method than of the variability implied by the data. As Beck, Farouki, and Hinds (1986) stress, polynomial surfaces selected to satisfy the data may include extraneous features because the potential complexity of polynomial surfaces often is not constrained fully by the data. This is particularly so when the data are scattered unevenly.

The terms modifying each z_i are the Lagrange polynomials. These are expressions in x that each have a value of one at their associated datum location, and zero at all other data locations, so the combination will interpolate the data exactly. Lagrange interpolation fits an $N-1$ degree polynomial function to N data, but it also may be seen as a weighted average of the data, where the weights for each interpolation are given by the Lagrange polynomials.

When seen as a weighted average, the reason that Lagrange interpolation is unsatisfactory becomes apparent, for the heights of these basis functions, at interpolation locations, increases as the distance to the data increases. This implies that the interpolated value between two data can be affected by a distant datum, and so the fitted function is not representative of the spatial variation in the data. Liszka (1984), for example, provides an illustration of Lagrange interpolation showing the strong tendency to overshoot where the slope changes. The interpolated curve incorporates excessively oscillatory behavior, and forms a more complex surface than may be expected. A commonplace solution for this problem, in other methods, is to limit the interpolation distance arbitrarily, so that the method becomes piecewise.

Interpolation

Lagrange interpolation has applications in finite element methods where Vandermonde matrices are used to develop serendipity functions (see Ball, 1980). Lee and Phillips (1987) discuss the Neville-Aitken recursive algorithm, appropriate for efficient computation of a Lagrange polynomial through multivariate data. The resulting surface interpolates the data, but can behave unexpectedly between data so cannot be recommended for topographical data.

Many other authors, such as Barnhill, Dube, and Little (1983), Busch (1985), Ellis and McLain (1977), Farin and Barry (1986), Gasca and Maeztu (1982), Gasca and Ramirez (1984), Giloi (1978), Hakopian (1982), Micchelli (1980), Rattray (1962), and Szabados (1986), have discussed Lagrange interpolation.

(b) Collocation

This general term refers to two surfaces that are coincident at specific locations. If one were describing the variation of a spatial function by a representative surface which agrees with the observations, this generated surface is said to collocate the data.

Several function fitting techniques, usually termed collocation methods, use a combination of basis functions having a distinctly different form. The height of Lagrange interpolants increases as the distance to the data increases. The collocation surface, in contrast, tends to a horizontal plane as the distance to the data increases, similar to IDWO surfaces (p. 112). This is because collocation surfaces outside the convex hull of the data are asymptotic to the mean level plane (p. 106), as IDWO surfaces are.

The form of collocation basis functions apparently developed from improvements to the decay function behavior of inverse distance. Before they are normalized, each inverse distance weight is infinitely large at zero distance, and decays smoothly to zero at infinite distance. Normalization makes the weight to be one at zero distance but, of course, it does not become zero at any finite distance. So schemes were developed to force the weight to vanish at a limiting distance.

Arthur (1965) specifies a decay-type function of Euclidean distance which has a maximum value of one at zero distance, and is positive for all distances less than some arbitrary limit. If d is the distance from the interpolation point to a datum and e is the arbitrary limit, then Arthur's function is

$$C(d) = 1 - d^2/e^2$$

This function was novel because, for a given interpolation point, the complement of the distance, rather than its inverse, provides the character of the decay function. Because Arthur's function is applied to bivariate data, it effectively provides a radially symmetric kernel of influence around each datum. The dome shape is a parabolic quadratic, and Arthur's surface was a linear combination of these basis functions, obtained by solving a system of linear equations.

Hardy (1971) introduced a method using multiquadric basis functions. The surface is a combination of a set of circular hyperboloids, each centered about a datum. Put into a comparable form with Arthur's decay function, and for arbitrary constant e (Hardy suggested $e=0.815$ times the average distance between data), Hardy's basis function is

$$C(d) = (d^2 + e^2)^{1/2}$$

and a related complementary form is

$$C(d) = 2 \cdot d + e - (d^2 + e^2)^{1/2}$$

Hardy's function behaves like an inverted decay-type function (the bowl shape is right side up), and so, effectively, it acts to increase the relative influence of other data as d becomes larger. Franke (1982b) compares the performance of these basis functions. Foley (1987a) tested the multiquadric approach for various choices of e. Hardy (1977) discusses and illustrates several decay, and inverted decay, basis functions.

Given observations, $z(P_i)$, at locations P_i, a system of simultaneous linear equations are set up and solved for the coefficients a_i. This is illustrated for four points, but any number may be used.

$$\begin{bmatrix} e_1 & C(P_1\text{-}P_2) & C(P_1\text{-}P_3) & C(P_1\text{-}P_4) \\ C(P_2\text{-}P_1) & e_2 & C(P_2\text{-}P_3) & C(P_2\text{-}P_4) \\ C(P_3\text{-}P_1) & C(P_3\text{-}P_2) & e_3 & C(P_3\text{-}P_4) \\ C(P_4\text{-}P_1) & C(P_4\text{-}P_2) & C(P_4\text{-}P_3) & e_4 \end{bmatrix} * \begin{bmatrix} a_1 \\ a_2 \\ a_3 \\ a_4 \end{bmatrix} = \begin{bmatrix} z(P_1) \\ z(P_2) \\ z(P_3) \\ z(P_4) \end{bmatrix}$$

where

$$C(P_i\text{-}P_j) = ((P_{ix}\text{-}P_{jx})^2 + (P_{iy}\text{-}P_{jy})^2 + e_j^2)^{0.5}$$

is a basis function of the distance on the x-y plane between P_i and P_j, modified by the arbitrary nonnegative constant e_j. In program FITFUNCT.BAS, set TB=5 or TB=8, in line 1100, for multiquadric collocation interpolation with either of the basis functions on the previous page. Values for e_j are placed in array RO, in line 1220.

The coefficients, a_i, whose sum is zero, are obtained by solving the system of linear equations, and these become the coefficients of the surface for an arbitrary location. If **X** is the interpolation point with Cartesian coordinates (x, y), and the coefficients, a_i, have been determined, then the interpolated value, $F(x, y)$, by the methods of collocation, multiquadric splines, or biharmonic splines (see next and p. 125), is

$$F(x, y) = a_1 C(X\text{-}P_1) + a_2 C(X\text{-}P_2) + a_3 C(X\text{-}P_3) + a_4 C(X\text{-}P_4)$$

Schut (1976) lists and describes several other kernel functions, that may be substituted for the basis function C. Mason (1984) uses power functions, and points out that any set of basis function may be used if they are linearly independent.

Hardy (1977) gives a comparison of multiquadric and covariance kernels, and a discussion on the summation of quadric and other forms of basis function. Hardy and Nelson (1986) report that a certain selection of basis function leads to a multiquadric-biharmonic surface. Briggs (1974) showed that a biharmonic spline surface has minimum curvature. Hardy (1990) gives a comprehensive review.

(c) Minimum Curvature Splines

Spline is a term used to evoke a property that distributes the change in slope as widely and evenly as possible; just as a rapier, if unrestrained, will flex in a wide curve rather than bend sharply. A surface that interpolates the data and has the least possible change in slope at all points is obtained by minimizing the total curvature. Formally, minimum curvature surfaces have been described as conforming to the variational criterion of smoothness obtained by minimizing the square of the Laplacian, or the quadratic variation of the gradient, which leads to the biharmonic or bilaplacian equation (see Langridge, 1984, p.293). Meinguet (1979, 1983) gives details of the mathematical theory of minimum curvature splines. Harder and Desmarias (1972) introduced minimum curvature spline interpolation, and report that it was developed originally for interpolating wing deflections in aircraft design. They derive the basis function, $d^2 \ln(d^2)$, and give a comparison with a quintic polynomial surface.

Franke (1982a) advocated the basis function $d^2 \log d$, and Sandwell (1987) gives the biharmonic Green function, $d^2 (\log d - 1)$, as a basis function, to obtain minimum curvature surfaces. Sandwell also shows how slope measurements can be used as data in a similar expression. A variety of basis functions are given by Ayeni (1979) who combines stepwise selection of trends with collocation.

This method can be applied to most data sets, but there is some difficulty if the range of heights is large; this appears as overshoot between data. Minimum curvature surfaces can be extrapolated and tend to a horizontal plane at some distance from the data set. This tendency can cause some distortion at the edge of the data set. However, as Schut (1976) noted, a systematic trend in the data can be simultaneously determined and expressed by a low order polynomial. Then by fitting the residuals with a minimum curvature surface, and adding it to the polynomial surface, the distortion is reduced.

For N data, $N+3$ simultaneous equations are set up as shown below. This system establishes any linear trend in the data, and builds the minimum curvature surface as departures from that trend which then is added to the surface.

$$\begin{vmatrix} 1 & x_1 & y_1 & 0 & C(P_1\text{-}P_2) & C(P_1\text{-}P_3) & C(P_1\text{-}P_4) \\ 1 & x_2 & y_2 & C(P_2\text{-}P_1) & 0 & C(P_2\text{-}P_3) & C(P_2\text{-}P_4) \\ 1 & x_3 & y_3 & C(P_3\text{-}P_1) & C(P_3\text{-}P_2) & 0 & C(P_3\text{-}P_4) \\ 1 & x_4 & y_4 & C(P_4\text{-}P_1) & C(P_4\text{-}P_2) & C(P_4\text{-}P_3) & 0 \\ 0 & 0 & 0 & 1 & 1 & 1 & 1 \\ 0 & 0 & 0 & x_1 & x_2 & x_3 & x_4 \\ 0 & 0 & 0 & y_1 & y_2 & y_3 & y_4 \end{vmatrix} * \begin{vmatrix} b_0 \\ b_1 \\ b_2 \\ a_1 \\ a_2 \\ a_3 \\ a_4 \end{vmatrix} = \begin{vmatrix} z_1 \\ z_2 \\ z_3 \\ z_4 \\ 0 \\ 0 \\ 0 \end{vmatrix}$$

Although the matrix shown is for only four data, any number of data may be used; however, round-off errors generally limit this method to less than 100 data. Each $C(P_i-P_j)$ is the value of $d^2 \log d$, where d is the distance on the x,y plane between P_i and P_j. The unknown values, a_i and b_j, are obtained by solving the system of equations. These are the coefficients of the surface for an arbitrary location, and the sum of a_i is zero. For any interpolation point, $X=(x,y)$, the interpolated value is

$$F(x,y) = b_0 + b_1 x + b_2 y + a_1 C(P_1\text{-}X) + a_2 C(P_2\text{-}X) + a_3 C(P_3\text{-}X) + a_4 C(P_4\text{-}X)$$

To obtain a minimum curvature surface with program FITFUNCT.BAS, set the parameter TB=3, in line **1100**. The minimum curvature distance function is on line **1060**. For approximating minimum curvature surfaces, see p. 128.

Enriquez, Thomann, and Goupillot (1983) describe the minimum curvature spline using $d^2 \log d^2$ and point out that if $d^{(3/2)}$ is used, it represents the best estimation of random functions. Swain (1976) solves iteratively the difference equations of minimum curvature given by Briggs (1974). Minimum curvature splines also have been discussed by Cheney (1986), Dyn and Levin (1982), Ebner (1984), Franke (1982a, 1985), Inoue (1986), Lenard (1985), Nielson and Franke (1984), Rentrop (1980), and Wahba (1981, 1986).

(d) Kriging

This is a term whose meaning includes the fitting of a representative surface using a criterion of minimum variance, or covariance, according to several somewhat different definitions. However, the generation of a kriged surface differs from a minimum curvature spline surface only in that the basis function of distance for kriging is a form of covariance kernel. The matrix is set up as follows

$$\begin{bmatrix} 1 & x_1 & y_1 & 0 & C(P_1\text{-}P_2) & C(P_1\text{-}P_3) & C(P_1\text{-}P_4) \\ 1 & x_2 & y_2 & C(P_2\text{-}P_1) & 0 & C(P_2\text{-}P_3) & C(P_2\text{-}P_4) \\ 1 & x_3 & y_3 & C(P_3\text{-}P_1) & C(P_3\text{-}P_2) & 0 & C(P_3\text{-}P_4) \\ 1 & x_4 & y_4 & C(P_4\text{-}P_1) & C(P_4\text{-}P_2) & C(P_4\text{-}P_3) & 0 \\ 0 & 0 & 0 & 1 & 1 & 1 & 1 \\ 0 & 0 & 0 & x_1 & x_2 & x_3 & x_4 \\ 0 & 0 & 0 & y_2 & y_2 & y_3 & y_4 \end{bmatrix} * \begin{bmatrix} b_0 \\ b_1 \\ b_2 \\ a_1 \\ a_2 \\ a_3 \\ a_4 \end{bmatrix} = \begin{bmatrix} C(P_1\text{-}X) \\ C(P_2\text{-}X) \\ C(P_3\text{-}X) \\ C(P_4\text{-}X) \\ 1 \\ x \\ y \end{bmatrix}$$

$C(P_i\text{-}P_j)$ is the chosen semivariogram basis function for the distance between data points P_i and P_j. The matrix shown is for only four data; for N data the matrix has $N+3$ rows. The coordinates, (x,y), of the interpolation point X, are included in the matrix, and the interpolation is

$$F(x,y) = b_0 + b_1 x + b_2 y + \sum a_i z_i$$

where the $z_i = z(P_i)$ are the observations. With this modification, the sum of a_i is one, rather than zero.

Interpolation

This modification is not implemented in the programs. However, use program FITFUNCT.BAS to obtain a kriged surface with the minimum curvature spline program by setting the parameter **TB**=3, in line **1100**, and substitute a user defined semivariogram for the distance function. The defining parameters of an exponential and a spherical semivariogram function are on lines **1070** and **1080**.

This may be done because, similar to collocation, kriging may be applied to the whole data set or to a running subset. Using a running subset increases the execution time considerably, of course, but allows a modification by solving the system of simultaneous equations directly for each interpolation location.

The surface obtained by using semivariograms as basis functions of distance can vary considerably, according the manner in which the semivariogram behaves near zero. If this function has a nonzero slope at zero, the surface will cusp at the data; this effect may be seen by using a "spherical" semivariogram. In order for the surface to have a continuous slope at the data, the semivariogram must have a zero slope at zero distance. Also, if the semivariogram functional value at zero is not zero, which is the so-called nugget effect, then the surface will not interpolate the data. Possibly a better way to krige such an approximation surface may be to use the experimental error bounds of the data (p. 128).

There are several heuristic methods to estimate the semivariogram parameters for a given data set, but, because of the probabilistic element, no rigorous method of deducing or deriving these parameters. Schagen (1979,1982) uses an estimated covariance basis function. Kratky (1978) discusses reflexive prediction in which the data are related by a prescribed covariance to the predicted values. Dermanis (1984) points to the basic similarities of collocation and kriging. Watson (1984) interprets smoothing splines as a special example of kriging. Davis (1981, p.1493) points out, "Kriging is an optimal interpolation [method] only if the correct form of the semivariogram can be deduced, ... ".

In general, the idea that the variability of a spatial function can be specified by a single succinct expression, possibly should not be expected to apply to the detailed shape of complex surfaces. A useful and considered discussion of covariance and semivariograms was given by Ripley (1981). Burrough (1986), Pouzet (1980), Schagen (1979, 1982), Schut (1974,1976), Watson (1972), Wong (1985), Virdee and Kottegoda (1984), among others, also have expressed qualified opinions about this method.

(e) Relaxation Surfaces

This is a general term for surfaces that are constructed by iteratively minimizing local curvature differences, and often referred to as LaPlacian surfaces. A set of grid nodes is superimposed on a region of data points. Then values of an interpolating surface are obtained by readjusting the heights at the nodes so that each node is an average of all its immediate neighbors, including data. Of course, the data themselves are not adjusted, and so the surface has a peak or pit at each datum.

The adjustment is a discrete approximation of a constant relationship between the rates of slope change in perpendicular directions. LaPlace's equation sets the sum of second partial derivatives to zero; for such a surface, the change in slope in perpendicular directions, at any location, is equal in magnitude and opposite in sign. For discrete grid nodes, this reduces to differences of differences.

To obtain relaxation surface values for an orthogonal grid, with fixed boundary nodes, supply initial values for each node, and then improve systematically these estimates by comparing the second differences in each direction. These differences indicate the direction and magnitude of the correction necessary to minimize curvature differences over the whole network. Using three adjacent data on a line, for example (x_1, z_1), (x_2, z_2) and (x_3, z_3), a second difference at x_2 is

$$z_2' = z_1 * w_0 + z_2 * (1-w_0-w_1) + z_3 * w_1$$

where $w_0 = (x_3-x_2)/(x_3-x_1)/4$ and $w_1 = (x_2-x_1)/(x_3-x_1)/4$.

For arbitrary initial values and systematic reapplication of this estimate in two directions, the values at the nodes of an orthogonal grid will relax toward a stable relaxation surface. These are quick computations, so the iterative nature of the method is not significant for execution times.

Renz (1982) illustrates divided differences, in the univariate situation, by smoothing the data according to the second divided differences. Kok and Begin (1981) discuss interpolation of scattered data by finite differences. This is an iterative approach toward a surface that approximates a differential equation, with the data as boundary conditions.

Tobler (1979a) applied relaxation techniques to scattered data by iteratively matching grid nodes to data by least squares planes and adjusting the node values to represent a finite equivalent of LaPlace's equation, or the biharmonic equation (see next). Rikitake, Sato, and Hagiwara (1987) discuss the relaxation method for solving Poisson and LaPlace differential equations over rectangular or equilateral triangular grids.

Cole (1968) proposed a relaxation method to iteratively approximate scattered data with piecewise quadratics. This is an improvement because LaPlacian surfaces have cusps at the data points, as Liszka (1984) illustrates. When these singularities at the data points are avoided by also averaging slopes at the grid nodes, this approach leads to fitting a biharmonic, minimum curvature, surface.

Crain (1970) discussed the second order equations leading to LaPlacian surfaces; these are smooth everywhere except at the data, where cones are formed. Briggs (1974) showed that the fourth order equations produce a minimum curvature surface; as smooth as possible, everywhere. Dyn and Wahba (1982) use an iterative method to obtain LaPlacian histosplines for volume-preserving interpolation of choropleth data. Lam (1983) reviews pycnophylactic, or volume preserving, LaPlacian interpolations.

Franke (1985a) develops a minimum curvature spline with a tension parameter to control overshoot in the vicinity of steep gradients. Franke (1985b) discusses Laplacian smoothing splines with cross validation.

Interpolation

(f) Approximation Surfaces

Although computer-generated surfaces almost never agree, exactly, with the data, a customary distinction is based on the magnitude of disagreement. If these differences are invisible at the selected scale, the surface is said to be an interpolation surface; in comparison, if disagreement with the data is conspicuous, the surface is said to be an approximation surface.

Approximation surfaces are considered to be preferable when the data are expected to have significant unknown observational error, or known experimental imprecision. Surface perturbations in approximate representative surfaces, due to individual data errors, are less apparent than they are in interpolating surfaces, but the broad behavior of the spatial variation is made clear.

The difference between the approximation surface and each datum, termed the residual, often is adopted as an expression of error at each datum. However, whether the computed residual is a measure of an unknown observation error, or of an allowable interpolation error, or of both, depends on the method and the spatial function from which the data were drawn.

The procedures for computing approximation surfaces may be divided into three groups. First, there are methods that determine residuals relative to a polynomial of specified degree by computing the parameters that satisfy a minimum sum of squared residuals criterion. Second, there are techniques that use additional information about data observation errors to select a surface satisfying the minimum curvature criterion. A third group of approximation procedures make use of additional information about both observation and interpolation errors, and satisfies either minimum squared difference, or minimum curvature, criteria.

(i) Polynomial Trend Surfaces

A polynomial surface, being a linear combination of power basis functions (p. 118), has its simplest form in the level plane produced by the statistical method (p. 106), which is zero degree. For any degree, g, the surface at any location (x, y) is a combination of power basis functions of location. An interpolated value, $F(x, y)$, when the coefficients, a_k, of the combination have been determined, is the summation over $k = 1 - m$

$$F(x,y) \sum a_k x^i y^j \qquad i, j, i+j = 0 - g, \quad m = (g+1)(g+2)/2$$

Although such a polynomial always can be selected to fit small data sets exactly, the surface exhibits the same unsatisfactory behavior between data shown by Lagrange methods (p. 119). It could be seen that what was needed was a lower order polynomial which, although it fit the data inexactly, behaved conservatively between observations. The idea was that by weakening the influence of individual observations, the interpolant would be regressed toward the shape of the measured spatial function.

The key to this approach was a criterion to control the misfit of the surface by requiring a minimum sum of squared differences between the surface and the data. This turned out to have a neat algebraic expression, and a convenient solution as a system of simultaneous linear equations (p. 88). The resulting surface was termed a regression surface, or trend surface, and it exhibited the broad regional trend in variation implied by the data.

In terms of classical statistics, the observations have a minimum mean squared difference from the regression function; in this sense, the selected function fits the data more closely than any other polynomial of the same degree. This idea, of course, realized many practical applications.

In the program FITFUNCT.BAS, nonlinear regression by least squares is turned on by setting **TB**=4, in line **1100**, and the degree of the approximating polynomial is set by **DG**, also in line **1100**. The residuals are put in array **RO**.

Cooper and Cross (1988) point out that the least squares method was applied first by Gauss in 1795, and described formally by Legendre in 1806. Gauss was concerned with observations from a surveying instrument, without functional reference to another variable. For univariate functional data, Riggs, Guarnieri, and Addelman (1978) attribute the first straightline approximation of a function, $f(x)$, by the least squares criterion, to Adcock in 1877. Merriam and Harbaugh (1963) have noted that this method has been advocated since Griswold and Munn in 1917. Many authors have proposed variations and extensions. Agocs (1951) records that the approximation approach also was used manually by "eyeballing" isolines that, locally, ignored the data in favor of a smoother surface.

Apparently Panofsky (1949) published the first computer contoured map. He fitted a third-degree polynomial as an approximation to wind and pressure fields, and concluded that isolines drawn from the polynomial surface were satisfactorily similar to contours drawn manually. This weather map, suggested by John von Neumann as a pilot study of "objective analysis" on a fledgling computer, was aimed at computer forecasting of weather by observation of regional trends in space and time. This approach illustrates one of the two early and useful roles of least squares regression in analysis of topographical data.

In the alternative role, determination of the regional trend allowed the subtraction of that trend, so that local anomalies could be isolated and interpreted. These anomalies were the subregions where all residuals are positive, or negative. Simpson (1954) noted the importance of using a low order polynomial so that local anomalies are necessarily revealed when the regional trend is subtracted. Of course, this admonition is relevant to all applications of nonlinear regression because higher order polynomials can create unacceptable artificial "anomalies" between data (see Lagrange interpolation, p. 119).

Grant (1957) advocated the word "trend" to describe the broad behavior of the variation in comparison to the local, short range, anomalies. He also held that positive and negative residuals were equally likely and, in this respect, was invoking a third role of least squares regression; the statistical interpretation of topographical heights as a symmetrically distributed random variable.

Gurnell (1981) suggested a trivariate trend surface to represent evapotranspiration by employing two locational coordinates, augmented by a linear term in the altitude coordinate. Neuman (1987) gives a tutorial paper on least squares fitted bivariate polynomials.

For additional discussions on trend surfaces, see Davis (1973), Unwin (1974), Ripley (1981), Burrough (1986), Jones, Hamilton, and Johnson (1986), and Sutterlin and Hastings (1986). Harbaugh and Merriam (1968) provide a good summary of trend analysis from the geological point of view.

(ii) Response Surfaces

Response surface is a general term for polynomial surface approximations of functions of two or more independent variables with a nonspatial connotation, such as pressure, temperature or frequency. Such data may be studied by treating them as topographical variables, and displaying their representative surface. Although the independent variables are not spatial coordinates, contouring the response surface gives a topographical expression to the behavior of the function. Because the variables are both quantitative and continuous, computation does not distinguish between the two forms of data, and the representative surface illustrates the variation.

The usual approach is to approximate the data with a low-order polynomial. A response surface usually is a least squares nonlinear regression using the criterion of minimum sum of squared differences. Sampson and Davis (1967) pointed out that trend surfaces are a variety of response surface. Since then, however, trend surfaces have become the more widely accepted term.

A few of the many authors who have demonstrated that the source of data does not prevent treating it as if it were topographical are mentioned next. Box (1954) explored the behavior of chemical reactions by fitting a cubic polynomial, derived from its Taylor series, to the multivariate observations by the criterion of least sum of squared differences. Brill, Gaunaurd, and Uberall (1981) use response surfaces to study and display the behavior of elastic wave scattering. Petersen (1985) gives an extensive coverage of response surfaces applied to replicate data sets of observations at the nodes of a regular grid. Steven (1984) discusses the smoothing effects of response surfaces based on minimizing surface area. Balaras and Jeter (1990) use response surfaces to model solar radiation.

(iii) Minimum Curvature Approximation Surfaces

Approximating surfaces using minimum curvature (p. 122), or kriging (p. 123), procedures are obtained easily and conveniently for a known, or estimated, precision in the observations. A constant value for the region, or a distinct value for each datum, may be used in place of the diagonal of zeros in the expressions on those pages. Then a minimum curvature surface is generated that approximates the variation in the data, while staying within these bounds. Cox (1984) provides a theoretical framework for applying such smoothing surfaces to data in higher dimensions.

Minimum curvature approximation is available in program FITFUNCT.BAS, by setting **TB**=3, in line **1100**, and putting approximation limits into array **RO**, in line **1220**, for each datum, instead of assigning zero or a constant value. The programs are set up to use random numbers to demonstrate this approximating effect. You may use the user defined function **RT#^2*LOG(RT#)**, in line **1060**, to determine the surface.

(iv) Objective Analysis

This term concerns a form of optimal linear interpolation, a class of methods where compensation is made for both observation and interpolation errors. In the ideal situation, the spatial functions representing the interpolation and observational errors are known. Then, whatever the interpolation mechanism that is used, the same representative surface must be inferred from the probabilistic information in conjunction with the observations.

In practice, the interpolation errors at the grid nodes are estimated by back-interpolation from the grid nodes to the observations, and the observational errors are estimated from replicate observations when these are available. Then a surface representing these interpolation errors is added to the grid values. The method is problematic when observation errors are not known. Davis (1985) refers to this approach as constrained optimal interpolation.

This method is useful where replicate observations are involved, such as weather records at a set of stations. As Goodin, McRae, and Seinfeld (1979) observe, a historical record of data values is necessary to estimate the observational error. So objective analysis or optimal (some term it optimum) interpolation has been used in meteorology for decades, and practical knowledge of the error functions have been accumulating.

Gilchrist and Cressman (1954) reported fitting quadratic polynomials, by least squares, to subsets of barometric data, with the standard deviations over time of the observational errors at each station, as a weighting factor. They also proposed that differences between successive maps be interpolated to provide an estimate where a datum is missing, and as a method of detecting erroneous data.

Bergthorsson and Doos (1955) devised a weighted average of the current, the forecast, and the mean observations at each station; the object was to obtain the most likely values, leading to an error estimate for the observations, and so allow objective analysis. Wunsch and Zlotnicki (1984) apply objective analysis to sea levels, as do Mazzega and Houry (1989).

When least squares regression is assumed to have normally distributed residuals, and used for objective analysis, the least squares regression surface for the data is added to the least squares regression surface for the residuals from the first regression. The resulting surface may be refined further by iteration, somewhat like relaxation surfaces (p. 124), and D'Autume (1979) refers to such a surface as being elastic; Sabin (1985) terms this method dynamic addition. Schagen (1982) takes a similar approach but uses an estimated covariance for the regression.

Interpolation

Franke (1985b) credits Gandin, in 1963, with the introduction of objective analysis to the meteorological literature, and Franke studies this approach in conjunction with LaPlacian smoothing splines (p. 125). For a problem similar to weather analysis, Yeske, Scarpace, and Green (1975) mapped the patterns of lake currents by using objective analysis to convert velocity data from moving markers to a geographically fixed velocity surface.

Franke (1985c) also has studied objective analysis by simulating both observation errors and interpolation errors, using several different interpolation schemes for the observation to grid, and grid to observation back-interpolation. He determined that optimal interpolation, with known spatial covariance function for the interpolator, and known standard deviation for the observation, outperformed the other methods. Unfortunately, neither of these statistical descriptors are known for most real data sets.

The problems of objective analysis of weather patterns are remain significant in spite of the variety and depth of the research efforts. For example, Connor (1988) reports on a strong weather anomaly that escaped detection by several large and highly developed computer contouring systems. The loss of life and property was extensive, but may have been less if the anomaly could have been inferred sooner. Navon (1986) gives a condensed overview of variational and optimization methods in meteorology.

This subsection has described ways in which analytical functions may be fitted to an entire set (less than a 100 data for personal computers when sets of simultaneous equations must be solved). In the next three numbered subsections, the data set is partitioned into subsets of 3, 4, or about 6, data, and functions are fitted in a piecewise sense. Some piecewise methods use related forms of the globally-fitted surfaces that were discussed in this subsection.

(3) Triangle-Based Methods.

Once a data set has been triangulated (p. 57), the spatial order of the data is available, and this allows adjacent triplets of data to be treated as subsets. The simplicity of this approach has made it appealing, and many computer algorithms have been designed to treat data triplets. The simplest of these combine triangular facets to form a continuous piecewise surface.

(a) Linear Facets.

The manual method of triangle-based linear interpolation, described on p. 109, is adapted easily to become a fast and economical method on the computer, giving a clear indication of pronounced trends and anomalies. In this approach, planar facets are fitted to each triangle. See the diagram (p. 110) of the sorted data set (p. 57). This forms a continuous representative surface that interpolates the data and shows the variation.

Triangle-based linear interpolation is useful for initial investigations and when smooth isolines are not important. It is entirely appropriate when the data are known to be obtained at ridge and ravine break points because then it expresses the erosion and drainage patterns (Watson, 1982). This technique also has applications for engineering design data (Oxley, 1985).

The altitude of the representative surface is calculated easily, for any point within a triangle, by a weighted average of the three observations, using the barycentric coordinates (p. 76) as weights for the associated datum. For example, at the centroid of the triangle the barycentric coordinates, w_i, are $1/3$, $1/3$, and $1/3$. The altitude of a plane, **L**, at that location is

$$L(x_c, y_c) = \sum 1/3 \, f(x_i, y_i)$$

where $f(x_i, y_i)$ is the observed height at the ith vertex and summation is over $i = 1 - 3$. This expression may be compared with the manual statistical method on p. 106. For any three other weights, w_i, whose sum is one,

$$L(x, y) = \sum w_i \, f(x_i, y_i)$$

is the altitude of the same plane at location (x, y), where

$$x = \sum w_i x_i \quad \text{and} \quad y = \sum w_i y_i$$

This plane forms a planar facet across the triangle and, together with facets on the other triangles, provides a continuous surface across the convex hull of the data.

The least satisfactory aspect of contouring by linear interpolation across triangles is the jagged appearance of the contours. This is caused by incorporating all the change in slope, required by the data, at the triangle edges and vertices. To distribute the change in slope more smoothly, gradient information is used to infer the convexity or concavity of a suitably smooth, curved, surface across each triangle. Linear facets are provided by TRIANGLE.BAS; set **CL**=0, in line **1100**.

(b) Nonlinear Triangular Patches.

Many schemes have been proposed that are intended to fit a smoothly curved surface over each triangle in such a way that the slope also changes smoothly from one triangle to the next. Including gradient information, either measured or estimated, in the interpolation procedure, allows the representative surface to be nonplanar within a triangle. Starting with estimated gradients at each datum, the various suggestions concern types of polynomial patches and ways of subdividing the triangles. Several of these approaches are mentioned under the next four roman headings.

Interpolation

(i) Polynomial Patches.

After linear patches, the most obvious choice for a nonlinear triangular patch is a quadratic polynomial. However, because of the low level of complexity in a 2nd degree equation, it is necessary to subdivide the triangle and apply the quadratics to subtriangles. Powell and Sabin (1977) apply quadratics across each of the subtriangles formed by dividing the triangle using the circumcenter and midpoints of each side. The Clough-Tocher subdivision, described by Percell (1976), forms three subtriangles by introducing a vertex anywhere in the interior of the triangular patch; then three polynomial patches are determined by the three data and the estimated gradients.

Lawson (1977) used the triangle centroid (p. 108) to form the Clough-Tocher subdivision. Cendes and Wong (1987), as well as Sablonniere (1987), suggest the incenter (p. 32) as a subdividing point of the triangle, and interpolation by bivariate quadratic subpatches with matching slopes by using Bezier coefficients (p. 147) applied to the three data and their estimated gradients. Lacombe and Bedard (1986) develop a so-called serendipity basis function (p. 170) for interpolating between univariate curves along the edges of subdivided triangles.

With the increased complexity provided by using a cubic polynomial, changes in the change of slope may be accommodated within a triangle. Cubic surfaces are capable of representing any amount of variability in a single valued spatial function over a triangle, although not always in an acceptable manner. Cubics are applied widely to rectangularly gridded data (see Rectangle-Based Methods, p. 137), but are less easily adapted to the irregular triangles of scattered data. For example, Cherenack (1984) discusses splines formed by bivariate cubic polynomials in Cartesian coordinates. These patches minimize strain energy in certain directions.

Dooley (1976) uses a piecewise cubic polynomial, expressed in barycentric coordinates, and with matching derivatives along triangle edges. Barnhill and Farin (1981) fit both cubic and quintic polynomials across triangles. They require first and second derivatives at the vertices of the triangle, and for the 5th-degree scheme they also require cross-boundary derivatives at points along the triangle edges; then a Bezier net (p. 147) of barycentric coordinates, is defined over the triangle.

Farin (1982b) describes cubic polynomials in barycentric coordinates for a regular array of triangles. These require a Bezier net to be set up and each triangle subdivided into nine subtriangles; then each subtriangle and the three adjacent subtriangles determine a Bernstein-Bezier patch of the generated surface. Herron (1985) uses cubic polynomial basis functions, in barycentric coordinates, to interpolate to altitude and slope at the vertices and slope across the edges.

Klucewicz (1978) uses the Barnhill-Gregory Boolean sum of functions defined on cardinal Hermite cubics, transformed from the standard right-angled triangle to a triangle of any proportions. Wang (1983) uses a quartic surface using gradient information at the vertices and across the edges of arbitrary triangles and based on cubic Hermite polynomials (p. 143).

Schut (1976) describes a method by Bauhuber, Erlacher, and Gunther, that first determines a cubic polynomial, along each edge of the triangle, and which satisfies the gradients at each pair of vertices. Within the triangle, the weighted average of these three cubic polynomials, each parallel to a side, becomes the interpolated value. The weights apparently are the barycentric coordinates.

A tricubic bivariate polynomial surface has a cubic univariate polynomial along each edge of the triangle (compare to bicubic polynomials, p. 143), and Mansfield (1980) recommends tricubic polynomials for interpolation to the data and partial derivatives at the vertices. Lowden (1985) presents a method using the Clough-Tocher three-part cubic polynomial patches for triangles.

Higher degree polynomial patches allow precise fitting of data and slopes along the patch boundaries; this has made them more popular, and studied widely. Alfeld, Piper, and Schumaker (1987) fit quartic polynomials by erecting a net of Bezier control points (p. 147) over a triangle in such a manner that control points along a triangle edge are coplanar with those of the adjacent triangle. This insures continuity of slope across triangle edges, which is a problem for many triangle-based methods.

Alfeld (1985) gives a method for piecewise interpolation of triangulated bivariate data by combining two univariate approaches, using barycentric coordinates. This method is analogous to tensor, or Cartesian product, methods (p. 141) for rectangularly gridded data. Alfeld, Piper, and Schumaker (1987) use quartic polynomials, and a Bezier net of control points, expressed in barycentric coordinates. Nielson (1979) describes a form of univariate interpolation along line segments joining a vertex and a side; notice that this is interpolation between a curve and a datum rather than between two or more data.

Akima (1978) fits a quintic polynomial over each triangle to suit estimated gradients at the vertices. This is a spline with eighteen degrees-of-freedom - three less than the full complement of a 5th degree surface - and is attained by also specifying that the quintic polynomial be expressible as a cubic function in a direction perpendicular to each triangle edge. The coefficients of the polynomial are selected to satisfy first and second derivatives at the vertices as well as match the three adjacent triangular patches. Segalman, Woyak, and Rowlands (1979) use a similar scheme for a finite element solution.

Preusser (1984a, 1984b) applies piecewise quintic polynomials to triangular patches; quintics require first and second partial derivatives at the vertices. Botkin and Bennett (1985) use quintic polynomials over triangles for finite element shape optimization (p. 170).

Alfeld and Barnhill (1984) obtain a surface with continuous second derivatives by using polynomials up to the 8th degree. They use so-called transfinite, univariate, quintic polynomials to interpolate value, slope, and rate of change in slope. A Boolean sum (p. 146) of three such operators is used to interpolate triangular patches, given a set of topographical data with first and second derivative information along a triangulated network. This is expressed conveniently in terms of barycentric coordinates.

Waggenspack and Anderson (1986) present a procedure for expressing a bivariate polynomial in standard power basis functions (p. 118) by the equivalent form in Bernstein basis polynomials (p. 148) in barycentric coordinates, for triangular regions. Ying (1982) describes interpolation over triangular elements such that given types of singularity may be induced at a vertex. Nielson and Mangeron (1981) give a method for bilinear interpolation on a right-angled triangular domain.

Eckstein (1989) compares triangular spline patches to inverse distance weighting. Lee and Phillips (1987) present an algorithm to evaluate Lagrangian polynomials over regular triangular arrays, analogous to the de Casteljau algorithm for Bernstein polynomials on the standard triangle, and they derive a forward difference formula.

(ii) Weighted Averages.

Triangle-based nonlinear weighted averages are an extension of linear facets (p. 130), but the sum of the weights, w_i, is variable rather than constant unity; in general it is either less, or more, than one. This approach depends upon methods for generating a set of weights that vary consistently, and provide a continuous surface with a continuous slope.

As an early example, Ojakangas and Basham (1964) interpolate at grid nodes by a distance weighted average (p. 114) of the two linear triangular facets that are formed by four data surrounding the interpolation point. This has similarities to the two-triangle method for rectangularly gridded data (p. 138), and has the same ambiguity; two triangles can be formed from four data in two different ways, leading to different results.

To control the manner in which the weights, w_i, differ from linearity, a gradient is estimated for each datum. This may be done by any of the methods discussed in Section 2.4, p. 85, and includes a variability estimate (p. 171) of the local roughness implied by the data that are adjacent to each datum.

McLain (1976) forms a weighted average of the three gradients at the triangle vertices. The weights are computed by a heuristic expression involving the distances between the interpolation point and the data, as well as the sublengths of triangle edges subdivided by perpendicular lines through the interpolation point.

The behavior of a weighted average of a triple of gradients may be understood by considering an extreme example; let the nearest gradient be given a weight of one while the other two gradients are given a weight of zero. This implies that everywhere inside the Voronoi polygon of a datum the surface is a plane; this is slant top polygonal interpolation (p. 108), and so the surface of each polygon is discontinuous with its neighbors.

Although the weight for a datum, or a gradient, must be one when interpolating at the datum, the weights of nonlinear weighted averages are otherwise unconstrained, but they must change smoothly to give a smooth surface. This is because the weights must be changed in a way that smooths and averages the gradient discontinuities between the data.

Although the sum of linearly weighted gradient planes (p. 108) provides a continuous surface, unsampled peaks often are overestimated in comparison with a linear facets (p. 130) by which unsampled peaks are underestimated. However, these two extreme surfaces provide bounds for an intermediate surface estimate. Then a conservative representative surface is assured because it is generated to lie between a linear interpolation and a projected gradient interpolation, as limiting surfaces. This may be done by blending these two surfaces.

(iii) Blended Surfaces.

Just as two inverse distance weighted surfaces were blended for a compound surface (p. 116), a combination of triangle-based surfaces may provide appropriate intermediate surface. Linear weighting of each triplet of data provides the foundation for a representative surface; then it is amended by adding proportions (see Blending Functions, p. 163) of the differences between it and the three gradient planes. This is provided by program TRIANGLE.BAS; set CL=1 in line **1100**.

When either of the two triangle-based surfaces, linear triangular facets or a weighted average of three gradients (for example, McLain, 1976), are used singly, then all the change in variation implied by the data is necessarily accommodated either along triangle edges in the first situation (Figure 2.1.3, p. 62), or along Voronoi polygon boundaries (p. 108) in the second situation. However, the combination of these two methods provides a conservative representative surface with variation distributed more evenly.

Linear triangular facets provide the basis for a representative surface that is obtained by blending in the gradient information in a manner that smooths all slope discontinuities. Such a surface always lies between the linear triangular facet and the linear gradient plane, so these effectively are bounds on the interpolation.

This approach to curvilinear interpolation is local in the sense that the estimated value within any triangle depends only on the measured values at its three vertices and the direct neighbors of those vertices that were involved in the gradient estimate. This typically consists of twelve data points, but more importantly, when a Delaunay triangulation is used, these are all the natural neighbors of the three data of the triangle.

Because the difference between the gradient plane and a linearly weighted triangle surface is zero at the data points, and because the blending function (p. 163) ranges between zero and one, with its first derivative being zero at these limits, the interpolated surface is tangent to the gradient plane. This causes the second derivatives of the blended surface to vanish at each data point.

Along triangle edges, the blended surface depends only on the pair of gradients at the vertices at either end of the edge, and the slope of the surface is a smooth blend between these two slopes. The surface also has a continuous slope perpendicular to an edge, however, because the slope influences to either side vanish smoothly.

Interpolation

A local measure of variability, for each datum in the data set, allows the interpolation to be ameliorated according to the local variability of the data. Each datum is evaluated relative to all the data with which it shares triangles. The roughness index (p. 171) used in the program TRIANGLE.BAS is formed, at line **2590**, from the vector sum of the normals to all the triangles that involve the datum.

The linear interpolation, gradient planes, and roughness indices are combined to provide a curvilinear interpolation of the data by using the roughness index as a parameter to regulate a spatial blending between the gradient planes and linear interpolation, that is,

$$F(x,y) = L(x,y) + \sum H(w_i, r_i) * (S_i(x,y) - L(x,y))$$

where $\mathbf{F}(x, y)$ is the triangle-based nonlinear interpolation at some point (x, y), $L(x, y)$ is the linear interpolation at (x, y), w_i is the local coordinate with respect to the ith vertex of the triangle containing (x, y), and r_i is the roughness index of the data point at that vertex. $S_i(x, y)$ is the height of the ith gradient plane at (x,y) and $H(w_i, r_i)$ is the blending function (p. 170). The summation is over $i=1 - 3$.

The blending of these two surfaces results in a smoothly varying, single-valued, interpolation, providing a representative surface that coincides with all data points. The slope is continuous because it is the sum of continuous functions. The surface displays any existing indication of unsampled highs or lows between the data points because it is sensitive to local trends in adjacent data and uses them to moderate the slope. This may mitigate the effects of an inappropriate triangulation if a sufficiently strong trend is inherent in the adjacent data.

The plausibility of this surface is indicated by its bounds. It lies between surfaces determined by weighted averages of the data and weighted averages of the gradients. Nevertheless, because the surface produced by any triangle-based method is triangulation dependent, an alternate triangulation will give a different surface. This can be a significant effect if the data are sparse relative to height variation.

(iv) Miscellaneous Triangle-Based Methods.

Many innovative adaptations, variations, and combinations, of interpolation methods for triplets of data have been published. Several of these are mentioned briefly.

Klucewicz (1978) applies the Barnhill-Gregory Boolean interpolant to triangular patches. Bohmer and Coman (1980), given derivative information along the edges of right-angle triangles, using a Hermite-Birkhoff Boolean sum (p. 146) of operators to obtain a representative surface. Barnhill and Farin (1981) use a quintic interpolant over triangles by Coons' method combined with Bezier patches. Segalman, Woyak, and Rowlands (1979) describe an 18-coefficient quintic triangular patch.

Nielson and Franke (1983) compare three forms of triangular patch; the Clough-Tocher subdivision used by Lawson (1977) and Lowden (1985), the triangular Coons' patch (Nielson, 1983), and the Madison Triangle (Barnhill, 1977). In addition to the data, these methods require slope information on the edges of the triangles. Surfaces for a test situation showed similar interpolation error patterns and, as well as the unsurprising error along the data set perimeter, there were anomalous interior errors arising apparently from the sampling pattern.

For comparison with a trivariate example, mention is made of Alfeld (1984) who describes an interpolant over tetrahedra formed by first developing a univariate interpolant along tetrahedral edges. These then are reinterpolated over the faces of the tetrahedra.

Nielson and Franke (1984) developed a tension spline for triangulated data sets. Gradient information from an inverse distance weighted least squares plane at each datum ameliorates the surface formed by the linear triangular facets, according to the tension parameter.

Herron (1985) characterizes interpolants for triangular patches, given partial derivatives at the vertices and the midpoints of each edge. Cherenack (1984) describes piecewise cubic splines over pairs of triangles. Grandine (1987) discusses B-splines (p. 149) on a simplex. Hobson (1972) explores surface roughness in terms of the normal vectors to triangles in a triangulated data set. Foley (1983) gives details of a recursive approach to Hermite interpolation (p. 143) of scattered data.

Percell (1976) describes Clough-Tocher triangular elements. The cubic Clough-Tocher element is a triangle which is subdivided into three triangles by any point within it, with values and gradients at the three vertices and normal derivatives for each edge. The quartic Clough-Tocher triangle, for the same subdivision, requires the values and gradients at the three vertices and one interior point, as well as one midpoint value and two normal derivatives for each edge.

Others who have discussed triangle-based interpolation include Akima (1978, 1984), Alfeld (1984, 1985), Alfeld and Barnhill (1984), Antoy (1983), Bideaux (1979), Bohmer and Coman (1980), Gold, Charters, and Ramsden (1977), Kashiyama and Kawahara (1985), Klucewicz (1978), Lacombe and Bedard (1986), Lawson (1977), Mansfield (1980), Nielson and Franke (1983), Nielson (1983), Nielson and Mangeron (1979), Oxley (1985), Powell and Sabin (1977), Renka and Cline (1984), Wang (1983), Watson and Philip (1984b), Ying (1982), and Yoeli (1977).

(4) Rectangle-Based Methods.

Rectangularly gridded data offers computational advantages because it allows an area-based (p. 79), local approach to interpolation by treating four, easily selected, data points at a time. This is convenient for array processing algorithms, and so allows fast implementations.

Interpolation

As well as being easy to select by fours, gridded data are stored economically because locational coordinates are implicit in the storage order, and so storage is reduced to one-third that required for scattered data.

The representative surface for a set of rectangularly gridded data can be formed, in patchwork-quilt fashion, by joining together many four-cornered patches. For each set of four mutually adjacent data from the grid, forming a rectangular prism or interstitial polygon, a rectangular surface patch is fitted. The resulting mosaic is an aggregate of, locally determined, surface pieces.

However, four data from the grid almost never will all lie on a plane, so the surface patch cannot be as simple as a planar rectangle. The obvious solution is to split the rectangle into two triangles by using either of its two diagonals. The two triangles then form a ridge or valley along the diagonal, and so accommodate the difference from a plane.

Then each triangle is contoured by linear interpolation using barycentric coordinates (p. 77). Level lines on a tilted triangle are determined from the algebraic solution of a linear equation for the intersection of each triangle edge and a level line. This is an application of binary local coordinates, as discussed on p. 76.

Unfortunately, this two-triangle surface patch is not unique because using the other diagonal will give a different pair of triangles and a different arrangement of contour lines. Nevertheless, because the two-triangle method is conceptually transparent and easy to program, it probably is the most frequently implemented method for initial attempts at computer contouring. The direction of the diagonal can be selected to be all left-handed, or all right-handed, according to the order in which the cell edges are traversed, or even randomly left and right. Yates (1987), for example, uses the two-triangle method, and contour lines are selected to have a clockwise direction of travel through the cell.

The two-triangle rectangle-based interpolation method is fast. This method provides an acceptable set of surface contours whenever the rectangular grid of data is sufficiently dense, relative to the variation in height, that the isolines do not seem angular. When the gridded data are less dense, several authors, such as Batcha and Reese (1964), have suggested establishing a fifth point in the cell center with each component variable set to the average of those at the surrounding four points. This subdivides the rectangle into four triangles.

Then each of these four triangles, formed by the central point and a pair of data on one side of the rectangular cell, may be contoured by linear interpolation using barycentric coordinates (p. 77). Although the interpolated surface has a discontinuous slope both at the grid nodes, and along the triangle edges, the four-triangle rectangle-based interpolation method is simple to implement, and avoids the inherent ambiguity of the two-triangle method.

The two-triangle and four-triangle subdivisions provide a surface that is formed of planar facets because one plane cannot accommodate the variation between four data. Methods that form single, smooth patches over the rectangle, are discussed next.

(a) Bilinear Patches.

Because each triangular subdivision has a planar surface with a different orientation, the patch formed by the four-triangle method is faceted, and the surface has sudden slope changes along each triangle edge. These slope differences can be melded and averaged to form a smooth surface over the rectangle, by making a weighted average of the four data using the rectangular coordinates (p. 79) as weights. This smooth patch agrees exactly with the four-triangle patch along the cell edges and at the central, fifth point.

A point on this surface has a height that is the weighted sum of the heights at the four data locations, each multiplied by the respective rectangular coordinate of the summation point. The altitude of the bilinear patch, at some point (x, y) within the rectangle, say $F(x, y)$, where $f(x_i, y_i)$ is the height at (x_i, y_i), summation is over $i=1 - 4$, and w_i is the ith rectangular coordinate of (x, y) with respect to the ith datum, is given by the expression

$$F(x, y) = \sum w_i \, f(x_i, y_i)$$

As discussed on p. 79, in general rectangular local coordinates change continuously between any two interpolation points; notice however, that, in directions parallel to the rectangle edges, the proportion between one pair of coordinates is constant and only the proportion between the other pair changes. This implies that the surface of the weighted average patch must change in a straightline fashion along directions parallel to the rectangle edges. Such a simply curved patch, formed with rectangular coordinates, may be thought of as being a set of straightlines over the rectangular grid cell, as shown in Figure 2.5.3. It is termed a ruled surface, or a lofted linear surface.

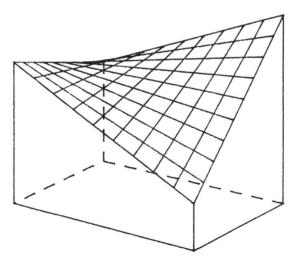

Figure 2.5.3 Rectangular hyperboloid interpolation

Interpolation

Figure 2.5.3 also illustrates a practical method of interpolating at some point, (x, y), within a rectangular cell. This method is another frequently implemented computer contouring algorithm (see Simons (1983), for example) and frequently taught in engineering courses (see Elfick and others, 1987). For simplicity, consider bilinear interpolation in the unit square.

First two preliminary univariate linear interpolations (p. 76) are made between the two pairs of data in the x direction, $(0,0)(1,0)$ and $(0,1)(1,1)$; then a final univariate interpolation between these two results in the y direction, $(x, 0)(x, 1)$, yielding the result at (x, y). Treating y first, and then x, gives an identical result for bilinear patch interpolation.

Figure 2.5.3 shows that an average of the data, weighted by rectangular local coordinates, produces a curved surface patch which has constant partial derivatives, and is a portion of a quadratic bivariate polynomial. In fact, it is a rectangular hyperboloid, and so, conveniently, it intersects the edges of a rectangular cell in straightlines, and also forms straightlines between corresponding points on opposite edges.

This ruled surface effect also may be demonstrated by holding either x or y constant in the general polynomial expression for a rectangular hyperboloid

$$F(x, y) = a_1 + a_2 x + a_3 y + a_4 xy$$

where the a_i are coefficients determined by the four data values.

To calculate the a_i for four data on a rectangle, (x_i, y_i, z_i), $i=1$ - 4, and where sigma indicates summation over these four data, set up the matrix, shown below, as a set of simultaneous equations and solve it for the coefficients of the quadratic.

$$\begin{pmatrix} 4 & \sum x_i & \sum y_i & \sum x_i y_i \\ \sum x_i & \sum x_i^2 & \sum x_i y_i & \sum x_i^2 y_i \\ \sum y_i & \sum x_i y_i & \sum y_i^2 & \sum x_i y_i^2 \\ \sum x_i y_i & \sum x_i^2 y_i & \sum x_i y_i^2 & \sum x_i^2 y_i^2 \end{pmatrix} * \begin{pmatrix} a_1 \\ a_2 \\ a_3 \\ a_4 \end{pmatrix} = \begin{pmatrix} \sum z_i \\ \sum x_i z_i \\ \sum y_i z_i \\ \sum x_i y_i z_i \end{pmatrix}$$

The solution of this system of simultaneous equations, the column vector (a_1, a_2, a_3, a_4), may be obtained by Gaussian elimination with pivoting, as implemented in the programs, starting at line **6690**.

Alternatively, by using a geometrical transformation, straightforward formulas may be derived for these coefficients. These formulas are applied and the results are transformed back to the original rectangle by the inverse tranformation. Let the rectangle be transformed to the unit square, then there are simple expressions for the a_i in terms of the z_i. When each of the corners of the unit square, in counter-clockwise order from $(0,0)$, are used in the equation for $F(x, y)$, shown above, the matrix of coefficients from the four equations, and its inverse, are

$$\begin{pmatrix} 1 & 0 & 0 & 0 \\ 1 & 1 & 0 & 0 \\ 1 & 1 & 1 & 1 \\ 1 & 0 & 1 & 0 \end{pmatrix} \qquad \begin{pmatrix} 1 & 0 & 0 & 0 \\ -1 & 1 & 0 & 0 \\ -1 & 0 & 0 & 1 \\ 1 & -1 & 1 & -1 \end{pmatrix}$$

The a_i are obtained by multiplying the inverse matrix with the column vector of observations, z_i, so these coefficients are

$$a_1 = z_1$$
$$a_2 = z_2 - z_1$$
$$a_3 = z_4 - z_1$$
$$a_4 = z_1 - z_2 + z_3 - z_4$$

The expression for $\mathbf{F}(x, y)$, on the previous page, describes a bilinear patch as a fitted function, using power basis functions and coefficients determined by the observations. In contrast, $\mathbf{F}(x, y)$, on p. 139, describes the same bilinear patch as a weighted average, using local coordinates and the observations. This indicates the close association between fitted functions and weighted averages; in this particular example the same surface is produced.

The bilinear patch method is important because it incorporates a conceptual approach that has been developed in several other, more sophisticated, algorithms. The idea is that the straightline intersections of the rectangular hyperboloid with the edges of the rectangular cell represent univariate interpolations between pairs of data. Then the bilinear surface at any point within the cell is a obtained from two univariate linear interpolations along a pair of sides, and a univariate linear interpolation between these two results. Notice that, although either pair of sides gives the same result in the bilinear situation, this is an exception; for interpolation with nonlinear functions, different surfaces are formed (p. 143) according to which pair of sides are used first.

This tactic, of forming a bivariate patch by combining univariate interpolators in perpendicular directions by matrix multiplication, is referred to as the Cartesian product, and also as the tensor product (see Cheney, 1986, for theoretical aspects of tensor products). Several authors have warned that tensor products of univariate splines are intrinsically univariate, and so not the most appropriate approach to bivariate functions.

Leberl (1975) compares bilinear interpolation with several other methods, and subjectively evaluates it as fast and economical of memory; Tobler (1979a) gives some examples. Sibson and Thompson (1981) use heights and directional derivatives of pairs of data to infer quadratic functions along each edge of the rectangular cell, and hence the value at a mid-point. Repeated application provides the rectangle with a biquadratic polynomial patch with continuous slope. When, instead of directional derivatives, data also are available at the mid-points of the cell edges, Stelzer and Welzel (1987) propose a biquadratic polynomial patch. The three data along each edge are sufficient to determine a univariate quadratic along each edge. Then these functions are treated like the linear functions in bilinear patches, resulting a cubic patch.

Bilinear interpolation, in program RECTANGL.BAS, is obtained with **TB**=6 and **CL**=0 in line **1100**. This approach makes the slope discontinuous along the grid lines. Set **CL**=1 for blended gradients and a smooth, bounded surface (see p. 135 and p. 157).

Interpolation

TABLE 4 Listing of Data Set A:\SHARPG.DAT

25	35				Number of data, number of contours
6.0	6.2	6.4	6.6	6.8	Thirty-five contour levels
7.0	7.2	7.4	7.6	7.8	.
8.0	8.2	8.4	8.6	8.8	.
9.0	9.2	9.4	9.6	9.8	.
10.0	10.2	10.4	10.6	10.8	.
11.0	11.2	11.4	11.6	11.8	.
12.0	12.2	12.4	12.6	12.8	.
1.001	1.001	3.998			Lower left corner of window and width
1.0	1.0	1.0	5.0		First (x, y), Grid spacing, Mesh size
8.43					Twenty-five height measurements
9.94				9.28	.
8.13				6.62	.
10.11				11.43	.
9.26				10.81	.
7.42				10.68	.
9.16				7.27	.
9.25				7.96	.
9.71				10.70	.
8.65				10.64	.
11.74				8.84	.
10.52				8.74	.
10.72				6.88	.
continued in next column					

The structure of data sets for RECTANGL.BAS is slightly different than for the other programs (p. 8) to take advantage of the grid. Instead of storing the locations of the data as (x, y) values, these are implicit in the grid spacing, mesh size, and location, of the first datum. The data list is folded only to save space.

Just as bilinear interpolation patches also can be thought of as weighted averages of the four data (p. 139), so can biquadratic patches. In the bilinear situation, the weights are the rectangular coordinates, which are the areas of the subrectangles determined by the interpolation point, expressed as fractions. Similarly, biquadratic interpolation patches can be thought of as weighted averages with the distinction that the weights are nonlinear functions of these fractions. For example, Junkins, Miller, and Jancaitis (1973) use weights that are quartic functions of rectangular coordinates. Seldner and Westermann (1988) use nonlinear functions of rectangular coordinates within trapezoidal cells.

Many authors, such as Ebner and Reiss (1984), have pointed out that when a separate bilinear, or biquadratic, patch is defined for each subset of four data, although the interpolated surface is smooth within the rectangular cell, the slope of the surface is discontinuous along grid lines. The surface formed by the aggregate of patches appears as a faceted surface because of this slope change between adjacent rectangles. For dense data this faceted appearance was not significant; for sparse data, matching of curvatures for contiguous patches was obviously required.

(b) Hermite Patches.

Hermite interpolation is a generic term that refers to methods that use gradients as well as altitudes. Many authors, such as Giloi (1978), have pointed out that the instability and ripples developed by Lagrange interpolation are avoided by Hermite interpolation techniques. There are, of course, several ways in which gradient information can be incorporated in the patch.

One way is an obvious extension to bilinear and biquadratic polynomial patches, achieved by requiring a univariate cubic polynomial to be fitted along each edge of the rectangle. Then, as for bilinear and biquadratic patches, the edge curves are combined, as a tensor product, to form a bicubic patch. The increased complexity of bicubic patches offers the advantage that slopes can be made to match across the rectangle boundaries, and this is achieved primarily by including gradient information in the construction of the patch.

The first aspect is the derivation and construction of the polynomial univariate curve along each cell edge. A univariate cubic polynomial, $F(x)$, and its first derivative, $F^1(x)$, have the forms

$$F(x) = a_1 + a_2x + a_3x^2 + a_4x^3$$
$$F^1(x) = a_2 + 2a_3x + 3a_4x^2$$

Because there are four unknown coefficients, a_i, in these expressions, four pieces of information are needed for a solution. These are the heights at the pair of corners, and the slope estimates at these points; name them z_1, z_2, r_1, and r_2.

If the edge is transformed to the unit interval, then again (cf. p. 140) there are simple expressions for the a_i in terms of the z_i and r_i. Four equations are formed when the data locations, (0) and (1), are substituted into the expression for $F(x)$ and $F^1(x)$. The matrix of values from the four equations, and its inverse which is termed the bicubic interpolation matrix, are

$$\begin{matrix} 1 & 0 & 0 & 0 \\ 0 & 1 & 0 & 0 \\ 1 & 1 & 1 & 1 \\ 0 & 1 & 2 & 3 \end{matrix} \qquad \begin{matrix} 1 & 0 & 0 & 0 \\ 0 & 1 & 0 & 0 \\ -3 & -2 & 3 & -1 \\ 2 & 1 & -2 & 1 \end{matrix}$$

The a_i are obtained by multiplying the inverse matrix, the matrix on the right, with the column vector, (z_1, r_1, z_2, r_2), so these coefficients are

$$a_1 = z_1$$
$$a_2 = r_1$$
$$a_3 = -3z_1 - 2r_1 + 3z_2 - r_2$$
$$a_4 = 2z_1 + r_1 - 2z_2 + r_2$$

The Cartesian product (see p. 141 and p. 145) of such cubic polynomials is a Hermite patch; this is a surface that matches four altitudes and four pairs of slopes at the corners of a rectangular patch, and it is often termed a piecewise bicubic spline. However, the slopes do not always match in the diagonal directions.

Interpolation

When a curve is termed a spline, it implies that the curve has minimum curvature for arc length, given the imposed restraints; that is, it is the smoothest curve possible for the data, in the sense of minimizing the mean squared sum of second derivatives. Birkhoff and Garabedian (1960) introduced the bicubic spline patch. They fitted a cubic polynomial along each edge of a grid cell in such a way that the slope is continuous between two adjacent cells, and curvature is minimal.

To generate a set of bicubic spline patches for a set of gridded data, gradient estimates at each node must be determined. So that the surface on each cell will match its neighbors, the partial derivatives are determined by solving tridiagonal systems of simultaneous equations for each grid line. DeBoor (1962), and Holroyd and Bhattacharyya (1970) give examples. This results in two partial derivatives and an altitude for each corner of each patch. Then a twelve-parameter quartic polynomial is fitted to each rectangle and its slope estimates. However, the curvature may not be minimal in diagonal directions.

De Boor (1962) described a bicubic spline interpolation method for fitting a rectangular patch using altitude and slope data at the four vertices. The slope data included partial derivatives in the directions of the rectangle edges and the second derivative at each node. This allows a sixteen-parameter bicubic patch. The coefficients of the polynomial are obtained by matrix multiplication of the altitude and three slope values, at each vertex, by the bicubic interpolation matrix.

In addition to the altitudes, z_i, at the four corners, the partial derivatives in the x-direction, r_i, and in the y-direction, s_i, and the second-order derivative in x and y, t_i must be estimated. Then, if the rectangle is transformed to the unit square, a straightforward expression follows for the sixteen coefficients, $a_{i,j}$, say in matrix **A**, in terms of the z_i, r_i, s_i and t_i. When each of the data, in counter-clockwise order from (0,0), are assembled in a matrix, postmultiplied by the bicubic interpolation matrix and premultiplied by its transpose, the result is the matrix **A**.

$$\begin{pmatrix} 1 & 0 & -3 & 2 \\ 0 & 1 & -2 & 1 \\ 0 & 0 & 3 & -2 \\ 0 & 0 & -1 & 1 \end{pmatrix} * \begin{pmatrix} z_1 & s_1 & z_4 & s_4 \\ r_1 & t_1 & r_4 & t_4 \\ z_2 & s_2 & z_3 & s_3 \\ r_2 & t_2 & r_3 & t_3 \end{pmatrix} * \begin{pmatrix} 1 & 0 & 0 & 0 \\ 0 & 1 & 0 & 0 \\ -3 & -2 & 3 & -1 \\ 2 & 1 & -2 & 1 \end{pmatrix} = \mathbf{A}$$

Then, using the coefficients of **A**, $\{a_{i,j}\}$, and summing over i and $j = 1 - 3$, the surface of the bicubic patch at any location (x, y), in the unit square, is

$$F_h(x, y) = \sum\sum a_{i+1, j+1} \, x^i y^j$$

The bicubic interpolation matrix often is termed the Hermite interpolation matrix, and has several forms resulting from the interchanging of rows or columns. Foley and Van Dam (1982) give a derivation for the Hermite interpolation matrix. Holroyd and Bhattacharyya (1970) reported an implementation of De Boor's (1962) bicubic method, and Davis and David (1980) adapt De Boor's method to large data sets.

Computer Methods

To see how this bicubic patch is generated, consider the univariate cubic Hermite polynomial basis functions, **h**, termed blending functions. These may be obtained from the Hermite matrix by premultiplying with the vector($1, w, w^2, w^3$), giving four cubic polynomials, on the interval 0 - 1.

$$\begin{aligned}
\mathbf{h}_{z1}(w) &= 1 \quad -3w^2 + 2w^3 = (1+2w)(1-w)^2 \\
\mathbf{h}_{r1}(w) &= \quad w \quad -2w^2 + w^3 = w(1-w)^2 \\
\mathbf{h}_{z2}(w) &= \quad 3w^2 - 2w^3 = \mathbf{h}_{z1}(1-w) \\
\mathbf{h}_{r2}(w) &= \quad -w^2 + w^3 = -\mathbf{h}_{r1}(1-w)
\end{aligned}$$

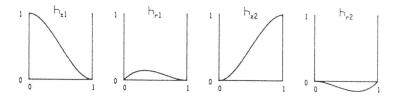

Figure 2.5.4 - Hermite cubic polynomials

These four functions, applied to their respective altitudes, z_i, and derivatives, r_i, for two data, add to give the Hermite univariate interpolant at any location, w, in the unit interval.

$$F_h(w) = z_1(1-3w^2+2w^3) + r_1(w-2w^2+w^3) + z_2(3w^2-2w^3) + r_2(-w^2+w^3)$$

In this expression, w ranges from 0 to 1, but binary coordinates (p. 76) may be used by applying the equivalent formulations for the second datum. Of course, when binary coordinates are used, interpolation can be within an arbitrary interval. Notice that $F_h(w)$ is a weighted average of the data, but now the weights, similar to the biquadratic patch on p. 142, are nonlinear functions of the local coordinates. This may be contrasted to the expression on p. 143, where $F(x)$, the same cubic curve, is a fitted function. This duality of expressions was also seen for bilinear patches, on p. 141.

If the vector $(1, u, u^2, u^3)$, also is premultiplied into the bicubic interpolation matrix, then the tensor product gives the surface of the bicubic patch at any location (w, u), in the unit square, as an alternative expression for $F_h(x, y)$ on p. 144.

Interpolation

$$F_h(w,u) = \begin{bmatrix} 1 & 0 & 0 & 0 \\ 0 & +u & 0 & 0 \\ -3u^2 & -2u^2 & +3u^2 & -u^2 \\ 2u^3 & +u^3 & -2u^3 & +u^3 \end{bmatrix} * \begin{bmatrix} z_1 & s_1 & z_4 & s_4 \\ r_1 & t_1 & r_4 & t_4 \\ z_2 & s_2 & z_3 & s_3 \\ r_2 & t_2 & r_3 & t_3 \end{bmatrix} * \begin{bmatrix} 1 & 0 & -3w^2 & +2w^3 \\ 0 & +w & -2w^2 & +w^3 \\ 0 & 0 & +3w^2 & -2w^3 \\ 0 & 0 & +w^2 & +w^3 \end{bmatrix}$$

Rogers and Adams (1976), and Rom and Bergman (1986), discuss the bicubic patch. Zwart (1973) discusses piecewise polynomial spline patches. Loikkanen (1985) fits a twelve-parameter quartic for a thin hybrid plate finite element. Gordon and Hall (1973) discuss transfinite interpolation between boundary curves. Jancaitis and Junkins (1973) use piecewise quartic polynomials to fit altitude and slope at the four vertices.

Harada and Nakamae (1982) show how a cubic polynomial may be fitted to three data, using the slope between successive data to estimate pseudo-data at both ends of the triplet. Dierckx (1987) has introduced bicubic surface splines, expressed in polar coordinates. Busch (1985) gives general Lagrange formulae for Hermite osculatory interpolation. Hessing and others (1972) fit a bicubic Ferguson patch with zero twist vectors. Kalkani (1977) considers the accuracy of contours compared to cubic spline approximations of the data.

(i) Coons' Patches.

The bilinear patch may be computed by a univariate linear interpolation between the two univariate linear interpolants along a pair of rectangle edges. Using the other pair of rectangle edges gives the surface of the same rectangular hyperboloid (p. 139), because these linear interpolants commute, and the ruled surface can be obtained by either approach.

However, if the interpolants along a pair of edges are not linear, but cubic, and a linear (or cubic) interpolation is made between them, the result will not be the same as a linear (or cubic) interpolation between cubic curves on the other pair of edges; nonlinear interpolants do not commute. The two surfaces obtained by these alternative approaches are termed lofted nonlinear surfaces, and Coons' patch is defined as the arithmetic sum of the two lofted surfaces minus the bicubic patch.

Coons' patch also may be seen to be a Boolean sum, as an alternative description of the combination of the two lofted surfaces and the bicubic patch. A Boolean sum of two contributing surfaces is a surface that, at each location within the rectangle, is the sum of the surface heights minus their difference.

Letting the bicubic patch be the Boolean intersection of the two surfaces lofted from the univariate cubic interpolants along each edge, then Coons' patch can be termed a Boolean sum. As Forrest (1973) points out, the Hermite interpolation polynomials are usually implemented, but the Boolean formulation allows a neat conceptual expression that may be extended to other methods.

Bos and Salkauskas (1988) discuss Boolean sums of surfaces. Barnhill (1977) provides details of univariate interpolants, tensor products, the bicubically blended Coons' patch, and lofting interpolants; Barnhill (1983) discusses Coons' patches. Farin (1982a, 1982b), generalize Coons' and Bernstein-Bezier methods.

Feng and Riesenfeld (1980) discuss Boolean sums of interpolants, the Cartesian or tensor product, and the decomposition of Coons' patch into four interpolants. Little (1983) describes Brown's square, an inverse area weighted average equivalent of a combination of Coons' lofting interpolants. Forrest (1973) gives a historical discussion of Coons' methods. Dodd, McAllister, and Roulier (1983) implement Gregory's patch, a variation of Coons' patch.

(c) Bezier Patches.

This is another curved rectangular patch, developed from the Cartesian product of Bezier's method of producing a cubic curve. Bernstein cubic polynomials are used as weights for the data, summed to form cubic curves along the grid lines. Bezier patches are used widely for interactive, heuristic design of smooth free-form surfaces, but seldom are applied to topographical data.

In contrast to Hermite cubic polynomials, which are the weights for two altitudes and two slopes, each univariate Bezier cubic curve requires four, equally spaced, collinear altitudes; sixteen data are required to fit a patch over a rectangle (having nine subrectangles). The four corner data are interpolated exactly, and the remaining data are only approximated. In each grid direction, the slopes of a Bezier curve at the corner data match the slopes of straightlines between each end point and its nearest intermediate datum.

Analogously to the bicubic matrix, there is a Bezier interpolation matrix.

$$\begin{matrix} -1 & 3 & -3 & 1 \\ 3 & -6 & 3 & 0 \\ -3 & 3 & 0 & 0 \\ 1 & 0 & 0 & 0 \end{matrix}$$

The equations of four univariate Bernstein polynomials, or blending functions, may be obtained by premultiplying the Bezier matrix by the vector $(w^3, w^2, w, 1)$.

$$\begin{aligned} b_{z1}(w) &= w^3 + 3w^2 - 3w + 1 = 1(1-w)^3 w^0 \\ b_{z2}(w) &= 3w^3 - 6w^2 + 3w \quad 0 = 3(1-w)^2 w^1 \\ b_{z3}(w) &= -3w^3 + 3w^2 \quad 0 \quad 0 = 3(1-w)^1 w^2 \\ b_{z4}(w) &= -w^3 \quad 0 \quad 0 \quad 0 = 1(1-w)^0 w^3 \end{aligned}$$

These four expressions are the cubic Bernstein polynomials, shown in Figure 2.5.5, on the next page. The univariate cubic Bezier curve, at any point w in the unit interval, is the sum

$$F_b(w) = z_1(1-w)^3 + z_2 3(1-w)^2 w + z_3 3(1-w)w^2 + z_4 w^3$$

This result will always lie within the quadrangle formed by the four altitudes; the Bezier curve is formed by their weighted average. Binary local coordinates, $1-w$ and w, may be applied by using the equivalent formulations.

Interpolation

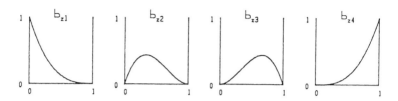

Figure 2.5.5 Bernstein cubic polynomials

If the vector, $(u^3, u^2, u, 1)$, also is premultiplied into the Bezier interpolation matrix, and the result transposed, then the Cartesian product with the sixteen height values gives the surface, $\mathbf{F}_b(w, u)$, of the Bezier patch at any location (w, u) within the data rectangle.

$$\mathbf{F}_b(w, u) = \begin{bmatrix} -u^3+3u^2-3u+1 \\ 3u^3-6u^2+3u \\ -3u^3+3u^2 \\ u^3 \end{bmatrix}^T * \begin{bmatrix} z_{1,1} & z_{1,2} & z_{1,3} & z_{1,4} \\ z_{2,1} & z_{2,2} & z_{2,3} & z_{2,4} \\ z_{3,1} & z_{3,2} & z_{3,3} & z_{3,4} \\ z_{4,1} & z_{4,2} & z_{4,3} & z_{4,4} \end{bmatrix} * \begin{bmatrix} -w^3+3w^2-3w+1 \\ 3w^3-6w^2+3w \\ -3w^3+3w^2 \\ w^3 \end{bmatrix}$$

Notice that the coefficients 1, 3, 3, and 1, in the right-hand expressions on the previous page, are set up to show that these are the coefficients of the four-class binomial probability distribution, and so give an indication that the Bernstein polynomials have a probabilistic interpretation. Goldman (1983b, 1986), and Piegl (1988), discuss these probabilistic aspects, and show how alternate forms of blending functions can be derived from the probabilistic model. Higher degrees of Bernstein polynomials can be constructed to provide a surface over a larger array of control points.

Enns (1986) gives implementations in BASIC for Bezier curves and B-spline curves (discussed on p. 149). A derivation of the Bezier matrix is given by Foley and Van Dam (1982) and they discuss the relationship between Hermite and Bezier matrices. Rogers and Adams (1976), Giloi (1978), and Newman (1979), discuss the derivation and use of Bezier patches. Brueckner (1980) discusses constructing a Bezier surface over a rectangle, starting from Bezier points over a triangle. Piegl (1984) introduces additional weights for improved control of the shape of the surface. Long (1987) proposed Bezier quartic patches instead of the usual cubic forms.

(d) B-Spline Patches

Univariate cubic B-splines are another method of interpolating in the unit interval, and similarly to Hermite and Bezier patches, their Cartesian product provides a bicubic B-spline patch. As with Bezier patches, B-spline patches require sixteen data, but are dissimilar in generating a surface over the rectangle formed by the four interior data only. This patch does not interpolate the data, but does provide an approximate surface that has second derivative continuity with contiguous patches. The bicubic B-spline patch interpolation matrix is

$$\frac{1}{6} * \begin{matrix} -1 & 3 & -3 & 1 \\ 3 & -6 & 3 & 0 \\ -3 & 0 & 3 & 0 \\ 1 & 4 & 1 & 0 \end{matrix}$$

Premultiplying this matrix by the vector $(w^3, w^2, w, 1)$, gives the four univariate cubic B-spline basis functions, **bs**, shown in Figure 2.5.6 on the next page, and the sum of these four kernels, for any w between zero and one, is one.

$$\begin{aligned}
\mathbf{bs}_{i-1}(w) &= (-w^3 + 3w^2 - 3w + 1)/6 = (1-w)^3/6 \\
\mathbf{bs}_i(w) &= (3w^3 - 6w^2 + 4)/6 \\
\mathbf{bs}_{i+1}(w) &= (-3w^3 + 3w^2 + 3w + 1)/6 = \mathbf{bs}_i(1-w) \\
\mathbf{bs}_{i+2}(w) &= (w^3)/6 = \mathbf{bs}_{i-1}(1-w)
\end{aligned}$$

Then for four collinear data, z_{i-1}, z_i, z_{i+1}, z_{i+2}, the univariate cubic B-spline curve, at any binary coordinate w in the interval between z_i and z_{i+1}, is the sum

$$F_{bs}(w) = (z_{i-1}(1-w)^3 + z_i(3w^3-6w^2+4) + z_{i+1}(-3w^3+3w^2+3w+1) + z_{i+2}w^3)/6$$

This univariate curve, like the Bezier curve, always will lie within the quadrangle formed by the four altitudes. B-spline curves usually are thought to be fitted functions; they are a compound of the four basis function, **bs**, shown in Figure 2.5.6. However, from this expression, the B-spline curve also can be seen to be a weighted average of the data, and the weights are functions of binary coordinates, w and $(1-w)$.

If the vector, $(u^3, u^2, u, 1)$, also is premultiplied into the cubic B-spline interpolation matrix, then the Cartesian product gives the surface of the bicubic B-spline patch at any location with rectangular coordinates, (w, u), within the four central data. This surface will always lie within the convex hull of the sixteen altitudes.

$F_{bs}(w, u) =$

$$\begin{pmatrix} (1-u)^3/6 \\ (3u^3-6u^2+4)/6 \\ (-3u^3+3u^2+3u+1)/6 \\ u^3/6 \end{pmatrix}^T \begin{pmatrix} z_{i-1,j-1} & z_{i-1,j} & z_{i-1,j-1} & z_{i-1,j-2} \\ z_{i,j-1} & z_{i,j} & z_{i,j-1} & z_{i,j-2} \\ z_{i+1,j-1} & z_{i+1,j} & z_{i+1,j-1} & z_{i+1,j-2} \\ z_{i+2,j-1} & z_{i+2,j} & z_{i+2,j-1} & z_{i+2,j-2} \end{pmatrix} * \begin{pmatrix} (1-w)^3/6 \\ (3w^3-6w^2+4)/6 \\ (-3w^3+3w^2+3w+1)/6 \\ w^3/6 \end{pmatrix}$$

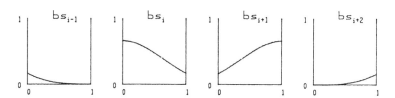

Figure 2.5.6 B-spline cubic basis functions

Jupp (1976) describes B-splines and attributes them to Schoenberg in 1946. Bohm, Farin, and Kahmann (1984) note that B-spline curves are a generalization of Bezier curves. Foley and Van Dam (1982) discuss the bicubic B-spline interpolation matrix, and give a comparative discussion of Hermite, Bezier, and B-spline interpolation forms, and their conversions.

Newman and Sproull (1979) describe B-spline local support for minimum curvature Cartesian product surface B-spline blending functions. Goldman (1983b, 1986) illustrates a link between geometry and probability exhibited by the Bezier and B-spline formulations. Monaghan (1985) discusses B-splines and other interpolating kernels. MacCallum and Zhang (1986) give the matrices for cubic and quartic B-splines. Wang (1981) discusses parametric B-spline cubic curves. Hu and Schumaker (1986) define complete splines as fitting derivative data at the boundaries. They also discuss tensor product bicubic B-splines and bivariate smoothing. Gardan and Lucas (1984) discuss Bezier and B-spline patches.

Barsky and Greenberg (1982) suggest an approach using a parametric uniform bicubic B-spline surface that interpolates the data. Solutions to sets of simultaneous equations provide control vertices, and these pseudodata force the surface to interpolate the actual data. Then a point on the surface is a weighted average of the sixteen B-spline control vertices.

Lord (1987) describes an iterative method for interpolating a topographical data set using B-splines. The control points are initially assigned the data values; then iteratively, for each datum that does not lie on the B-spline surface, move that control point away from that datum along the normal to the surface, until the surface is acceptably close to the data points. It is not clear how quickly this approach will converge.

Beta-splines are a generalization of B-splines and they provide another form of surface shape control. Two parameters, to control bias and tension, are introduced to give more flexible control of the tautness of the B-spline, and make it easier to manipulate the surface shape. Goodman and Unsworth (1986) discuss these Beta-splines. Barsky and De Rose (1985) present a special example of the Beta-spline involving only one parameter. This allows more economical computing with only slightly less flexibility.

Nielson (1986) reviews v-splines; these are another economical alternative to Beta-splines. Foley (1987b) gives a discussion of weighted v-splines. Alia and others (1987) introduce the univariate angular spline, a parametric spline whose parameter is an angle. It is not clear how, or even if, this may be extended to bivariate data.

Tzimbalario (1980) considers cardinal discrete splines using exponential Euler discrete polynomials; these splines are determined by differences instead of derivatives.

Stoer (1982) discusses Clothoidal splines, or Cornu-spirals, whose curvature is a linear function of arc length, and gives a procedure for a minimum curvature univariate Clothoidal spline. It is not clear how these splines may be applied to bivariate data.

(e) Taylor Interpolants

The expansion of a Taylor series fixes the relationships between the derivatives of a function, and this expansion may be extended to bivariate functions. If the value and derivatives are known at a nearby location, then the interpolation is an evaluation of the Taylor expansion around the data.

Several authors have used this approach to develop an interpolating surface. However, in the strict sense of the word, this method is an extrapolator rather than an interpolator because it can be applied outside the convex hull of the data and uses only one datum.

Cavaretta, Micchelli, and Sharma (1980) discuss a general method of extending univariate schemes to higher dimensions using a multivariate analog of the Taylor polynomial. Liszka (1984) proposes a method based on a Taylor expansion at each datum in a subset; then the interpolation coefficients are obtained by solving a set of simultaneous equations. Navlakha (1984) applies the Taylor expansion to trivariate data and includes an estimator for partial derivatives. Lenard (1985) uses directional differences and the Taylor expansion to develop a quintic polynomial spline.

Cavaretta, Micchelli, and Sharma (1980) also describe Kergin's extension of univariate Taylor expansions to higher dimensions. Micchelli (1980) points out that Kergin interpolation is a natural multivariate version of Lagrange interpolation. Lorentz and Lorentz (1983) discuss Hakopian interpolation, Kergin interpolation, and interpolating to a function and its derivatives (as opposed to interpolating function values).

(f) Tension Patches

Bicubic interpolation of the data, even using estimated gradients, may not provide sufficient control over the behavior of the surface between the data; certain data configurations lead to a surface that exhibits pronounced overshoots or ripples, although it interpolates altitude and slope at each datum.

This unrestrained behavior seems to be most obvious when the data are evenly spaced but highly variable, and many authors have proposed methods for constraining the representative surfaces generated for gridded data.

Control of the surface shape between the data requires both a control mechanism and a concept, or definition, of preferred shape. Additional degrees of freedom for computational control are obtained by including extra parameters in the surface generating function, but approaches for establishing and justifying a standard for surface behavior are less obvious. The basic assumption, of course, is that the "ideal" surface for a given data set has a shape that is not more complex than necessary to suit the altitudes and slopes of the data. This is termed shape-preserving, and it involves spatial properties of the surface such as monotonicity, convexity, nonnegativity, and tension.

Tension is a popular method of controlling surface shape. For example, rectangular patches can be generated by a Cartesian product of exponential splines. These splines, termed L-splines, are splines under tension, so they are less smooth than tensionless splines; the change in slope is most pronounced at the data locations. The tension parameters may be varied between the limiting situations of cubic splines and linear splines. For topographical data, these two extremes may be visualized as a minimum curvature surface or a taut elastic sheet.

Surfaces generated by tension splines are controlled by a set of tension parameters. Wever (1988) discusses the determination of these parameters. Appropriate values are obtained by the solution to a set of simultaneous equations. Rentrop (1980), for example, describes procedures for computing univariate exponential tension splines, using sinh and cosh functions. Tornow (1982) introduces a univariate exponential spline for irregularly spaced data.

More direct constraints for the generated surface, than those implied by tautness, are obtained by piecewise planes supported by subsets of data. Wherever a set of these planes are convex, the generated surface is required to be convex.

Ubhaya (1987) discusses an approximation surface that is quasiconvex in the sense that it is constrained by the upper convex hulls of subsets of data. Scott (1984) obtains convexity by making a search for sets of triangles that can be interpolated by a convex set of facets. Loh (1981) defines local convexity for B-spline surfaces.

At a finer scale, the differences between adjacent data establish a set of local slope estimates. An interpolated surface is said to be shape-preserving if its local behavior is constrained by the behavior of the slope estimates, involving monotonicity and convexity. Roulier (1980) gives definitions for these concepts, which may be used to quantify the tautness of generated surfaces.

McAllister and Roulier (1981) present an interpolatory quadratic spline with monotonicity and convexity consistent with that of the data. Mettke (1983) constructs a compromise Hermite spline interpolation for monotone and convex functions.

Bacchelli-Montefusco (1987) discusses shape-preserving tension splines with an interactive adjustment procedure. This is a desirable option for created surfaces, and could be useful for interpreted surfaces. Schmidt and Hess (1987) describe nonnegative rational quadratic splines. Nonnegative surfaces are required to represent spatial functions that are positive or zero everywhere, such as the thickness of a stratigraphic unit. Roulier (1987) uses a shape-preserving quadratic spline that maintains any convexity exhibited by the data. McLaughlin (1983) classifies the data into segments, according to these notions of shape, for alternative interpolatory treatment.

Approaches that are oriented specifically toward data of the single-valued, topographic form usually are referred to as monotonic because, over any sufficiently short interval, the surface is either nondecreasing or nonincreasing, never both as as a multivalued surface may be. Carlson and Fritsch (1985) describe a monotonic algorithm using bicubic Hermite interpolation. Beatson and Ziegler (1985) discuss a quadratic monotonic interpolation.

Similar to Beta-splines, rational splines provide the effect of splines under tension. These splines are defined as a ratio of polynomial expressions, and have some advantages over nonrational forms when interpolation is required for a function with singularities. Their application to surfaces usually involves the parametric form and tensor product. There seems to have been little application of rational curves and surfaces to topographic data for exploratory purposes.

Miller (1986) gives a general discussion of Bezier and B-spline rational surfaces. Farin (1983) describes a rational Bezier curve with weights, and Piegl (1986) introduces a recursive algorithm for rational Bezier interpolation. Forrest (1980) discusses rational cubic curve constructions that combine both conic section and parametric cubic curve forms. Farin (1983) describes recursive evaluation of a rational Bezier curve which has a constant cross ratio. It is invariant under projective transformation, and can interpolate conic sections. Abhyanker and Bajaj (1987) discuss algorithms for rational algebraic curves and surfaces.

Gregory (1986) discusses a rational spline solution to shape-preserving interpolation. Mason (1984) describes rational interpolation functions and recommends them whenever a function with an infinite range is required. Tiller (1983) reviews nonuniform, rational B-splines.

Piegl (1984) generalizes the nonrational Bezier curve and compares it to rational curves. Beck, Farouki, and Hinds (1986) note that the degree of parametric rational polynomial surfaces can be high, leading to extraneous features. Piegl (1986) presents recursive algorithms for rational curves and surfaces. Piegl (1988) introduces a rational Bezier/B-spline scheme with embeded Hermite/Coons interpolants.

Piegl and Tiller (1987) present a nonuniform, rational B-spline for freeform and analytic curves. Schmidt and Hess (1987) discuss nonnegative rational quadratic splines. Sederberg, Anderson, and Goldman (1985) discuss the transformation and intersection of rational polynomial curves. Gal-Ezer and Zwas (1988) compare rational and polynomial interpolation.

Press and others (1986) discuss the use of polynomial ratios for Padé interpolation. Other, more theoretical, work on rational interpolation includes Berrut (1988), Cuyt and Verdonk (1985), Cuyt (1987), Cuyt and Wuytack (1987), Graves-Morris and Jones (1976), Graves-Morris (1983, 1984a, 1984b), and Larkin (1967).

(g) Fourier Surfaces

Some methods require rectangularly gridded data, not to treat the data as rectangles in a piecewise manner, but for the bidirectional regularity of data spacing. These procedures require periodicity of the observations to develop and express the periodic aspects of a representative surface.

Fast Fourier Transforms (FFT) (after Cooley, Lewis, and Welch, 1967b) are applied easily to topographic data. A representative surface is developed by applying two one-dimensional Fourier series, as a Cartesian product (p. 141), giving a surface that is a linear combination of sine and cosine functions.

Although two-dimensional Fourier series have been widely applied as tools for interpolation, extrapolation, and approximation, of topographical data, results have been unsatisfactory for several reasons arising generally from the complexity of the observed phenomenon.

The double Fourier functions are periodic in two directions, so a periodic representative surface is generated; this always is possible even when the data source is not obviously periodic. However, when it is not known whether the spatial function from which the data are drawn contains periodicity, a periodic representative surface can be misleading.

Another drawback for the well known FFT algorithm is that the number of data is required to be a power of 2. Yfantis and Borgman (1981) introduce six other FFT algorithms which require the number of data to be powers of 3, powers of 5, products of powers of 2 and powers of 3, and other such products.

Complex numbers, involved in the FFT, require twice the storage and computational time that real numbers use. There is an alternative, the Fast Hartley Transform (FHT) which uses only real numbers so is twice as fast while using only one-half the storage. Bracewell (1983), Paik and Fox (1988), and O'Neill (1988), have discussed Fast Hartley Transforms.

Ferrari and others (1986) discuss another alternative to FFT which uses a B-spline filter; they report that this approach is an order of magnitude faster than the bivariate FFT. Jupp (1976) also discusses B-splines in comparison with FFT. Hayes (1986) discusses periodic B-spline surfaces. Tartar, Freeman, and Hopkins (1986) use a Fourier series to interpolate a density for points on a line.

For other papers about Fourier surfaces, see Agterberg (1969), Ayeni (1982), Davis (1973), Dimitriadis, Tselentis, and Thanassoulas (1987), Esler and Preston (1967), Franklin (1987), Georgiou and Khargonekar (1987), Good (1958), Harbaugh and Sackin (1968), James (1966), Jones, Hamilton, and Johnson (1986), Kok and Blais (1987), Kratky (1981), MacGillivray, Hawkins, and Berjak (1969), Monro (1985), Papo and Gelbman (1984), Rayner (1967), Ripley (1981), Wren (1972), Young, Pielke, and Kessler (1984), and Whitten (1967).

(5) Neighborhood-Based Interpolation

This is the most general interpolation method available, generating a stable and robust surface for all forms of topographical data. It incorporates several desirable aspects that are analogous to properties of traditional and well-known methods. When describing the construction of this estimate, reference is made to these established concepts to reveal the relationships between neighborhood-based interpolation and other methods.

Two, three, or four local coordinates (p. 75) have been used in many interpolation methods designed for certain patterns of data, such as triangulated or rectangularly gridded data. To be general, however, a method for contouring topographical data must employ a variable number of local coordinates, to suit the configuration of the data. Any alternative approach runs the risk of diluting the information in the data when a variable number of local data surround the interpolation point.

(a) Linear Interpolation

This is the simplest form of neighborhood-based interpolation and it uses natural neighbor local coordinates (p. 81) as weights, just as interpolation by linear triangular facets use barycentric coordinates (p. 130) and bilinear rectangular patches use rectangular coordinates (p. 139). However, this form of interpolation does not generate slope discontinuities anywhere between the data, and contour lines will be smooth, with no sharp corners.

Local coordinates are a set of proportions, fractions of a whole. So the sum of these natural neighbor coordinates, at any location, is one, although the number of natural neighbors may change from location to location. The coordinate associated with a given datum varies smoothly between one, at the datum, and zero where the interpolation point ceases to be a natural neighbor of the datum.

Neighborhood coordinates are used as weights for a weighted average of the subset of data which are natural neighbors of the interpolation location. This results in a surface that is smooth everywhere except at the data locations. Neighborhood-based linear interpolation surfaces form cone-like peaks or pits at the data; the surface is smooth because the required change in slope is concentrated around the data locations.

Interpolation

The surface developed by neighborhood-based linear interpolation may be compared with that obtained by triangle-based linear interpolation or rectangle-based bilinear interpolation. When neighborhood-based linear interpolation is applied to three data, the interpolated value at the triangle centroid is the arithmetic average of the three data, and at each vertex the surface surface has the same altitude as that datum. This surface is similar to the plane that is formed by triangle-based linear interpolation, but the surface is curved above, or below, the plane along the triangle edges.

When neighborhood-based linear interpolation is applied to four data at the vertices of a rectangle, the interpolated value at the rectangle centroid is the arithmetic average of the four data, and at each vertex the surface has the same altitude as that datum, as does rectangle-based bilinear interpolation.

Neighborhood-based linear interpolation of four data appears similar to the isometric view of a bilinear surface in Figure 2.5.3 on p. 139, but the surface is not a rectangular hyperboloid, and usually does not contain a straightline. This is because, unlike rectangular local coordinates, the four neighborhood coordinates are positive everywhere inside the circumcircle of the data, so the surface is curved along rectangle edges. More importantly, neighborhood-based linear surfaces will have a continuous slope everywhere between data, whereas linear triangle, or bilinear rectangle, based surfaces have abrupt changes of slope along lines between data.

Neighborhood-based linear surfaces have an abrupt change of slope at the data points. A cone-shape is formed in the surface, just as if each datum were forcing the surface to be nonflat, and the change in slope due to that datum is concentrated around its location. Figure 2.5.7 shows a perspective view of neighborhood-based linear surface that illustrates the cones at the data and the taut regions between the data.

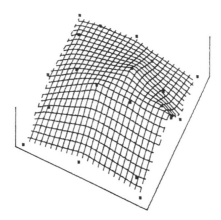

Figure 2.5.7 Neighborhood-based linear interpolation

To obtain a neighborhood-based linear surface with program CON2R.BAS or NEIGHBOR.BAS, set the parameter **CL**=0, in line **1100**.

(b) Nonlinear Interpolation

The cone-shape in the linear surface, at each datum, indicates the amount of change in slope, due to that datum, that must be accommodated in a representative surface that is smooth everywhere, including the data points. The gradient estimates and variability indices (p. 157) provide the information that governs this distribution of slope change.

(i) Gradients

Because a neighborhood-based linear surface does not have a slope at each datum, an estimate must be made and adopted as the slope of a nonlinear representative surface, at that datum. Estimation of gradients is discussed on p. 85; each estimate is based only on the data which are natural neighbors of the estimation point.

(ii) Blending Gradients

The gradients are blended in with the linear interpolation, just as was done for IDWO on p. 116, and for triangle-based interpolation on p. 135. The representative surface agrees with both the altitude, and the estimated gradient, at each datum. In effect, a linear interpolation of the altitudes is combined with the projected gradients as a form of Boolean sum (p. 146) for the regions between data.

The linear surface provides the basis, and a proportion of each gradient is added to or subtracted from it; this, effectively, distributes the difference in slopes between the estimated gradient and the linear surface over a small region around each datum, and is achieved by varying the influence of the gradient, from one at the datum to zero where it ceases to be a neighbor. More importantly, this approach causes the generated surface to be bounded by the conservative linear surface and the gradients established by local trends. Anomalous unsampled peaks in the spatial function may be inferred from the local slopes as an amendment to the linear surface.

Three parameters are used to control the gradient blend. Two of these are available for user adjustment of the generated surface, and are discussed on p. 170; the third parameter is the local variability index, mentioned next, which allows more flexibility when the heights of adjacent data are highly variable or strongly inconsistent.

(iii) Variability Indices

Local control of surface tautness requires an estimate of local variability such as the outlier index (p. 173) or the roughness index (p. 171). Then the final interpolation depends upon the altitudes, the estimated slopes, and the estimated local variability, for each of the data that are natural neighbors of the interpolation location. The influence of each datum is approximately proportional to its associated natural neighbor coordinate under these constraints.

Interpolation

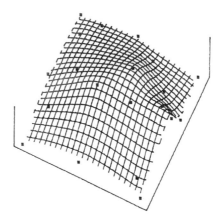

Figure 2.5.8 Neighborhood-based blended interpolation

Figure 2.5.8 shows a neighborhood-based curvilinear surface that interpolates the data exactly and is smooth everywhere within the convex hull of the data. A tighter or puffier surface is obtained by varying the blending parameters.

To obtain a neighborhood-based curvilinear surface with program CON2R.BAS or NEIGHBOR.BAS, set the parameter CL=1, in line 1100. Parameters BI and BJ, in line 1100, control the blending functions, and these are discussed on p. 170.

Sibson (1980a, 1981) introduced natural neighbor coordinates, and reported neighborhood-based nonlinear interpolation using estimated spherical quadratic gradients. There have been few accounts of applications of this method. Moffat and others (1986) made representative contours of a stratigraphic surface using natural neighbor coordinates. Laslett and others (1987) discuss natural neighbor linear interpolation blended with spherical quadratic gradients; the blending weights are selected to give an exact fit for data from a spherical surface. These authors point to the differences between distance-based and neighborhood-based interpolated surfaces for unevenly scattered data. Watson and Philip (1987) demonstrated neighborhood-based interpolation of a set of traverse data. They also showed that the consistency of this method allowed map-wide, low amplitude, features to be extracted from satellite data.

A few authors have treated neighborhood-based interpolation without using neighborhood coordinates. Pal (1980) introduced a polygonal surface patch with any number of sides, and using cubic Bezier curves along each side. The resulting surface is a hybrid, using both intrinsic and parametric polynomials. This approach can be considered to be neighborhood-based interpolation because it is local, and allows a variable configuration of data. Kennedy and Tobler (1983) consider a related problem of neighborhood averaging among a set of polygons. They use a weighting by adjacent polygon edge lengths, which is a distance-based approach.

Goodchild and Lam (1980) use area weighted averages. Bayer (1985) discusses the generalization of linear triangle-based interpolation to any convex polygon, that is, linear neighborhood-based interpolation. He obtains a linear surface over the polygon by first transforming the polygon to the unit circle, where a harmonic surface function is determined by applying Fourier series, and then transforming back to the original polygon. The resulting surface appears as a taut membrane stretched between the data at the vertices of the polygon.

TABLE 5 Synoptic Tabulation of Computer Interpolation Methods

Method	Assumption	Characterization	Disadvantage	Page
Distance-Based	---	---	---	113
IDWO	distance dependent	approximates	flat spots	114
IDWG	distance dependent	proximal polygons	flat spots	116
Fitted Function	---	---	---	118
Lagrange	analytical	classical	overshoots	119
Collocation	characteristic curve	smooth	matrix inversion	120
Min. Curv.	characteristic curve	smoothest	matrix inversion	122
Kriging	characteristic curve	probabilistic	matrix inversion	123
Relaxation	dense data	iterative	data bounded	124
Approximation	data errors	regression	generalization	126
Triangle-Based	---	(area-based)	---	130
Linear	correct triangulation	faceted	triangulation dependent	130
Nonlinear	correct triangulation	uses gradients	triangulation dependent	131
Rectangle-Based	---	(area-based)	---	137
Bilinear	dense data	faceted	double-univariate	139
Hermite	dense data	uses gradients	matrix inversion	143
Bezier	dense data	probabilistic	control points	147
B-Spline	dense data	probabilistic	control points	149
Taylor	analytical	n derivatives	matrix inversion	151
Tension	preferred shape	shape-preserving	monotonic	152
Fourier	periodic	trigometric	double-univariate	154
Neighborhood-Based	---	(area-based)	---	155
Linear	no assumptions	local data	bounded	155
Nonlinear	no assumptions	uses gradients	slow	157

DISCUSSION

Neighborhood-based linear interpolation with blended gradients may be the most general and flexible scheme. It can be applied to all patterns of data, and a stable surface always is obtained. Because it is a weighted average of local data, any size of data set may be treated in time proportional to the number of data. The ability to adjust the tautness of the representative surface allows scope both to study a wide variety of spatial functions, and to study a particular data set by varying the tautness. Because it is a blend of surfaces, rather than the result of a single interpolator, the generated surface is a conservative estimate of the data variability.

Interpolation

Although practical problems of data insufficiency usually make impossible a true, or ideal, surface to represent the sampled spatial function, the interpolation method employed should be expected to have certain general properties of ideal interpolation (p. 103). Neighborhood-based interpolation may be judged against those criteria.

To give assurance that it is not degrading the information in the data, an ideal interpolation method should be able to generate a representative surface for any configuration, pattern, or density variation, of data, to within prescribed limits, as neighborhood-based interpolation does. In comparison, other interpolation methods are unable to develop an acceptable representative surfaces, or else generate spurious features, especially for extremely erratic data, and so indicate their partiality. Some interpolation methods require regularly spaced data, but in some situations measurements cannot be obtained at the nodes of an assigned grid for practical reasons. Variations in the density of data, across a region, also may be desirable from an economic point of view. However, methods that use a subset of data selected by distance behave unsatisfactorily when the data are scattered unevenly over the region. When data density is anisotropic, such as a set of seismic traverses, fixed distance or fixed area subsets cannot be used.

An ideal interpolation method must produce a continuous and smooth surface, or else the method is demonstrably inconsistent. Blended gradients provide a surface with a continuous first derivative and a second derivative that vanishes at each datum. In contrast, some methods that are based on running subsets produce surfaces with discontinuities at locations where a datum is added or dropped from the subset; such methods are unable to provide a single-valued surface. This also implies that the surface is unrepresentative, for the data set as a whole, on at least one side of the discontinuity.

An ideal method must be locally based in order that the representative surface be stable; this prevents widespread changes in the surface resulting from local perturbation, and so accommodates some observational error. Some methods are global methods in the sense that the height of the generated surface at any location depends on all the data and not just nearby data. This global dependence is unsatisfactory whenever there is a possibility that any given datum is erroneous, because that error is propagated throughout the region.

An ideal interpolation method must allow adjustment of surface tautness to suit the data; otherwise one cannot be sure that the style of the interpolator is not imposed on the data. Blended gradients allow a two-parameter adjustment of surface tension. Some methods, such as those based on minimum curvature criteria, are limited to surfaces of fixed tautness. Methods that provide surfaces of fixed form, or style, may be suitable for certain types of topographical data. For example, physical functions such as gravity are considered to vary inversely with the square of distance, and may be well represented by a polynomial surface. On the other hand, altitude data from a mountainous region may be poorly represented by such a polynomial surface. However, most topographical data are samples of a spatial function with an unknown variability.

In these terms, neighborhood-based interpolation with blended gradients provides representative surfaces that are the most stable, flexible, and general, in the sense that it incorporates the properties of an ideal interpolator. Figure 2.5.9 shows isometric views of four interpolated surfaces. Clockwise from upper left, these are IDW with blended gradients, triangle-based linear facets, neighborhood-based with blended gradients, and a minimum curvature spline surface. The neighborhood-based blended surface may be as smooth as the spline surface, or may give tighter curves where this provides a more conservative surface; in addition, it is not limited by size of data set.

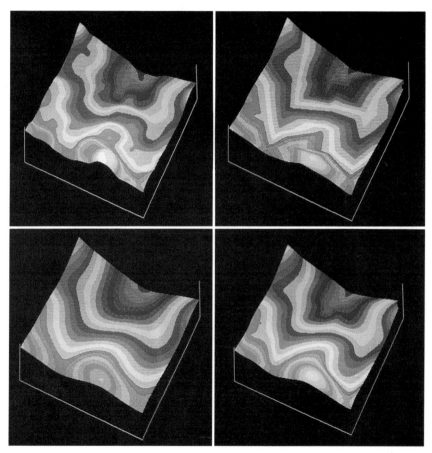

Figure 2.5.9 Comparison of four surfaces

SECTION 2.6

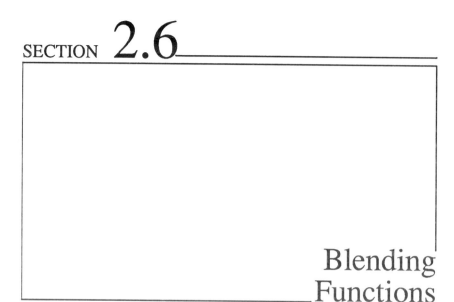

Blending Functions

SECTION SUMMARY

Blending functions control the shape of a representative surface. They fall into two distinct groups: global blending functions are applied to the entire data set, and local blending functions are applied to a subset of the data.

INTRODUCTION

An interpolating surface is determined easily at each datum by using the known altitude and estimated gradient. Blending functions are numerical bridges, providing arches between known locations, and so determining the shape of a representative surface in the unknown regions between the data. They control interpolation by governing the extent to which the influence and effect of individual data are projected into the surrounding surface. Blending functions also are termed basis, bell, haystack, Prussian helmet, or bridging, functions.

Blending functions classify neatly into two general types, local and global. Global functions are those that may be nonzero anywhere in Euclidean space, such as a tilted plane. Local functions are those that may nonzero only within a specified boundary; these functions are centered on the data locations and are zero everywhere outside their boundary. This distinction determines the two approaches to synopsis of topographical data, fitted functions and weighted averages, mentioned on p. 104.

GLOBAL BLENDING FUNCTIONS

These functions are defined by expressions that may be evaluated anywhere in Euclidean space, and the functional value may be nonzero anywhere. This implies that global blending functions allow both interpolation and extrapolation. A data set may be fitted precisely, or only approximately, according to the amount, and configuration, of the data. A surface is formed by a weighted average of global blending functions from a particular family. The weights usually are given by the solution to a system of simultaneous equations and, being constant, may be applied to give a surface value at any location. Of course, any combination of blending functions with infinite domains also has an infinite domain.

The use of global blending functions may be recognized, in an interpolation method, by the characteristic that the presence or absence of a particular datum may have an effect on the generated surface anywhere within the region. When the influence of a given datum can only perturb a limited region, the blending functions are local (discussed on p. 167).

(1) Power Functions

These monomial blending functions are straightforward powers of the location coordinates, the usual x and y, and recognizable as the components of a typical polynomial. Some examples are $1, x, y, xy, x^2, y^2$, and so on.

The first example, the number one, is the simplest blending function; it can be thought of as a zero-degree polynomial having only a single, constant, term. This function is used in the statistical method (p. 106) giving a level plane, at a height that is the arithmetic average of the heights at the data locations. This is the situation where the representative surface is a constant $(1/N)$ times the zeroth power of location, that is, one. The statistical method provides a surface which, some would say, is a poor representation or approximation, and this example may seem trivial to the point of vacuity. However, it does illustrate the way in which the power functions express the global influence of the data.

More complex surfaces can interpolate, as well as approximate, the data, and are built by using higher degree power functions. The weights usually are established by solving a system of k simultaneous equations when k functions are used. For example, a combination of the monomial functions $1, x, y, xy, x^2, y^2, x^2y, xy^2, x^3$ and y^3 requires $k=10$ equations and is a third degree polynomial.

Such representative surfaces are built by combining "tiles" whose upper surfaces, and so their thickness, vary as powers of geographical coordinates. Each tile may be thought of as a prescribed variation in the influence attributable to each datum. Although third or fourth degree polynomials often give satisfactory approximations to a surface representing the data (p. 126), if higher degree blending functions are used to interpolate, their behavior between adjacent data can seem to be unreasonable (see Lagrange interpolation, p. 119).

Power functions are linearly independent, meaning that none can be equaled by a sum of proportions of the others. Other, more complicated, power functions can be used that also are orthogonal, which implies that the inner product of any two of them is zero.

Grant (1957) pointed to the special advantage of using orthogonal polynomials; then the behavior of the surface between data is constrained by the orthogonality condition. These blending functions also can be expressed in terms of ratios of polynomials (p. 153), but are computed less easily. Mason (1984) discusses possible choices of blending functions and recommends Chebyshev polynomials.

(2) Trigometric Functions

A certain combination of selected members from the family of power functions can be used as a surface that approximates a topographical data set, and the trigometric functions, sine, cosine, and tangent, along with their inverses, may be used in a similar way.

Trigometric functions have an obvious appropriateness when the data are expected to exhibit periodic patterns in altitude variability. The most well known of these, the double Fourier series fit, has an extensive literature, and is discussed on p. 154. A major disadvantage of trigometric blending functions is that an apparent, but not necessarily justified, periodicity may be induced in the behavior of the representative surface between data.

(3) Nonfinite Kernels

Blending functions that concentrate the influence of a datum in the subregion immediately around it, can be thought of as a kernel function in the sense that this central subregion has most significance for the representative surface. A finite kernel function has a zero value everywhere outside a specified limit, and these are discussed on p. 167. Nonfinite kernels do not have a specific limit or boundary, and the influence of a given datum extends, although weakly, throughout the region.

(a) Global Inverse Distance

The most obvious examples of nonfinite kernel blending functions are the global inverse distance weighting schemes, mentioned on p. 112, and discussed on p. 114 and following. With respect to a given interpolation point, these methods establish a weight for each datum, although most weights are vanishingly small, and the interpolated altitude is just a globally-based weighted average of the data.

Blending Functions

When the influence of a datum is made to decrease as its distance from an interpolation location increases, the inverse relationship causes the datum's influence to fall off, somewhat similar to a decay function. For N topographical data, this blending function produces N radially symmetric kernels, each centered on a datum, and nonzero everywhere.

Two difficulties emerge when the inverse distance is used to provide a weight that decreases as the distance increases. First, as the distance reduces toward zero, the inverse approaches infinity, and an infinitely large weight is not computable. Obviously, the appropriate weight for a datum, when interpolation is at the datum location, is one. Second, as the distance increases, the inverse remains positive, and the influence of a given datum never becomes zero.

The first problem is solved by collecting all the inverse distances for the set of data, and converting these to proportions of unity. This is done by dividing each inverse by the sum of all the inverses, and involves two conventions: (i) a ratio of infinite numbers is equal to one, and (ii) any finite number divided by an infinite number is equal to zero. Then the sum of all the weights is one, and when interpolation occurs at a datum location, the weight for that datum is one, while all other weights are zero. See the expressions for IDWO on p. 112 and p. 114.

The second problem is often handled by additional "cut-off" conditions, and is discussed in the subsection on local blending functions, on the next page.

Other distance weighting schemes are exponential. Goodin,, McRae, and Seinfeld (1979) discusses distance weighting schemes which behave similarly to inverse distance, and are variations of the expression $w_i = \mathbf{EXP}(-d_i)$, where d_i is the distance to the ith datum from the interpolation point.

(b) Minimum Curvature

These distance-based blending functions provide a surface which, at any point, has the least curvature possible, consistent with the restraints of the data. When the generated surface is required to meet such a minimum curvature criterion, which minimizes the integral of the squared second derivatives, the blending function kernel is defined in Hilbert space.

Various authors have given the expressions $d^2 \log d^2$, $d^2 \log d$, and $d^2 (\log d - 1)$, where d is Euclidean distance. The mathematical details are outside the scope of this book, but may be seen in papers by Meinguet (1979, 1983), Franke (1985), and Villalobos and Wahba (1987). Interpolation is discussed on p. 122.

(c) Gaussian

A Gaussian kernel causes the influence of a datum to fall off as a normal distribution, the well-known bell curve. This approach is most appropriate when it is known that the data come from a spatial function with probabilistic properties. Enriquez, Thomann, and Goupillot (1983) discuss the function $d^{(3/2)}$, and suggest that it represents the best estimation of random functions.

LOCAL BLENDING FUNCTIONS

The kernels of influence for local blending functions are finite, and designed to give a datum a zero influence outside of the local subregion, so insuring that the contribution of any datum to the generated surface is only local. This implies that the use of local functions, in an interpolation scheme, may be recognized by the characteristic that the presence or absence of a particular datum has no effect on the generated surface beyond certain limits around the location of the datum. This is very useful to inhibit the propagation of errors.

Local blending functions allow a piecewise approach to interpolation. For each interpolation point, the blending function is computed separately for each datum in the local subset, so the influence of a datum is regulated only by the local coordinate of the interpolation point. This implies that a single combination of influences, applicable to the whole region, cannot be obtained because the combination of local data changes with location. In general, each interpolation requires the generation of a new set of weights, and possibly for a changed subset of data.

Although it may seem, at first, that using local blending functions requires much more computation than using global functions, when the influence contributed by a datum has a finite nonzero domain, each datum only needs to be processed for interpolation locations within the range of its blending function; this is because its influence is zero for other locations. So, since interpolation with local blending functions involves only a subset of the data for a given interpolation location, this approach is only slightly slower than interpolation using global functions. The advantage is that the generated surface, as a whole, may be more complex, and so provide a more detailed portrayal of the variability in the topographical data.

(1) Explicit Local Functions

The simplest local blending function has the profile of a box, or square wave function, with a flat and level top, providing a surface at the height of the datum and with a discontinuous change to zero at the limits of the range. This is the function used for constant value proximal polygon interpolation (p. 107), where the surface is made up of flat top, proximal polygonal prisms, and appears as a set of terraces and cliffs. In this situation the representative surface is discontinuous at the boundary of each polygon.

To remedy this, two major modifications can be made. (i) If the blending function is specified to decrease gradually as the interpolation point moves toward the limit, the representative surface can be continuous, although the slope will be discontinuous. (ii) When the local blending function also is specified to have a zero first derivative as it becomes zero, the slope is continuous, and so the representative surface can be continuous with a continuous slope.

As an example, boxcar functions with a circular boundary are used for manual contouring of density; for a particular instance, the number five contour, on such a manual density map, is the locus of points traced out by the center of a circle that always contains five data while it moves over the region.

The obvious disadvantage of the boxcar blending function, in such an approach, is the discontinuous surface that is generated. This suggests that some kind of bell-shaped function, which would taper off more gradually, could allow continuity in the generated surface by matching along boundaries. It often is desirable to match the slope along the boundaries also. To do this, the blending function must have zero derivatives both at the datum and at the limits of influence. The maximum of the function occurs at the datum and its effect approaches zero at the limits of its range. Such bell-shaped blending functions, give a continuous slope to a surface so there are no sudden changes of slope.

Although local blending function are univariate, their bivariate effect is compounded by the plan view of the influence kernel, which may be circular, elliptical, rectangular, a Voronoi polygon, or a union of natural neighbor circumcircles (p. 171).

(a) Local Inverse Distance

When unbounded functions, such as inverse distance weighting (p. 165), are used as local influence functions, their contribution is local only because they are applied to a subset of data, and so have a truncated nonzero domain. A datum has a positive inverse distance weight until it is dropped from the running subset; then its influence becomes zero and a discontinuity occurs in the representative surface.

This problem is often handled by additional functions which are incorporated to abbreviate the influence more smoothly, and even early variations of inverse distance weighting included adapted functions to limit the domain within which the weight was nonzero, see Goodin, McRae, and Seinfeld (1979). This usually is a circular domain. For example, the weight for the ith datum may be

$$w_i = ((R^2 - d_i^2)/(R^2 + d_i^2))^p$$

where R is maximum radius of influence, d_i is the distance to the ith datum and p is an arbitrary power.

These weights must be normalized to make them sum to one before they are applied to the data; this is done by dividing each weight by the sum of all the weights.

Feng and Riesenfeld (1980) discuss blended gradients in terms of Boolean sums of component functions with circular and square support regions. When the data are much more dense in one direction than in the other, an elliptical boundary for the blending function can be useful (for example, Smith, 1968). Generally, the radial symmetry of circularly-based influence ascribed by these functions is unrealistic for scattered data, or even gridded data that has much variability.

(2) Parametric Local Functions

Contouring of topographical data is seldom done with the parametric blending functions because their greater flexibility makes them more suitable for created surfaces than exploratory surfaces. However, interpolation, using these functions, is discussed with the rectangular patch methods, because their algorithmic similarities offer insights into computer methods in general. For the same reason, parametric blending functions are mentioned briefly in this section for comparison with other blending functions.

The parametric blending functions are defined on the interval zero to one. They are applied by using a transformation between the actual interval and the unit interval. This is useful, and makes parametric blending functions widely applicable, particularly for creating closed surfaces such as solid objects.

(a) Hermite Polynomials

These blending functions were discussed, on p. 143, from the point of view of their use for bicubic Hermite interpolation. They provide the smoothest possible surface over the interval in the sense that it is formed by a Cartesian product of minimum curvature piecewise cubic polynomial splines. Many variations have been used; Cadete (1987) discusses a form of cubic Hermite polynomial blending function that is expressed as a ratio of polynomials.

(b) Bernstein Polynomials

These polynomials are the blending functions for Bezier patches (p. 147), and they have an interesting probabilistic interpretation. These functions reflect the ways in which a set of colored balls may be drawn from an urn. Goldman (1983b, 1986) describes the Bernstein polynomials in terms of these urn models, and shows how such models can be used to generate other blending functions. For example, Timmer (1980) introduces a variation of parametric cubic curves that approximate the data, but more closely than Bezier patches, or B-splines. Timmer uses

$$\begin{aligned} t_{z1}(w) &= (1-2w)(1-w)^2 \\ t_{z2}(w) &= 4w(1-w)^2 \\ t_{z3}(w) &= 4w^2(1-w) \\ t_{z4}(w) &= w^2(2w-1) \end{aligned}$$

(c) B-splines Basis Functions

These functions are used as weights on the control vertices for B-spline approximation (p. 149). Like the Bernstein polynomials, they have a probabilistic interpretation (see Goldman, 1983b, 1986).

(d) Serendipity Functions

These basis functions interpolate between univariate curves, and seldom are applied to topographical data. Ball (1980) and Zlamal (1973) describe these blending functions, obtained from the inverses of Vandermonde matrices.

(3) Compound Exponential Function

This blending function is termed a compound exponential because it is composed of two exponential functions, joined together. For any value of w between zero and one, and any value of r greater than one, a graph of the exponential function w^r is an increasing curve with an increasing slope. The slope is zero at $w=0$, and it increases smoothly until it is r at $w=1$. A second graph of this function may be rotated by 180° and positioned so that the two slope values of r coincide. This forms an S-shaped curve with zero slope at both ends.

The proportions between the top and bottom of the "S" may be adjusted by a nonlinear dilation of the function. Three such curves are illustrated in Figure 2.6.1 below.

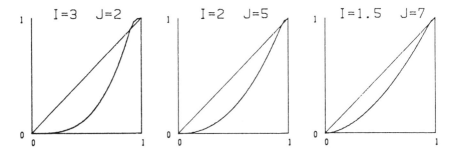

Figure 2.6.1 Exponential blending functions

The purpose of this blending function, similar to others mentioned in this Section, is to constrain and moderate the influence of each datum. It uses a local coordinate (p. 75) and an estimate of local variability (p. 157) for each datum in the interpolation subset. However, this function also uses two assigned parameters, referred to as **I** and **J**, to give the user control over the maximum slope and the lateral displacement of the maximum slope in the S-shaped curve. This determines both the amount and the placement of gradient influences. The effect is that the resulting surface can varied from taut, with peaks at each datum, to puffy and rounded.

The compound exponential blending function is defined on the unit interval, zero to one. The value returned by this function, for a given local coordinate, w, of an interpolation point, and a local variability estimate, r, is applied as a weight to the difference between the linear interpolation and the gradient plane (p. 157).

This function specifies the proportion of the gradient that will be appended to the linear surface. Specifically, this proportion is expressed as

$$H(w,r) = [(\{2[w^{(J*r)}]\}(I+r))/2](J_*r)^{-1} \text{ if } w \leq 0.5$$
$$= 1-[(\{2-2[w^{(J*r)}]\}(I+r))/2](J_*r)^{-1} \text{ if } w > 0.5$$

where the assigned parameters, **I** and **J**, control the amount, and distribution, of gradient influence in the generated surface. Larger values of **I** give more influence to the gradient and larger values of **J** increase the peakedness of the surface at the datum.

The representative surface is obtained by computing this proportion for each datum in the interpolation subset, relative to the interpolation point. The surface is a linearly weighted average of the data plus a nonlinearly weighted average of the gradient contributions. The values of r, the local variability index, are determined for each datum as either the roughness index or outlier index. The boundary of the influence kernel is the boundary of the set union of all the natural neighbor circumcircles that involve each datum.

Compared to interpolation methods discussed earlier, in which the surface was formed from a weighted average, and the weights were either the local coordinates of the interpolation point or a function of these coordinates, the surface here is formed from a weighting derived as a function of the local coordinates and a function of the variability indices. This allows the the surface to conform to the local trends in the data even when these trends are asymmetric and anisotropic.

(a) Roughness Index

The roughness index is a method of assessing the local variability of the data, and this is important because such roughness affects the reliability of interpolation. The roughness index is based on the natural neighbors of each datum; each contiguous pair of these neighbors in conjunction with the datum, defines a natural neighbor (Delaunay) triangle. Each datum has a local variability as determined by these triangles.

To construct this roughness index, so that it is comparable between data sets, the data must first be normalized in the sense that they are transformed to lie in a unit cube, with the z coordinate being full range and the more extensive of the x and y coordinates being full range. Then if the triangles formed by the natural neighbors of a datum, with the measured heights, are treated as vectorial structures in three-dimensional space, cross products (p. 33 and p. 94) for each triangle can be calculated using any two sides of the triangle.

For each triple of data {A,B,C}, with coordinates A=(x_1, y_1, z_1), B=(x_2, y_2, z_2), and C=(x_3, y_3, z_3), the cross product is a perpendicular vector, (x_N, y_N, z_N), where

$$x_N = (y_2-y_1)*(z_3-z_1) - (y_3-y_1)*(z_2-z_1)$$
$$y_N = (z_2-z_1)*(x_3-x_1) - (z_3-z_1)*(x_2-x_1)$$
$$z_N = (x_2-x_1)*(y_3-y_1) - (x_3-x_1)*(y_2-y_1)$$

Blending Functions

Then the area of the triangle enclosed by {A,B,C}, which is one-half the length of the normal vector (x_N, y_N, z_N), is given by the expression

$$\text{Area} = ((x_N^2 + y_N^2 + z_N^2)^{1/2})/2$$

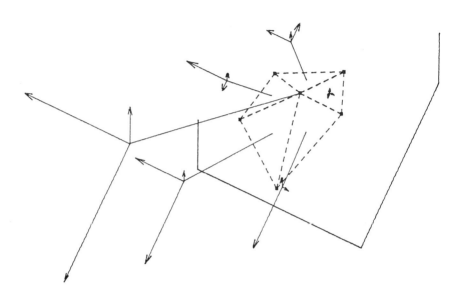

Figure 2.6.2 Roughness index

The three-dimensional cross product vector is perpendicular to the triangle face, and with a length proportional to the area of the triangle. However, the vector may have a negative z-value, which is the height component, and so must be converted into the opposite, positive, sense. If z_N is less than zero then multiply all three components, x_N, y_N, and z_N, by negative one.

At each data point, the vector sum of cross products (each taken in the positive direction with respect to the height) of all the triangles that it shares can be accumulated. The plane perpendicular to this vector sum can be thought of as a gradient plane through the data point in the sense that it is a spatial average of slopes and sizes of the surrounding triangles. This implies that computation of the roughness index does not require the direct manipulation of areas; the ratio of the normals gives the same value because the length of each normal is proportional to the area of its triangle.

The complement of the ratio of the area of all the triangles surrounding a given data point divided into the area of these triangles as projected onto the gradient plane of the data point is taken as a measure of local roughness for each data point.

Specifically, if r is the roughness index, then

$$r = 1 - A_P/A$$

where A is the area of a triangle group and A_P is the area of that group as projected parallel to the vector sum of cross products. Then r is zero for a triangle group that lies entirely on some plane and approaches one for a triangle group with a large variation. This unitless parameter is local because it is based only on those data that share triangles with the given datum.

The roughness index can serve as a indication of the dependability of the interpolated surface; high roughness indices show that there is lack of consistency among the data to be interpolated, while conversely, low indices mean that adjacent data are concordant, and the spatial function is sufficiently well known to allow confident interpolation.

A measure of local variation provides an important controlling parameter for interpolation procedures because it allows automatic adjustment of weighting and blending operations to suit the data. The roughness index is used this way in the programs. When cross product gradients are computed, starting at line **2460**, the roughness index is computed and stored in **P1#(I0%,6)**. Cross product gradients and roughness indices may be computed for all the data by setting SL=1, in line **1100**.

Fisher's circular standard deviation may be computed for the set of cross product vectors, as an alternative estimate of local variability (Philip and Watson, 1982b), although comparisons of the roughness index and Fisher's method are inconclusive because the typical sample size (six) of vectors (triangles) per datum is too small for reliable results from such statistical methods. The roughness indices for a data set, or data subset, also may be used to estimate an average spatial roughness parameter for a comparison between subregions.

Hobson (1967, 1972) considered three aspects of surface roughness (p. 88), estimated by computing normals, and areas, of square and triangular patches; the distribution of roughness was displayed by contour maps. Turner and Miles (1967) implemented Hobson's (1967) measure of roughness, using unit normal vectors to groups of three data.

(b) Outlier Index

It is not unusual for a particular datum to be atypical of the other data in the set. Even more often, a datum may be typical of the whole data set but not be typical of its natural neighbor subset. This is the datum that influences the shape of the interpolated surface to a greater extent than any of its natural neighbors, so it is important to identify such a datum.

The amount by which any datum outlies its natural neighbors can be estimated by the difference between it and the surface implied by those data alone. This is a form of cross-validation. Wahba and Wendelberger (1980) define other methods for cross-validation of an interpolated surface.

Blending Functions

When a neighborhood-based gradient is computed (p. 98), a linear interpolation is made using neighborhood coordinates of the datum location. This is a point on a surface formed by a linearly weighted average of heights of the natural neighbors of the datum, and in this sense shows their joint height at the location of the datum whose gradient is being estimated.

The difference between the height at a datum and a weighted average of its neighbor's heights gives a signed magnitude estimate of the change that will have to be incorporated in the interpolated surface due to that datum. If r_i is the outlier index, then

$$r_i = f(x_i, y_i) - L(x_i, y_i) = f(x_i, y_i) - \sum w_j f(x_j, y_j)$$

where the $f(x_j, y_j)$ are heights of natural neighbors of the datum at (x_i, y_i), and $L(x_i, y_i)$ is the linear neighborhood-based surface at (x_i, y_i). Summation is over j from 1 to N natural neighbors of datum i. Figure 2.6.3 illustrates how a datum may be an outlier with respect to its natural neighbors; the difference in height, either positive or negative, is the outlier index.

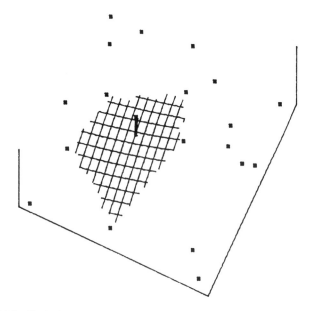

Figure 2.6.3 Outlier index

If a datum is on the perimeter of the data set, it may not be fully bounded by natural neighbors, and may be outside the convex hull of the remaining data. In that situation, because neighborhood-based interpolation can only be done within the convex hull of the remaining data, the outlier index is not available. The roughness index is an alternative.

Contours of the outlier indices of a data set reveal subregional differences by the pattern of values. Outlier indices are predominantly negative where the topography has the form of a basin or syncline, while they are positive for domes and anticlines. Closely spaced contours of outlier indices indicate that the data are erratic, that is, they are insufficiently dense for the scope of variation in their heights.

Automatic recognition of such subregional differences allows the interpolation parameters to be adjusted. For example, in a subregion where the outlier indices are close to zero the interpolated surface may undulate smoothly through the data while large indices develop a more taut surface. In particular, the tautness may be adjusted around each datum according to its outlier index.

This is achieved by using the outlier index as a parameter modulating the blending function as applied to the gradient of the associated datum. Because the blending function controls the extent to which the linear interpolation is mitigated by the gradient estimate, the surface is made nearly linear in the immediate vicinity of a datum with a large positive or negative outlier index. The overshoot, that may occur with unmodulated interpolators, may be diminished or avoided entirely.

Starting at line **2170** in the programs, the outlier index is computed and stored in **P1#(I0%,6)** at line **2450**. Natural neighbor linear interpolation gradients and outlier indices may be computed for all the interior data by setting **SL=0** and **CL=1**, in line **1100**.

SECTION **2.7**

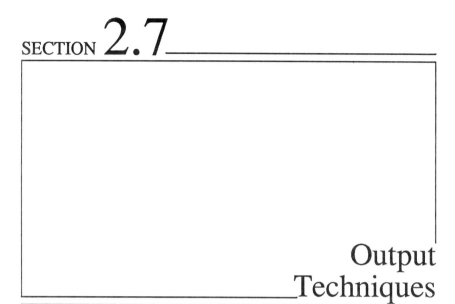

Output Techniques

SECTION SUMMARY

This Section covers a variety of topics involved in presentation of the analysis: display media, output databases, rectangular and triangular grids, following contours, isometric views, interval filling, color-filled isometrics, orthogonal profiles, stereograms, numerical output, integration, and volume-products.

INTRODUCTION

Pictorial and numerical output provide the culmination of the procedures and mechanisms discussed in previous sections. The techniques in this section are applied to the interpolated grid, the output database. It has been thought that computer contouring, as a subject, is limited to the topics covered in this section, but generating the representative surface requires those preliminary steps.

DISPLAY MEDIA

The pictures produced by computer contouring, which are images of the output database, are usually displayed on a monitor screen, and a screen dump to a dot matrix printer can provide hardcopy of the screen picture. A major alternative is a pen-on-paper plotter.

The first significant difference between these media is resolution. Because plotters usually have a higher resolution than monitor screens or dot matrix printers, smoother curves can be drawn. The "staircase" effect is not apparent in curves drawn by plotter because small increments are less apparent when a line is drawn diagonally.

The second difference is that most monitor screens are set up for matrix indices, whereas plotters operate in Cartesian coordinates. By convention, a pixel on a monitor is addressed as a row-column pair and a point on a plotter as an x-y pair. Notice that the order is reversed; x, as in column, is the abscissa while y, as in row, is the ordinate. This can cause confusion.

Both monitor screens and pen-plotters have their abscissa, or x value, arranged so that values of x increase toward the right. The ordinates are different; on plotters the ordinate is arranged so that values of y increase toward the top of the page, while on monitors, following matrix conventions, the ordinate increases toward the bottom of the screen. The location of origin, (0,0), usually is in the lower left corner of plotters and the upper left corner of monitor screens.

A straightline is drawn, on either monitor screen or pen-plotter, by specifying the position of its ends. The possible positions are any monitor pixel or plotter coordinate pair, and the position label is the pair of integers, (X%,Y%), which is their counting order.

Contour construction is based on these straightline segments, and curves are formed when many short, connected, segments are combined to appear as a curved line. There are two output factors that control the smoothness of the curved lines on a display. The first is the mesh size of the output database, and the second is the step size of the display medium. One has to accept the step size restriction, and appropriate mesh size is a trade-off between the smoothness of the drawn curve and computing time.

Notice that the smoothness referred to here, that of the full length of isoline extracted from the output database, is conceptually distinct to the potential smoothness of the representative surface, as discussed in terms of precision on p. 55 and following.

OUTPUT DATABASE

Before contours can be constructed for a topographical data set, interpolation is performed at each node of a regular grid. This grid of interpolated values is the output database, and it is a discrete version of the representative surface generated for the data. Output techniques portray the surface solely on the basis of this output database.

Conventionally, rectangular grids are used for the output database. The surface is estimated at each node where two series of orthogonal profiles intersect. Then isoline segments within each group of four interpolation points are computed and plotted.

However, considerable advantage lies in using a triangular mesh (see next), both in terms of economy in execution and storage as well as ease in extracting isoline segments. Constructing a display of a surface is done in a piecewise fashion using an array of surface patches. These patches are interstitial polygons defined by a regular grid of interpolation points. Usually, a data set is interpolated at the nodes of a square grid, and the positions of contour lines are estimated within each cell by a second interpolation method. As discussed in the section on manual methods on p. 106, linear methods of interpolation on rectangular grids, although economical, are not satisfactory unless the grid nodes are dense.

TRIANGULAR GRIDS

Figure 1.5.1, on p. 22, shows a triangular grid. Triangular output database grids are an alternative that may be preferred for three reasons.

(1) Economy

Triangular grids require approximately 30% fewer nodes than square grids, for an equal maximum interpolation distance, and so provide the most efficient storage of interpolated surfaces. If a measure of interpolation confidence at a point were to be based on the maximum distance to known values (equivalent to the radius of the largest circle that can fit between the nodes), triangular grids are distinctly more economical, as demonstrated on p. 22 and following.

(2) Unambiguity

The interstitial polygons formed by a rectangular mesh, as an output database, present an ambiguous contouring situation because a single plane will not fit all four nodes, and triangular subdivision may be done in two ways (see p. 138). This saddle problem does not arise with triangular grids, because a triangular interstitial polygon offers no ambiguities to be resolved in approximating the generated surface by a plane, and hence an isoline segment.

(3) Isotropy

Increased interpolation isotropy is obtained from a triangular output grid. This implies that, because the grid nodes are equally spaced in three directions, any interpolation point has a more uniform context, and a isoline segment depends upon a more equidirectional array of nodes than are available for output databases with equal spacing in only two directions.

The only array of nodes that are more isotropic than a triangular grid is a uniformly random array. Such a configuration is probabilistic and not really comparable to a regular grid, but it has an average sample spacing that is probably equal in all directions, and so is the ultimate in isotropy. In this sense a triangular grid is more isotropic than a square grid.

The programs use a triangular mesh of interpolation points, generated at line **3840** and following. The size of the mesh is set by the parameter **GR** in line **1100**; the larger values, the finer the mesh. See p. 39 for details of increasing the storage for fine meshes with **GR** greater than 40.

Stearns (1968), Philip and Watson (1982a, 1983), Hamilton and Tasker (1984), Tasker (1985), and Whitehorse and Phillips (1985), have discussed the advantages of triangular grids.

FOLLOWING CONTOURS

The extraction, termed threading, order of the straightline segments that make up the isolines may be either (i) in raster sequence across the output window, or (ii) contiguously along each isoline level in turn. These segments then are drawn in the order of extraction; with the raster approach, the segments may be sorted into contiguous order if the isolines are to be drawn by a plotter with minimal pen-up and pen-down action.

Notice that any attempt to redraw, or smooth, a series of these segments to form a more cosmetic curve, after they have been extracted from a representative surface, is to abandon the initial representative surface; isolines then may seem to contradict the observations. This redrawing, referred to as threading by Sabin (1985) who correctly labels it misleading and dangerous, can cause isolines to cross or occur on the wrong side of the data.

(1) Extraction In Raster Sequence

In the first situation, the interstitial polygons of the output database are visited systematically, cell after cell along a row and then row after row, and all the contour intersections in each rectangle or triangle may be drawn immediately, until the picture is complete. This method draws each segment of each contour level as it is computed, and the entire picture is composed of these unsorted short line segments.

If it is desired to draw each contour in its entirety as one continuous curve, the short segments must be sorted and stored until the contour line is fully known. This is not complicated, but requires a considerable amount of storage even for a moderate number of contour levels and a moderate output mesh size. Generally, it is not practical with less than a megabyte of storage available. Anderson (1983) proposes a quadtree storage method, and discusses quadtree manipulation.

(2) Extraction In Contour Sequence

In the second situation, contours can be drawn by beginning at any point and following the contour until it is complete, or ends at the perimeter of the output database grid. Usually, this is done by scanning adjacent cells, and this requires a bookkeeping system to insure that all cells have been checked for all possible contour segments. Cottafava and Moli (1969), and Snyder (1978), have discussed such schemes.

For either technique, there are two approaches to selecting the end points for the next segment. The surface across the interstitial cell is assumed to be a plane, or a hyperboloid, and the linear intersections of the contour level with the cell edges are used as end points.

Alternatively, for a higher degree polynomial surface, because it has an analytical expression, the equation for the intersection of the polynomial surface with a level plane at the contour height is available. This expression may be followed across the cell with a variable step size, and so govern the smoothness of the resulting isoline.

Batcha and Reese (1964), Bowden (1983), Briggs (1974), Crain (1970), Davis (1973), Dayhoff (1963), Deakin and Pluckett (1984), Eyton (1984), Freiberger and Grenander (1977), Garfinkel (1962), Kok and Begin (1981), Lodwick and Whittle (1970), Lowden (1985), MacCallum and Zhang (1986), Petrie and Kennie (1987), Rhind (1975), Satterfield and Rogers (1985a, 1985b), Schut (1976), Suffern (1984), Watson (1982, 1983), have discussed contour extraction.

ISOMETRIC VIEWS

The conventional plan view of level curves, depicting a representative surface, may be difficult to interpret when the highs and lows are not labeled. The difference in elevation between two adjacent isolines is not obvious in the plan view, so hills are confused easily with valleys.

Of course, an isometric view of the same set of level curves would make the hills and valleys obvious, and such a view can be generated on the computer by rotating the line segments in three dimensions. Compared to isometric views, perspective views have the property that all pairs of parallel lines seem to meet in the far distance. This is a necessary effect to make an oblique view of a general scene realistic. Daulton (1987) discusses the construction of perspective views.

However, perspective views of objects that are predominantly in the foreground may be portrayed in an isometric view that approximates a perspective view. This allows some saving in computation time.

Figure 1.2.1, on p. 11, is an isometric view of a set of isolines. Although this view approximates the natural perspective view, a major difference is that lines that should be hidden can be seen. This form of isometric view is termed a "wireframe" isometric.

Output Techniques

Isometric views are obtained by rotating the isoline set in three-dimensional space. This always can be done with two 2-dimensional rotations; for example, one about a line parallel to the z-axis, and one about a line parallel to the x-axis. The combination can result in any required orientation.

For an example of isometric rotation, suppose that the two points in three-dimensional space, (x_1, y_1, z) and (x_2, y_2, z), are the end points of a line segment on the isoline at level z. The desired isometric view has, for example, a positive rotation of 30° about a line parallel to the z-axis, the azimuth angle, and a negative rotation of 15° about a line parallel to the x-axis, the tilt angle. Both axes of rotation pass through the centroid of the topographical data set, but the rotational axes could be located anywhere.

The first rotation transforms the coordinates of each end point, (x_1, y_1, z) and (x_2, y_2, z), to (x_{1a}, y_{1a}, z) and (x_{2a}, y_{2a}, z), respectively, and leaves both z-values unchanged. Label the centroid (x_c, y_c, z_c). The first transformation is

$$x_{1a} = (x_1 - x_c) * \cos(30) - (y_1 - y_c) * \sin(30) + x_c$$
$$y_{1a} = (x_1 - x_c) * \sin(30) + (y_1 - y_c) * \cos(30) + y_c$$
$$x_{2a} = (x_2 - x_c) * \cos(30) - (y_2 - y_c) * \sin(30) + x_c$$
$$y_{2a} = (x_2 - x_c) * \sin(30) + (y_2 - y_c) * \cos(30) + y_c$$

The second rotation transforms the points, (x_{1a}, y_{1a}, z) and (x_{2a}, y_{2a}, z) to (x_{1a}, y_{1b}, z_{1b}) and (x_{2a}, y_{2b}, z_{2b}), respectively. This rotation leaves x_{1a} and x_{2a} unchanged.

$$y_{1b} = (y_{1a} - y_c) * \cos(-15) - (z - z_c) * \sin(-15) + y_c$$
$$z_{1b} = (y_{1a} - y_c) * \sin(-15) + (z - z_c) * \cos(-15) + z_c$$
$$y_{2b} = (y_{2a} - y_c) * \cos(-15) - (z - z_c) * \sin(-15) + y_c$$
$$z_{2b} = (y_{2a} - y_c) * \sin(-15) + (z - z_c) * \cos(-15) + z_c$$

Such transformations are executed in the programs, at lines **6530** and **6540**, and the angles are set by **AZ** and **TL** in line **1100**. Because only the transformed values for x and y are required for plotting, the expressions for z_{1b} and z_{2b} are not used, but are shown here for completeness. To obtain an isometric view from the programs, set **DP=1** on line **1100**. Set **DP=0** for a plan view.

INTERVAL FILLING

As well as by drawing isolines, the set of level curves may be displayed by filling the contour interval between the isolines with solid colors, or a pattern of motifs. When a set of colors are assigned to the set of height intervals between the contour levels, the isolines, themselves, need not be drawn because their position is indicated by the boundary between two adjacent colors. This produces a picture in which the change in the surface is implied by the change in colors. The color code identifies the subregion according to the height interval of the surface (see isochor maps, on p. 50).

Notice that although a gradation in height from one isoline to the other is implied by the surface that the isolines were extracted from, the process of interval filling has categorized the surface into a set of classes. This effectively asserts that the subregion between two isolines is an equivalence class. A set of these isochors (places of equalness) is implicit in each set of level curves. The interval filling operation has translated the topographical data into an isochor map.

Color filling is accomplished in the programs, at line **5870**, by linearly interpolating each interstitial triangle from the output database, at pixel locations. With an alternative approach, each pixel could be colored by interpolating the representative surface directly. This effectively makes the output database the same as the pixel array, and so this will increase the number of interpolations, and the running time, required for a picture.

Color filling is turned on by setting **CR**=1 in line **1100**, and **CR**=0 turns color filling off. Color filling is available only in the programs with output to the monitor screen, in directory SCRN.

COLOR-FILLED ISOMETRIC VIEWS

Interval filling also can be applied to isometric views (see Figure 2.5.9 on p. 161) and offers a second advantage. When solid color interval filling is used in a screen display, the "wireframe" nature of isometric isolines is avoided. Hidden lines are hidden simplistically by working from background to foreground, and so overwriting the hidden portions. Other, more complex, approaches to hidden line removal are discussed by Hilbert (1981), Spillers and Law (1987), and Boese (1988).

By rotating the set of line segments as they are extracted from a generated surface, an isometric view of the isolines can be drawn. However, when color is filled into the region between isometric isolines, more or less color must be applied because the observed area has changed as a result of the change in view point; the individual pixel positions cannot be rotated without some unfilled pixels appearing within some isochors.

The appropriate approach requires rotating the surface of the interstitial triangle before the isolines are extracted. This allows the projected area of the triangle to decrease or increase, according to the rotation, and then a complete set of pixels can be assigned for each triangle.

In the programs, starting at line **5880**, color filled isometric views are obtained by rotating each interstitial triangle from the output database into its isometric position. The pixels "covered" by the rotated triangle then are colored according to the height of the equivalent location on the unrotated triangle.

This technique produces a linear interpolation within the triangle for each pixel covered by the rotated interstitial triangle. If the output data base is sufficiently dense, the resulting isochors will appear to be smoothly curved.

ORTHOGONAL PROFILES

A surface may be displayed in isometric view, or in perspective, by a fine mesh of lines that lies like a fishnet over the surface. This crosshatching mesh is made up of two perpendicular series of profiles through the surface. When the surface is flat and level, the mesh appears as a set of squares. However, when the surface is sloping, the squares are distorted in a manner that implies the change in the surface.

Two approaches to the generation of such a display are possible. Usually the representative surface is interpolated at the nodes of a square grid, then the horizontal and vertical rows of the grid are drawn in isometric view. The square grid should be fine enough so that a straightline from one interpolation point to the next produces an acceptable smooth curve along the profile.

The other approach, used in the programs, determines the trace of a square grid on each of the interstitial triangles from the output grid. This is considerably slower, grid cell for grid cell, but is compensated in part by the need for fewer interpolation points in the output grid.

With this method, the individual segments of line in the profile usually do not begin or end at the apparent mesh nodes, so the profile appears more smooth than a similar size interpolation grid for the first method. Examples of orthogonal profiles have been published by Watson (1983), Todd (1987), and Watson and Philip (1988). Orthogonal profiles are produced by the programs when **DP=2**, on line **1100**. Set **DP=0** for a plan view.

STEREOGRAMS

A stereogram is a pair of pictures, such as would be made by two cameras held side-by-side; the scene is the same but the points of view are slightly different. The interocular angle, of 4° to 6°, is sufficient to create the illusion of three-dimensional depth when the pair of pictures are viewed simultaneously, but only one picture is viewed by each eye. Obviously, there are two ways in which this may be done.

A conventional stereogram requires parallel lines of sight as each eye views a picture. Generally stereo glasses are required to view parallel stereograms because not many people can direct their eyes into parallel lines of sight.

A cross-eyed stereogram is a parallel stereogram with the views interchanged. Cross-eyed stereograms allow the lines of sight to cross in the natural manner except that the stereogram lies beyond the crossover point.

When a single object, such as a finger, is held beyond the crossover point of the lines of sight, two images of the finger are perceived. When a double object, such as a stereogram, is held beyond the intersection of the lines of sight, three images can be perceived. The central image gives an illusion of three-dimensional depth because it is composed of both pictures viewed simultaneously.

Cross-eyed stereograms are useful on the personal computer screen, as well as the plotter. Stereograms are produced by the programs when **SR=1**, on line **1100**, using an interocular angle of 5.7° which is equivalent to .1 radian. Set **SR=0** for single pictures. Stereograms are produced only for isometric isolines and isometric orthogonal profiles.

Gay (1983) describes the relationships between vertical and horizontal scale for stereo pairs. Eyton (1984) discusses parallel stereograms and gives an example; these stereograms may be viewed in the cross-eyed manner, but, in that situation, they exhibit reverse relief. Todd (1987) gives several parallel stereograms for stratigraphical surfaces.

NUMERICAL OUTPUT

Pictorial displays are a fundamental method of comprehending a topographical data set because they give a visual impression of the topographical variability exhibited by the data. Numerical assessment of a data set gives an auxiliary method to understanding a spatial function, usually with a firm estimate of precision.

The conventional approach to numerical assessment of topographical data is to produce a histogram of altitudes from a systematic sampling of the representative surface. An ogive (hypsometric curve) is a summation of a histogram and is illustrated in Figure 1.9.1 on p. 42. The histogram and ogive provide a synopsis of the data set.

Another method of assessing the data is by their topographical quantities; an assessment of the volume under any bounded portion of a surface that represents the data. This provides an excellent basis for comparison between the surfaces generated by various interpolation schemes. More importantly, the volume estimate is useful for practical purposes, and often is the bottom line that foretells the economic success or failure of a project. The pictorial display gives visual assurance and lends plausibility to the final number.

(1) Precise Volumes.

In the programs, volume estimates are obtained directly from the triangles in the output database. The area of each triangle by the mean height of its vertices is accumulated as the triangles are checked for contours. Then this sum is divided by the area of all the triangles to obtain an average height for the output window. Notice that if the convex hull (p. 102) of the data does not fill the output window, then the average height applies only to the region covered by the convex hull. Also, notice that the precision of this estimate is dependent on the size of the triangles; the more and smaller the triangles are, the more precise is the estimate of volume.

By reducing the size of the display window (p. 13), or increasing the mesh size (**GR** in line **1100**), or both, volumes can be estimated to high precision for any of the interpolated surfaces. Execution time will increase or decrease as the square of change in mesh size or change in display window size.

As well as computing an average height of the representative surface for the whole data set, average heights may be computed for blocks, or regular subregions, within the data set. Such block averages, discussed on p. 43, provide a numerical and graphical insight into the spatial distribution of height values, as opposed to the numerical distribution, and so illustrate the broad behavior of the spatial function (see Figure 1.9.2 on p. 43). Block averages have been discussed by Unwin (1981).

Hall (1979) provides an algorithm for integrating polynomial surfaces, and notes that this is a measure of goodness of fit for polynomial approximation surfaces. Integration over a triangle has been discussed by Barnhill and Little (1984a), Doncker and Robinson (1984), Hall (1979), and Hughes (1959).

(2) Volume Products.

Interpolation provides the heights of a representative surface at the grid nodes to generate the output database. When the heights of the representative surface for another spatial function, for the same region, also are known at those nodes, the sum, difference, product, or ratio of the two representative surfaces may be obtained by applying the arithmetic operations to the heights, node by node, in raster fashion.

For example, if the grade and thickness of a certain ore body were obtained by drilling at several locations, average grade for the region is obtained by multiplying the grade and thickness surfaces, and then dividing by the volume under the thickness surface.

Of course, here again, the precision of the results will depend on the density of the output grid. Round-off errors can accumulate quickly when surfaces are combined. The ultimate numbers never can be given a greater precision than the input data, but one should suspect even that precision for results from surface combinations.

(3) Isted's Formula.

An interesting formula used in piecewise calculations of spatial averages of ore-deposits is the Isted formula, or so-called percentage method (Reedman, 1979; Mendelsohn, 1980). The problem is to calculate average grade for the volume of a triangular prism, the vertical edges of which are three parallel drill holes, with ore thickness T_1, T_2, T_3, and average grade for each intersection of G_1, G_2, G_3, as in Figure 2.7.1 on the next page.

Numerical Output

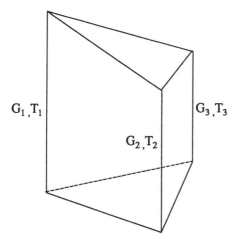

Figure 2.7.1 Isted volume-product

The spatial average of grade is considered to be the incremental grade times volume product, divided by volume. Isted's Formula for average grade, G_A, is

$$G_A = \tfrac{1}{4}((\textstyle\sum T_i G_i)/(\sum T_i) + \sum G_i)$$

where the summations are over $i = 1 - 3$. This expression gives the exact integral of the product of grade and thickness, assuming linear interpolation of each function over the triangle, and divides this product by the volume of the thickness prism. Watson and Philip (1986a, 1986b) provide a derivation of this expression, and a FORTRAN implementation. That program selects triplets of natural neighbor drillholes and computes the average grade for any number of drillholes.

SECTION 2.8

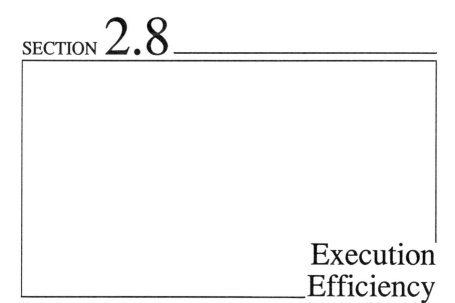

Execution Efficiency

SECTION SUMMARY

Execution time and storage requirements for various sorting, gradient estimation, interpolation, and display, procedures are discussed in this section.

SPATIAL SORTING

The spatial order of a topographical data set on a regular grid is, of course, implicit in the order of the indices, so spatial sorting is not necessary. For scattered and traverse data, proximal sorting is required to determine all adjacency relationships in the data. Natural neighbor order expresses these relationships, and it is determined by the set of natural neighbor circumcircles (p. 59).

The execution time to compute these circumcircles is controlled by the least efficient step in the procedure, which is $O(N^2)$ for the programs. However, sorting procedures of $O(N \log N)$ and even $O(N)$ are available; they are not implemented in the programs because their greater number of operations makes them slower, except for large amounts of data, and because their greater use of working storage makes them less easily accommodated on personal computers.

Sorting a set of bivariate topographical data requires approximately five percent of the execution time when gradient estimation, interpolation, and display construction, are considered.

GRADIENT ESTIMATION

Estimating a gradient for each datum requires, first, the selection of a subset of adjacent data, and second, computing some form of average slope for that subset. Of course, subset selection is easy with rectangularly gridded data. For the datum in the ith row and jth column, the nearby data are determined by subtracting or adding increments to i and j.

For scattered data, the conventional approach has been to select all the data within a fixed distance, or the nearest fixed number of data. However, when the data are scattered unevenly, some subsets will have too much data, too little data, or a directionally unbalanced group of data. Such subsets, generally, will not provide satisfactory gradient estimates because of inconsistencies with neighboring estimates. Subsets composed only of the natural neighbors of a datum avoid such inconsistencies. Selecting natural neighbor subsets is fast because it is done as a spatial sort.

For traverse data, natural neighbor subset selection is necessary because selecting a fixed number of data or all the data within a fixed distance almost invariably provides unacceptable gradient estimates, for the same reasons as for scattered data.

The average slope implied by a subset of topographical data, can be obtained in the programs by any of five methods. Set parameter **SL**, on line **1100**, equal to one of the integers from 0 to 4 (not in sequence) to obtain any of the following estimated gradients.

(1) Linear Gradients.

The gradient subset, the datum and its natural neighbors, is approximated by a plane. The orientation of the plane is selected to minimize the sum of squared differences between the data and the plane. The slope of this plane is adopted as the estimated gradient through the datum. This estimator is the fastest for either gridded or ungridded data. Set **SL=2**, on line **1100**.

(2) Quadratic Gradients.

The gradient subset is approximated by a quadratic surface. The parameters of the second degree polynomial are selected to minimize the sum of squared differences between the data and the surface. The slope of this surface at the datum is adopted as the estimated gradient.

Quadratic gradient estimates are fast but not always satisfactory. Possibly this is due to the great amount of flexibility in this method. The selected quadratic tends to accommodate, rather than summarize, the orientation influences of the datum's neighbors. Set **SL=4**, on line **1100**.

(3) Minimum Curvature Spline Gradients.

The datum and its set of neighbors are fitted by a minimum curvature spline. This is achieved by solving for parameters that minimize the curvature of a "thin plate" spline through the datum and its neighbors. It is the slope of this surface at the datum which is adopted as the estimated gradient. This estimator is moderately fast, but usually is satisfactory. Each gradient requires the solution of k simultaneous equations, where k is the number of data in the gradient subset. Set SL=3, in line **1100**.

(4) Cross Product Gradients.

This estimator determines an average orientation for the set of natural neighbor triangles formed by a datum and its neighbors. Although it depends on a spatial sorting of a datum and its natural neighbors, cross product estimation is fast and reliable. However, this is not a good approach for gridded data because the triangulation process arbitrarily rejects some diagonal neighbors of each datum. Set **SL**=1, in line **1100**, for cross product gradients.

(5) Natural Neighbor Linear Gradients.

This estimator is unusual in that it estimates the slope at a datum without using that datum except to select neighbors. The local coordinates of the datum with respect to the natural neighbors are used to compute three interpolations near the datum. These three values indicate the slope of the natural neighbor interpolation surface at the location of the datum *in the supposed absence of the datum*. This slope is used as the gradient estimate. Natural neighbor local coordinates are not available for the data on the perimeter of the data set because a complete set of natural neighbors are necessary to obtain these coordinates; for these data, cross product gradients are used.

Natural neighbor linear gradient estimates are slow, but generally produce the most satisfactory gradients. Set **SL**=0, in line **1100**.

INTERPOLATION EFFICIENCY

There are two general approaches to interpolation, fitted functions and weighted averages (p. 105). Function fitting requires the initial task of parameter calculation, usually by solving a system of simultaneous linear equations. This preinterpolation step uses time that increases in proportion to the third power of the number of parameters. However, the subsequent interpolation at the nodes of the output grid is comparably fast.

Execution Efficiency

Execution time for computing parameters of a fitted function is not significant for less than about 50 data and becomes impractical for more than about 100 because ill-conditioning of systems of simultaneous equations usually occurs. As well as potential ill-conditioning of the matrix, excessive computation time and round-off errors insure that function fitting is practical only for small data sets.

Weighted averages are the alternative to fitted functions; Tempfli and Makarovic (1979) conclude that interpolation time efficiency for a single interpolation is related primarily to the size of the interpolation subset. Generally this is small for natural neighbor subsets and, of course, total time is proportional to the number of interpolation points.

Interpolation efficiency includes the preinterpolation steps of gradient estimation and spatial sorting. Sorting (p. 189) scattered or traverse data uses time that increases in proportion to the square of the number of data. Sorting time is not significant for less than about 100 data and becomes impractical for several thousand data.

Gradient estimation (p. 189) for weighted averages uses time that increases in proportion to the number of data. Gradient estimation methods vary somewhat in their time use per datum. Cross product gradients are the fastest, but most of the execution time is used in identifying subsets.

The weighted average interpolation methods, like the gradient estimation methods, differ in their time per interpolation. Additionally, however, the interpolation time to produce the output database depends on output grid size as well as the interpolation procedure that must be executed for each node of the output grid. This implies that a dense output database grid, and a computationally expensive interpolation method, such as fitted functions on running subsets of data, can take many hours on a personal computer. On the other hand, a coarse mesh size for the output grid and a linear, area-based, weighted average can produce a display in minutes.

The principal topographical features will be obvious in both displays but the extra computation time provides a more appealing portrayal. Interpolation potentially is the most time consuming step in computer contouring. The price of fast execution is a less plausible output database and less smooth contours.

DISPLAY EFFICIENCY

The execution time for drawing a display is controlled by the output database grid size. This determines the number of interstitial polygons that must be checked for contour segments. Again, there is a trade-off between fast execution and smooth curves.

There are two approaches to color filling. The faster method linearly interpolates each interstitial triangle of the output grid at each pixel that the triangle "covers". The resulting surface is planar between the nodes of a triangular grid. These nodes are interpolated by any of the programmed methods.

As an alternative, not implemented in these programs, the output grid may be selected to correspond to a set of screen pixels. In this situation, the color of each pixel is determined directly by a single interpolation. Although less pronounced, again there is a trade-off between fast execution and smooth curves.

Construction of stereograms is treated differently for screens than for plotters. Producing stereo pairs on the screen only requires that any contour segment be drawn, then displaced, rotated and drawn again; the two views develop in parallel.

With this approach, plotters have too much pen-up time, as the pen travels between the two views. Producing two views in parallel requires far more than twice the time. The alternative is to draw one view, rotate the whole output database, and draw the second view. Although more computation is involved, plotting time only is double that for a single view.

The programs in SCRN produce stereograms in parallel while the programs in PLOT produce them serially.

Some Published Programs

SUMMARY

A list of published programs and subroutines for topographical data analysis and display.

Antoy (1983) FORTRAN program to contour traverse data.
Barrodale and others (1983) FORTRAN subroutine for spectrum analysis.
Baumann (1978) FORTRAN program using inverse distance weighting.
Bourke (1987) BASIC subroutine to contour gridded data.
Braile (1978) FORTRAN subroutine for inverse distance weighting.
Bregoli (1982) BASIC subroutine for line printer plotting.
Davis and David (1980) FORTRAN program for bicubic splines.
Devereux (1985) FORTRAN program using inverse distance weighting.
Dierckx (1980) FORTRAN subroutines for cubic splines.
Dimitriadis, Tselentis, and Thanassoulas (1987) BASIC subroutines for Fourier analysis.
Esler and Preston (1967) FORTRAN program for power spectrum.
Eyton (1984) FORTRAN subroutines for raster contouring.
Inoue (1986) FORTRAN subroutines for cubic splines.
Holroyd and Bhattacharyya (1970) FORTRAN bicubic splines.
James (1966) FORTRAN program for Fourier analysis of scattered data.
Kane and others (1982) FORTRAN program for inverse distance weighting.
Liszka (1984) FORTRAN subroutine for Taylor interpolation.

Some Published Programs

Mason (1984) BASIC subroutines for splines.
Oldknow (1987) BASIC subroutines for Bezier splines.
Rogers and Adams, 1976, BASIC subroutines for surfaces.
Sampson and Davis (1967) FORTRAN response surface.
Swain (1976) FORTRAN program for minimum curvature.
Tartar, Freeman, and Hopkins (1986) FORTRAN subroutines for Fourier analysis.
Watson (1982) FORTRAN program for linear interpolation.
Watson (1983) BASIC program for linear interpolation in stereo.
Watson (1986) FORTRAN subroutine for triangular prism volumes.
Yarnal (1984) FORTRAN programs for gridded data.
Yates (1987) FORTRAN subroutines for linear interpolation.
Yfantis and Borgman (1981) FORTRAN subroutine for Fast Fourier Transform.
Yeo (1984) FORTRAN program for linear interpolation.

Glossary Of Contouring Terms

Anisotropic - The property of having a preferred direction, not equidirectional

Anomaly - A local phenomenon involving untypically high or low values, or unusual local shape

Barycentric Coordinates - These are local coordinates that relate a designated location to three reference points (see p. 76)

Bivariate - This refers to functions of two independent variables, such as a surface

Bounded - An estimate, or a region, is bounded when it has fully defined limits to its size

Break Lines - Ridge lines or drainage lines, characterized by a sharp change in surface slope

Cartesian Coordinates - These are global coordinates that relate a designated location to a reference frame of orthogonal axes. Cartesian coordinates on the plane increase to the east and to the north

Chorochromatic Map - A color-patch map where each region is color coded (see p. 50)

Chorography - The description of the features of a region

Choropleth - A value associated with a specific region, literally a place-quantity datum

Choropleth Map - A map of the boundaries of regions which are each homogeneous in their value of interest, otherwise termed an isochor map

Col - A saddle-like surface, such as a pass in a ridgeline or between two mountain peaks

Contouring - Generating a graphical expression of the shape and form of a representative surface from knowledge of its elevations at particular locations

Convex Hull - For a bivariate data set, this is the maximum region enclosed by all straightlines between pairs of data points (see p. 102)

Delaunay Triangulation - A contiguous tiling of triangles, each of which has a circumcircle that does not enclose the vertices of any other triangle. Also termed a Delaunay Tessellation (see p. 62)

Dirichlet Tessellation - A contiguous tiling of Voronoi proximal polygons (see p. 62)

Extrapolation - Estimating the height of a topographical surface outside the convex hull of the data set

Global - Involving the entire data set rather than a subset of the data

Gradient - The direction and value of the steepest slope at a given location

Grid, also **Graticule** - A regular network or mesh, having a configuration of nodes that have constant spacing in each direction

Gridding - Interpolating a data set at the nodes of a grid

Hachures - Graphical expressions indicating steepness and direction of slope of a surface, or categorical value of a region

Interpolation - Estimating the height of a topographical surface at a given location within the convex hull of a data set

Isarithms, also **Isorithms** - Lines, or regions, of equal numerical value

Isobars - Lines of equal atmospheric pressure

Isobaths - Lines of equal depth of water

Isocheims, also **Isochimes** - Lines of equal winter temperature

Isochors - Regions of equal value

Isochromes - Lines, or regions, of equal color

Isochrones - Lines of equal time

Isoclines - Lines of equal slope

Isocrymes - Lines of equal coldest winter temperature

Isogones - Lines of equal angles

Isograms - Lines of equal numerical value

Isohalines - Lines of equal salinity

Isoheights - Lines of equal height

Isohels - Lines, or regions, of equal sunlight

Isohyets - Lines, or regions, of equal rainfall

Isohypses - Lines of equal elevation

Isolines - Lines of equal numerical value

Isonephs - Lines, or regions, of equal duration of cloudiness

Isopachs - Lines, or regions, of equal thickness

Isopags - Lines, or regions, of equal duration of ice cover

Isophotes - Lines of equal illumination

Isopleths - Lines, or regions, of equal quantity

Isopluvials - Lines, or regions, of equal rainfall

Isopycs - Lines, or regions, of equal density

Glossary Of Contouring Terms

Isosteres - Lines of equal density of air
Isotachs - Lines of equal windspeed
Isotheres - Lines of equal summer temperature
Linear Interpolation - Interpolation with weights that are confined to sum to one The interpolated surface will not extend above or below all of the data Compare to **Nonlinear Interpolation**
Local - Involving a subset of adjacent data rather than the entire data set
Local Coordinates - These are coordinates that relate a designated location to a number of reference points (see p. 75)
Model - A mathematical construction intended to be analogous to, and representative of, a physical system or function
Natural Neighbors - All the data that share empty circumcircles with the datum and such that no data lie within any circle. The Voronoi polygons of these data have at least one point of contact with the Voronoi polygon of the datum (see p. 59)
Neighborhood - The region shared by the neighbors of a datum
Node - A grid point, at the intersections of grid lines
Nonlinear Interpolation - interpolation with weights that are not confined to sum to one. The interpolated surface may extend above or below any of the data
Polygon - A shape on the plane that is bounded by straightlines. Any shape with a curved boundary may be approximated with a polygon having many short sides
Proximal Order - The spatial adjacency relationships among multivariate data
Proximal Polygon - A polygon defining a region associated with a regional center
Sorting, Natural Neighbor - Generating a list of the natural neighbor relationships of each datum within a data set
Tessellation - A mosaic, or tiling, of contiguous, space-filling, polygons. A triangulation is a tessellation that contains only triangles
Topography - The description and expression of a surface. This may concern the actual height of a physical surface, or, more abstractly, the variation in functional values of any continuous bivariate phenomenon
Triangulation - A set of triangles obtained by generating an exhaustive set of nonintersecting lines connecting each datum to other nearby data
Voronoi Tessellation - A set of proximal polygons having vertices which are the centers of the natural neighbor circumcircles (see p. 62)
Window - A conceptual frame that enclosed some or all of the data

References

Abhyanker, S.S., and Bajaj, C., 1987, Automatic parametrization of rational curves and surfaces II: cubics and cubicoids: Computer-aided Design, v. 19, no. 9, p. 499-502.

Agocs, W.B., 1951, Least squares residual anomaly determination: Geophysics, v. 16, p. 686-696.

Agterberg, F.P., 1969, Interpolation of areally distributed data: Colorado School of Mines Quart., v. 64, p. 217-237.

Agterberg, F.P., 1982, Recent developments in geomathematics: Geo-Processing, v. 2, p. 1-32.

Ahuja, N., 1982, Dot pattern processing using Voronoi neighbourhoods: IEEE Trans. on Pattern Analysis and Machine Intelligence, v. PAMI-4, p. 336-343.

Akima, H., 1978a, A method for bivariate interpolation and smooth surface fitting for irregularly distributed data points: ACM Trans. on Mathematical Software, v. 4, no. 2, p. 148-159.

Akima, H., 1978b, Algorithm 526 - bivariate interpolation and smooth surface fitting for irregularly distributed data points: ACM Trans. on Mathematical Software, v. 4, no. 2, p. 160-164.

Akima, H., 1984, On estimating partial derivatives for bivariate interpolation of scattered data: Rocky Mountain Jour. Math., v. 14, no. 1, p. 41-52.

Alfeld, P., 1984, A discrete C^1 interpolant for tetrahedral data: Rocky Mountain Jour. Math., v. 14, no. 1, p. 5-15.

Alfeld, P., 1985, Multivariate perpendicular interpolation: SIAM Jour. Numerical Anal., v. 22, no. 1, p. 95-106.

Alfeld, P. and Barnhill, R.E., 1984, A transfinite C^2 interpolant over triangles: Rocky Mountain Jour. Math., v. 14, no. 1, p. 17-39.

Alfeld, P., Piper, B. and Schumaker, L.L., 1987, An explicit basis for C^1 quartic bivariate splines: SIAM Jour. Numerical Anal., v. 24, no. 4, p. 891-911.

References

Alia, G., Barsi, F., Martinelli, E. and Tani, N., 1987, Angular spline: a new approach to the interpolation problem in computer graphics: Computer Vision, Graphics and Image Processing, v. 39, p. 56-72.

Anderson, D.P., 1983, Techniques for reducing pen plotting time: ACM Trans. on Graphics, v. 2, no. 3, p. 197-212.

Antoy, S., 1983, Contour plotting for function specified at nodal points of a mesh based on a set of irregular profiles: Computers & Geosciences, v. 9, no. 2, p. 235-244.

Arthur, D.W.G., 1965, Interpolation of a function of many variables: Photogrammetric Engineering and Remote Sensing, v. 31, no. 2, p. 348-349.

Arthur, D.W.G., 1973, Interpolation of a function of many variables II: Photogrammetric Engineering and Remote Sensing, v. 39, p. 261-266.

Ayeni, O.O., 1979, Optimum least squares interpolation for digital terrain models: Photogrammetric Record, v. 9, p. 633-644.

Ayeni, O.O., 1982, Optimum sampling for digital terrain models: a trend towards automation, Photogrammetric Engineering and Remote Sensing, v. 48, no. 11, p. 1687-1694.

Bacchelli-Montefusco, L., 1987, An interactive procedure for shape preserving cubic spline interpolation: Computers & Graphics, v. 11, no. 4, p. 389-392.

Bacchelli-Montefusco, L., and Casciola, G., 1984, Using interactive graphics for fitting surfaces to scattered data: IEEE Trans. on Computer Graphics and Applic., v. 4, no. 7, p. 43-45.

Balaras, C.A., and Jeter, S.M., 1990, A surface fitting method for three dimensional scattered data: Intern. Jour. Numerical Methods in Engineering, v. 29, p. 633-645.

Ball, A.A., 1980, The interpolation function of a general serendipity rectangular element: Intern. Jour. Numerical Methods in Engineering, v. 15, p. 773-778.

Barnhill, R.E., 1977, Representation and approximation of surfaces, *in* Rice, J.R., ed., Mathematical software III: Academic Press, New York, p. 69-120.

Barnhill, R.E., 1983a, A survey of the representation and design of surfaces: IEEE Trans. on Computer Graphics and Applic., v. 3, p. 9-16.

Barnhill, R.E., 1983b, Computer aided surface representation and design, *in* Barnhill, R.E., and Boehm, W., eds., Surfaces in computer aided geometric design: North-Holland, Amsterdam, p. 1-24.

Barnhill, R.E., and Farin, G., 1981, C^1 Quintic interpolation over triangles: two explicit representations: Intern. Jour. Numerical Methods in Engineering, v. 17, p. 1763-1778.

Barnhill, R.E., Dube, R.P., and Little, F.F., 1983, Properties of Shepard's surfaces: Rocky Mountain Jour. Math., v. 13, p. 365-382.

Barnhill, R.E., and Little, F.F., 1984a, Adaptive triangular cubatures: Rocky Mountain Jour. Math., v. 14, no. 1, p. 53-75.

Barnhill, R.E., and Little, F.F., 1984b, Three- and four-dimensional surfaces: Rocky Mountain Jour. Math., v. 14, no. 1, p. 77-102.

Barnhill, R.E., and Stead, S.E., 1984, Multistage trivariate surfaces: Rocky Mountain Jour. Math., v. 14, no. 1, p. 103-118.

Barr, R.C., Gallie, T.M., and Spach, M.S., 1980, Automated production of contour maps for electrophysiology: Computers and Biomedical Research, v. 13, p. 142-153.

Barrodale, I., Delves, L.M., Erickson, R.E., and Zala, C.A., 1983, Computational experience with Marple's algorithm for autoregressive spectrum analysis: Geophysics, v. 48, no. 9, p. 1274-1286.

Barsky, B.A., 1986, Parametric spline curves and surfaces: IEEE Trans. on Computer Graphics and Applic., v. 6, p. 33-34.

Barsky, B.A., and De Rose, T.D., 1985, The beta2-spline: a special case of the beta-spline curve and surface representation: IEEE Trans. on Computer Graphics and Applic., v. 5, 46-58.

Barsky, B.A., and Greenberg, D.P., 1982, Interactive surface representation system using a B-spline formulation with interpolation capability: Computer-aided Design, v. 14, no. 4, p. 187-194.

Batcha, J.P., and Reese, J.R., 1964, Surface determination and automatic contouring for mineral exploration, extraction and processing: Colorado School of Mines Quart., v. 59, p. 1-14.

Baumann, P.R., 1978, Iso: a FORTRAN IV program for generating isopleth maps on small computers: Computers & Geosciences, v. 4, p. 23-32.

Bayer, U., 1985, Pattern recognition problems in geology and paleontology: Springer-Verlag, Berlin, 229 p.

Beatson, R.K., and Ziegler, Z., 1985, Monotonicity preserving surface interpolation: SIAM Jour. Numerical Anal., v. 22, no. 2, p. 401-411.

Beck, J.M., Farouki, R.T., and Hinds, J.K., 1986, Surface analysis methods: IEEE Trans. on Computer Graphics and Applic., v. 6, p. 18-36.

Bengtsson, B- E., and Nordbeck, S., 1964, Construction of isarithms and isarithmic maps by computers: BIT, v. 4, p. 87-105.

Bentley, J.L., Weide, B.W., and Yao, A.C., 1980, Optimal expected-time algorithms for closest point problems: ACM Trans. on Mathematical Software, v. 6, p. 563-580.

Bergthorsson, P., and Doos, B.R., 1955, Numerical weather map analysis: Tellus VII, v. 3, p. 329-340.

Berrut, J-P., 1988, Rational functions for guaranteed and experimentally well-conditioned global interpolation: Computers Math. With Applic., v. 15, no. 1, p. 1-16.

Berry, J.K., 1987a, A mathematical structure for analysizing maps: Environmental Management, v. 11, p. 317-325.

Berry, J.K., 1987b, Computer-assisted map analysis: potential and pitfalls: Photogrammetric Engineering and Remote Sensing, v. 53, no. 10, p. 1405-1410.

Bezier, P.E., and Sioussiou, S., 1983, Semi-automatic system for defining free-form curves and surfaces: Computer-aided Design, v. 15, no. 2, p. 65-71.

Bideaux, R.A., 1979, Drill-hole management and display, in Computer methods for the 80's in the mineral industry: Amer. Inst. Mining, Metall., and Petroleum Engineers, New York, p. 155-162.

Birkhoff, G., and Garabedian, H.L., 1960, Smooth surface interpolation: Jour. Math. Physics, v. 39, p. 258-268.

Birkhoff, G., and Gordon, W.J., 1968, The draftsman's and related equations: Jour. Approximation Theory, v. 1, p. 199-208.

Bisseling, R.H., Kosloff, R., and Kosloff, D., 1986, Multidimensional interpolation and differentiation based on an accelerated sinc interpolation procedure: Computer Physics Commun., v. 39, p. 313-332.

Blumenstock, D.I., 1953, The reliability factor in the drawing of isarithms: Annals Assoc. Am. Geographers, v. 43, p. 289-304.

Boehm, W., 1980, Inserting knots into B-spline curves: Computer-aided Design, v. 12, no. 4, p. 199-201.

Boehm, W., 1986a, Multivariate spline methods in CAGD: Computer-aided Design, v. 18, no. 2, p. 102-104.

Boehm, W., 1986b, Curvature continuous curves and surfaces: Computer-aided Design, v. 18, no. 2, p. 105-106.

Boese, F.G., 1988, Surface drawing made simple - but not too simple: Computer-aided Design, v. 20, no. 5, p. 249-258.

Bohm, W., 1981, Generating the Bezier points of B-spline curves and surfaces: Computer-aided Design, v. 13, no. 6, p. 365-366.

Bohm, W., 1983, Subdividing multivariate splines: Computer-aided Design, v. 15, no. 6, p. 345-352.

Bohm, W., Farin, G., and Kahmann, J., 1984, A survey of curve and surface methods in CAGD: Computer Aided Geometric Design, v. 1, p. 1-60.
Bohmer, K., and Coman, Gh., 1980, On some approximation schemes on triangle: Mathematica, v. 22, no. 2, p. 231-235.
Boots, B.N., 1973, Some models of the random subdivision of space: Geografiska Annaler, v. 55, p. 34-48.
Boots, B.N., 1974, Delaunay triangles: an alternative approach to point pattern analysis: Proc. Assoc. Am. Geographers, v. 6, p. 26-29.
Boots, B.N., and Murdock, D.J., 1983, The spatial arrangement of random Voronoi polygons: Computers & Geosciences, v. 9, no. 3, p. 351-365.
Bos, L.P., and Salkauskas, K., 1988, Comment on the representation of splines as Boolean sums: Jour. Approximation Theory, v. 53, p. 155-162.
Botkin, M.E., and Bennett, J.A., 1985, Shape optimization of three-dimensional folded-plate structures: AIAA Jour., v. 23, no. 11, p. 1804-1810.
Bourke, P.D., 1987, A contouring subroutine: Byte, v. 12, p. 143-150.
Bowden, K.G., 1983, A fast contouring algorithm for potential arrays: The Matrix Tensor Quart., v. 33, no. 3, p. 43-47.
Bowyer, A., 1981, Computing Dirichlet tessellations: The Computer Jour., v. 24, no. 2, p. 162-166.
Box, G.E.P., 1954, The exploration and exploitation of response surfaces: Biometrics, v. 10, p. 16-60.
Bracewell, R.N., 1983, Discrete Hartley transform: Jour. Optic. Soc. Am., v. 73, no. 12, p. 1832-1835.
Braile, L.W., 1978, Comparison of four random to grid methods: Computers & Geosciences, v. 4, p. 341-349.
Brassel, K.E., and Reif, D., 1979, A procedure to generate Thiessen polygons: Geographical Analysis, v. 11, no. 3, p. 289-303.
Bregoli, L.J., 1982, A BASIC plotting subroutine sophisticated plotting with your MX-80: Byte, v. 7, no. 3, p. 142-156.
Briggs, I.C., 1974, Machine contouring using minimum curvature: Geophysics, v. 39, no. 1, p. 39-48.
Brill, D., Gaunaurd, G., and Uberall, H., 1981, The response surface in elastic wave scattering: Jour. Appl. Physics, v. 52, no. 5, p. 3205-3214.
Brostow, W., Dussault, J.P., and Fox, B.C., 1978, Construction of Voronoi polyhedra: Jour. Computational Physics, v. 29, p. 81-92.
Brueckner, I., 1980, Construction of Bezier points of quadrilaterals from those of triangles: Computer-aided Design, v. 12, no. 1, p. 21-24.
Bugry, R., 1981, Computer contouring packages: an historical view: Bull. Can. Petroleum Geol., v. 29, no. 2, p. 209-214.
Burrough, P.A., 1986, Principles of geographical information systems for land resources assessment: Clarendon Press, Oxford, 193 p.
Busch, J.R., 1985, Osculatory interpolation in R^n: SIAM Jour. Numerical Anal., v. 22, no. 1, p. 107-113.
Butland, J., 1979, Surface drawing made simple: Computer-aided Design, v. 11, no. 1, p. 19-22.
Cadete, M.O.R., 1987, A note on piecewise blending function interpolation applied to networks of curves in R^3: Jour. Computational Appl. Math., v. 17, p. 291-298.
Cain, J.C., and Neilon, J.R., 1963, Automatic mapping of the geomagnetic field: Jour. Geophys. Research, v. 68, no. 16, p. 4689-4696.
Carlson, R.E., and Fritsch, F.N., 1985, Monotone piecewise bicubic interpolation: SIAM Jour. Numerical Anal., v. 22, no. 2, p. 386-400.
Carlson, T.R., Erickson, J.D., O'Brian, D.T., and Pana, M.T., 1966, Computer techniques in mine planning: Mining Engineering, May, p. 53-56, 80.

Catsaros, N., 1987, A new method of interpolation and numerical integration: Computer Physics Commun., v. 43, p. 339-346.
Cavaretta, A.S., Micchelli, C.A., and Sharma, A., 1980, Multivariate interpolation and the Radon transform: Math. Zeit., v. 174, p. 263-279.
Cavendish, J.C., Field, D.A., and Frey, W.H., 1985, An approach to automatic three-dimensional finite element mesh generation: Intern. Jour. Numerical Methods in Engineering, v. 21, p. 329-347.
Cayley, A., 1859, On contour and slope lines: Philosophical Magazine, v. 18, p. 264-268.
Cendes, Z.J., and Shenton, D.N., 1985, Adaptive mesh refinement in the finite element computation of magnetic fields: IEEE Trans. on Magnetics, v. 21, p. 1811-1816.
Cendes, Z.J., and Wong, S.H., 1987, C^1 quadratic interpolation over arbitrary point sets, IEEE Trans. on Computer Graphics and Applic., v. 7, p. 8-16.
Chang, G., and Wu, J., 1981, Mathematical foundations of Bezier's technique: Computer-aided Design, v. 13, no. 3, p. 133-136.
Cheney, E.W., 1986, Multivariate approximation theory: selected topics: SIAM, Philadelphia, 68 p.
Cherenack, P., 1984, Conditions for cubic spline interpolation on triangular elements: Computers Math. With Applic., v. 10, no. 3, p. 235-244.
Clarke, K.C., 1985, A comparative analysis of polygon to raster interpolation methods, Photogrammetric Engineering and Remote Sensing, v. 51, no. 5, p. 575-582.
Cline, A.K., and Renka, R.L., 1984, A storage-efficient method for construction of a Thiessen triangulation: Rocky Mountain Jour. Math., v. 14, no. 1, p. 119-139.
Cole, A.J., 1968, Algorithm for the production of contour maps from scattered data: Nature, v. 220, p. 92-94.
Connor, S., 1988, It was a dark and stormy night: New Scientist, March 3, p. 36-37.
Conolly, H.J.C., 1936, A contour method of revealing some ore structures: Econ. Geology, v. 31, p. 259-271.
Cooley, J.W., Lewis, P.A.W., and Welch, P.D., 1967a, Historical notes on the fast Fourier transform: IEEE Trans. on Audio Electroacoustics, v. AU-15, no. 2, p. 76-79.
Cooley, J.W., Lewis, P.A.W., and Welch, P.D., 1967b, Application of the fast Fourier transform to computation of Fourier integrals, Fourier series, and convolution integrals: IEEE Trans. on Audio Electroacoustics, v. AU-15, no. 2, p. 79-85.
Cooper, M.A.R., and Cross, P.A., 1988, Statistical concepts and their application in photogrammetry and surveying: Photogrammetric Record, v. 12, no. 71, p. 637-663.
Cortey, N.E., 1982, Interpolation of arbitrary spaced points by closed surfaces: Computers & Graphics, v. 6, no. 1, p. 19-21.
Costantini, P., 1988, An algorithm for computing shape-preserving interpolating splines of arbitrary degree: Jour. Computational Appl. Math., v. 22, p. 89-136.
Cottafava, G., and LeMoli, G., 1969, Automatic contour map: Commun. ACM, v. 12, no. 7, p. 386-391.
Cox, D.D., 1984, Multivariate smoothing spline functions: SIAM Jour. Numerical Anal., v. 21, no. 4, p. 789-813.
Crain, I.K., 1970, Computer interpolation and contouring of two-dimensional data: a review: Geoexploration, v. 8, p. 71-86.
Crain, I.K., and Bhattacharyya, B.K., 1967, Treatment of non-equispaced two-dimensional data with a digital computer: Geoexploration, v. 5, p. 173-194.
Craven, P., and Wahba, G., 1979, Smoothing noisy data with spline functions: Numerische Mathematik, v. 31, p. 377-403.
Cuyt, A., 1987, A recursive computation scheme for multivariate rational interpolants: SIAM Jour. Numerical Anal., v. 24, no. 1, p. 228-239.
Cuyt, A.A.M., and Verdonk, B.M., 1985, Multivariate rational interpolation: Computing, v. 34, p. 41-61.

References

Cuyt, A., and Wuytack, L., 1987, Nonlinear methods in numerical analysis: North-Holland, Amsterdam, 278 p.
Dahmen, W., 1980, On multivariate B-splines: SIAM Jour. Numerical Anal., v. 17, no. 2, p. 179-191.
Dahmen, W., 1981, Approximation by linear combinations of multivariate B-splines: Jour. Approximation Theory, v. 31, p. 299-324.
Dartt, D.G., 1972, Automated streamline analysis utilizing "optimum interpolation": Jour. Appl. Meteorology, v. 11, p. 901-908.
Daulton, T., 1987, Three-dimensional perspective plotting: Byte, v. 12, no. 14, p. 307-314.
D'Autume, G. De M., 1979, Surface modelling by means of an elastic grid: Photogrammetria, v. 35, p. 65-74.
David, M., and Blais, R.A., 1968, Discussion on practical aspects of computer methods in ore reserve analysis, *in* Ore reserve estimation and grade control: Can. Inst. Min. Metall., Special v. 9, p. 114-115.
Davis, J.C., 1973, Statistics and data analysis in geology: John Wiley & Sons, New York, 550 p.
Davis, J.C., 1975, Contouring algorithms, Proc. 2nd Intern. Symp. on Computer-assisted Cartography (AUTO-CARTA II); U.S. Dept. of Commerce, Bureau of the Census and Am. Congresses on Survey and Mapping, Washington, D.C., p. 352-359.
Davis, J.C., 1981, Statistical techniques in petroleum exploration: Commun. Stat.-Theor. Meth., v. A10, no. 15, p. 1479-1503.
Davis, J.C., 1987, Contour mapping and SURFACE II: Science, v. 237, p. 669-672.
Davis, M.W.D., and David, M., 1980, Generating bicubic spline coefficients on a large regular grid: Computers & Geosciences, v. 6, p. 1-6.
Davis, R.E., 1941, Elementary plane surveying: McGraw-Hill, New York, 464 p.
Davis, R.E., 1985, Objective mapping by least squares fitting: Jour. Geophys. Research, v. 90, no. C3, p. 4773-4777.
Dayhoff, M.O., 1963, A contour-map program for X-ray crystallography: Commun. ACM, v. 6, no. 10, p. 620-622.
Deakin, J.J., and Puckett, G.A., 1984, Mine plan drafting: ACIRL Coal Res., v. 18, 59 p.
De Boor, C., 1962, Bicubic spline interpolation: Jour. Math. Physics, v. 41, p. 212-218.
DeMontaudouin, Y., Tiller, W., and Vold, H., 1986, Applications of power series in computational geometry: Computer-aided Design, v. 18, no. 10, p. 514-524.
Dermanis, A., 1984, Kriging and collocation - a comparison: Manuscripta Geodaetica, v. 8, p. 159-167.
Devereux, B.J., 1985, The construction of digital terrain models on small computers: Computers & Geosciences, v. 11, no. 6, p. 713-724.
Devijver, P.A., and Dekesel, M., 1982, Insert and delete algorithms for maintaining dynamic Delaunay triangulations: Pattern Recognition Letters, v. 1, p. 73-77.
Dierckx, P., 1980, An algorithm for cubic spline fitting with convexity constraints: Computing, v. 24, p. 349-371.
Dierckx, P., 1981, An algorithm for surface-fitting with spline functions: IMA Jour. Numerical Anal., v. 1, p. 267-283.
Dierckx, P., 1986, An algorithm for fitting data over a circle using tensor product splines: Jour. Computational Appl. Math., v. 15, p. 161-173.
Dimitriadis, K., Tselentis, G.-A., and Thanassoulas, K., 1987, A BASIC program for 2-D spectral analysis of gravity data and source-depth estimation: Computers & Geosciences, v. 13, no. 5, p. 549-560.
Dodd, J.R., Cain, J.A., and Bugh, J.E., 1965, Apparently significant contour patterns demonstrated with random data: Jour. Geol. Educ., v. 13, p. 109-112.
Dodd, S.L., McAllister, D.F., and Roulier, J.A., 1983, Shape-preserving spline interpolation for specifying bivariate functions on grids: IEEE Trans. on Comput. Graphics and Applic., v. 3, no. 9, p. 70-79.

Doncker, E.D., and Robinson, I., 1984, An algorithm for automatic integration over a triangle using nonlinear extrapolation, ACM Trans. on Mathematical Software, v. 10, no. 1, p. 1-16.

Dooley, J.C., 1976, Two-dimensional interpolation of irregularly spaced data using polynomial splines: Physics Earth and Planetary Interiors, v. 12, p. 180-187.

Drury, S.A., 1987, Image interpretation in geology: Allen & Unwin, London, 243 p.

Dwyer, R.A., 1987, A faster divide-and-conquer algorithm for constructing Delaunay triangulations: Algorithmics, v. 2, p. 137-151.

Dyn, N., and Levin, D., 1982, Construction of surface spline interpolants of scattered data over finite domains: RAIRO Numerical Anal., v. 16, no. 3, p. 201-209.

Dyn, N., and Wahba, G., 1982, On the estimation of functions of several variables from aggregated data: SIAM Jour. Math. Anal., v. 13, no. 1, p. 134-152.

Ebner, H., and Reiss, P., 1984, Experience with height interpolation by finite elements: Photogrammetric Engineering Remote Sensing, v. 50, no. 2, p. 177-182.

Eckstein, B.A., 1989, Evaluation of spline and weighted average interpolation algorithms: Computers & Geosciences, v. 15, no. 1, p. 79-94.

Elfick, M.H., Fryer, J.G., Brinker, R.C., and Wolf, P.R., 1987, Elementary surveying (7th ed.): Harper & Row, Sydney, 472 p.

Ellis, T.M.R., and McLain, D.H., 1977, A new method of cubic curve fitting using local data: ACM Trans. on Mathematical Software, v. 3, no. 2, p. 175-178.

Enns, S., 1986, Free-form curves on your micro: Byte, v. 11, no. 13, p. 225-230.

Enriquez, J.O.C., Thomann, J., and Goupillot, M., 1983, Applications of bidimensional spline functions to geophysics: Geophysics, v. 48, no. 9, p. 1269-1273.

Esler, J.E., and Preston, F.W., 1967, FORTRAN IV program for the GE625 to compute the power spectrum of geological surfaces: Kansas Geol. Survey Computer Contrib., v. 16, p. 1-10.

Eyton, J.R., 1984, Raster contouring: Geo-Processing, v. 2, p. 221-242.

Farin, G., 1982a, A construction for visual C^1 continuity of polynomial surface patches: Computer Graphics and Image Processing, v. 20, p. 272-282.

Farin, G., 1982b, Designing C^1 surfaces consisting of triangular cubic patches: Computer-aided Design, v. 14, no. 5, p. 253-256.

Farin, G., 1983, Algorithms for rational Bezier curves: Computer-aided Design, v. 15, no. 2, p. 73-77.

Farin, G., 1986, Piecewise triangular C^1 surface strips: Computer-aided Design, v. 18, no. 1, p. 45-47.

Farin, G., and Barry, P.J., 1986, Link between Bezier and Lagrange curve and surface schemes: Computer-aided Design, v. 18, no. 10, p. 525-528.

Farwig, R., 1986a, Rate of convergence of Shepard's global interpolation formula: Math. Computation, v. 46, p. 577-590.

Farwig, R., 1986b, Multivariate interpolation of arbitrarily spaced data by moving least squares methods: Jour. Computational Appl. Math., v. 16, p. 79-93.

Faugeras, O.D., Lebrasme, E., and Boissonn, J.D., 1990, Representing stereo data with the Delaunay triangulation: Artificial Intelligence, v. 44, no. 1/2, p. 41-87.

Field, D.A., and Smith, W.D., 1991, Graded tetrahedral finite-element meshes: Intern. Jour. Numerical Methods in Engineering, v. 31, no. 3, p. 413-425.

Feng, D.Y., and Riesenfeld, R.F., 1980, Some new surface forms for computer aided geometric design: The Computer Jour., v. 23, no. 4, p. 324-331.

Ferguson, D.R., 1986, Construction of curves and surfaces using numerical optimization techniques: Computer-aided Design, v. 18, no. 1, p. 15-21.

Ferrari, L.A., Sankar, P.V., Sklansky, J., and Leeman, S., 1986, Efficient two-dimensional filters using B-spline functions: Computer Vision, Graphics and Image Processing, v. 35, p. 152-169.

References

Finney, J.L., 1979, A procedure for the construction of Voronoi polyhedra: Jour. Computational Physics, v. 32, p. 137-143.

Fisher, H.T., 1982, Mapping information: Abt Books, Cambridge, Mass., 384 p.

Fletcher, G.Y., and McAllister, D.F., 1986, Natural bias approach to shape preserving curves: Computer-aided Design, v. 18, no. 1, p. 48-52.

Foley, J.D., and Van Dam, A., 1982, Fundamentals of interactive computer graphics: Addison-Wesley, Reading, Mass., 664 p.

Foley, T.A., 1983, Full Hermite interpolation to multivariate scattered data, in Chui, C.K., Schumaker, L.L., and Ward, J.D., eds., Approximation Theory IV: Academic Press, New York, 785 p.

Foley, T.A., 1984, Three-stage interpolation to scattered data: Rocky Mountain Jour. Math., v. 14, no. 1, p. 141-149.

Foley, T.A., 1987a, Interpolation and approximation of 3-D and 4-D scattered data: Computers Math. With Applic., v. 13, no. 8, p. 711-740.

Foley, T.A., 1987b, Interpolation with interval and point tension controls using cubic weighted v-splines: ACM Trans. on Mathematical Software, v. 13, no. 1, p. 68-96.

Foley, T.A., and Nielson, G.M., 1980, Multivariate interpolation to scattered data using delta iteration, in Cheney, E.W., ed., Approximation theory III: Academic Press, New York, p. 419-424.

Forrest, A.R., 1973, On Coons and other methods for the representation of curved surfaces: Computer Graphics and Image Processing, v. 1, p. 341-359.

Forrest, A.R., 1980, Recent work on geometric algorithms, in Brodlie, K.W., ed., Mathematical methods in computer graphics and design, Academic Press, New York, p. 105-121.

Fortune, S., 1987, A sweepline algorithm for Voronoi diagrams: Algorithmica, v. 2, p. 153-174.

Franke, R., 1982a, Smooth interpolation of scattered data by local thin plate splines: Computers Math. With Applic., v. 8, no. 4, p. 273-281.

Franke, R., 1982b, Scattered data interpolation: tests of some methods: Math. Computation, v. 38, p. 181-200.

Franke, R., 1985a, Thin plate splines with tension: Computer Aided Geometric Design, v. 2, p. 87-95.

Franke, R., 1985b, Laplacian smoothing splines with generalized cross validation for objective analysis of meteorological data: Naval Postgraduate School Tech. Rept., 33 p.

Franke, R., 1985c, Sources of error in objective analysis: Mon. Wea. Rev., v. 113, no. 2, p. 260-270.

Franke, R., 1987, Recent advances in the approximation of surfaces from scattered data, in Chui, C.K., Schumaker, L.L., and F.I. Utreras, F.I., eds., Topics in multivariate approximation: Academic Press, New York, p. 79-98.

Franke, R., and Nielson, G., 1980, Smooth interpolation of large sets of scattered data: Intern. Jour. Numerical Methods in Engineering, v. 15, p. 1691-1704.

Franke, R., and Nielson, G.M., 1983, Surface approximation with imposed conditions, in Barnhill, R.E., and Boehm, W., eds., Surfaces in computer aided geometric design: North-Holland, Amsterdam, p. 135-146.

Franklin, S.E., 1987, Terrain analysis from digital patterns in geomorphometry and landsat MSS spectral response: Photogrammetric Engineering and Remote Sensing, v. 53, no. 1, p. 59-65.

Freiberger, W., and Grenander, U., 1977, Surface patterns in theoretical geography: Computers & Geosciences, v. 3, p. 547-578.

Frenkel, Y., and Gill, D., 1975, An algorithm for contouring random data without gridding: Israel Jour. Earth Sciences, v. 24, p. 56.

Fryer, J.G., 1979, Automated mapping and surveyors: The Australian Surveyor, v. 29, no. 5, p. 325-330.

Fuchs, H., Kedem, Z.M., and Uselton, S.P., 1977, Optimal surface reconstruction from planar contours: Commun. ACM, v. 20, no. 10, p. 693-702.
Gale, N., and Halperin, W.C., 1982, A case for better graphics: the unclassed choropleth map: The Am. Statistician, v. 36, no. 4, p. 330-336.
Gal-Ezer, J., and Zwas, G., 1988, Computational aspects of rational versus polynomial interpolations: Intern. Jour. Math. Educ. Sci. Technol., v. 19, no. 4, p. 567-579.
Ganapathy, S., and Dennehy, T.G., 1982, A new general triangulation method for planar contours: Computer Graphics, v. 16, no. 3, p. 69-75.
Gardan, Y., and Lucas, M., 1984, Interactive graphics in CAD: Kogan Page, London, 256 p.
Garfinkel, D., 1962, Programmed methods for printer graphical output: Commun. ACM, v. 5, no. 9, p. 477-479.
Gasca, M., and Maeztu, J.I., 1982, On Lagrange and Hermite interpolation in R^k: Numerische Mathematik, v. 39, p. 1-14.
Gasca, M., and Ramirez, V., 1984, Interpolation systems in R^k: Jour. Approximation Theory, v. 42, p. 36-51.
Gay, S.P., 1983, Vertical scale and vertical exaggeration in three-dimensional contour maps of physical data: Geophysics, v. 48, no. 6, p. 792-793.
Georgiou, T.T., and Khargonekar, P.P., 1987, Spectral factorization and Nevanlinna-Pick interpolation: SIAM Jour. Control Optim., v. 25, no. 3, p. 754-766.
Gilchrist, B., and Cressman, G.P., 1954, An experiment in objective analysis: Tellus, v. 6, no. 4, p. 309-318.
Giloi, W.K., 1978, Interactive computer graphics: Prentice-Hall, Englewood Cliffs, 354 p.
Godwin, A.N., 1979, Family of cubic splines with one degree of freedom, Computer-aided Design, v. 11, no. 1, p. 13-18.
Gold, C.M., Charters, T.D., and Ramsden, J., 1977, Automated contour mapping using triangular element data structures and an interpolant over each irregular triangular domain: Computer Graphics (Proc. SIGGRAPH '77), v. 11, p. 170-175.
Goldman, R.N., 1983a, Subdivision algorithms for Bezier triangles: Computer-aided Design, v. 15, no. 3, p. 159-166.
Goldman, R.N., 1983b, An urnful of blending functions: IEEE Trans. on Computer Graphics and Applic., v. 3, no. 10, p. 49-54.
Goldman, R.N., 1986, Urn models and beta-splines: IEEE Trans. on Computer Graphics and Applic., v. 6, no. 2, p. 57-64.
Good, I.J., 1958, The interaction algorithm and practical Fourier analysis, Jour. Roy. Tat. Soc. B, v. 20, no. 2, p. 361-372.
Goodchild, M.F., and Lam, N.S.-N., 1980, Areal interpolation: a variant of the traditional spatial problem: Geo-Processing, v. 1, p. 297-312.
Goodin, W.R., McRae, G.J., and Seinfeld, J.H., 1979, A comparison of interpolation methods for sparse data: application to wind and concentration fields: Jour. Appl. Meteorology, v. 18, p. 761-771.
Goodman, T.N.T., and Unsworth, K., 1986, Manipulating shape and producing geometric continuity in [beta]-spline curves: IEEE Trans. on Computer Graphics and Applic., v. 6, no. 2, p. 50-56.
Gordon, W.J., and Hall, C.A., 1973, Transfinite element methods: blending-function interpolation over arbitrary curved element domains: Numerical Math., v. 21, p. 109-129.
Gordon, W.J., and Wixom, J.A., 1978, Shepard's method of "metric interpolation" to bivariate and multivariate interpolation: Math. Computation, v. 32, p. 253-264.
Grandine, T.A., 1987, The computational cost of simplex spline functions: SIAM Jour. Numerical Anal., v. 24, no. 4, p. 887-890.
Grant, F., 1957, A problem in the analysis of geophysical data: Geophysics, v. 22, no. 2, p. 309-344.

References

Graves-Morris, P.R., 1983, Vector valued rational interpolants I: Numerische Mathematik, v. 42, p. 331-348.

Graves-Morris, P.R., 1984a, Vector-valued rational interpolants II, IMA Jour. Numerical Anal., v. 4, p. 209-224.

Graves-Morris, P.R., 1984b, Symmetrical formulas for rational interpolants: Jour. Computational Appl. Math., v. 10, p. 107-111.

Graves-Morris, P.R., and Jones, R.H., 1976, An analysis of two variable rational approximants: Jour. Computational Appl. Math., v. 2, no. 1, p. 41-48.

Green, P.J., and Sibson, R., 1978, Computing Dirichlet tessellations in the plane: The Computer Jour., v. 21, no. 2, p. 168-173.

Green, P.J., and Silverman, B.W., 1979, Constructing the convex hull of a set of points in the plane: The Computer Jour., v. 22, p. 262-266.

Gregory, J.S., 1986, Shape preserving spline interpolation: Computer-aided Design, v. 18, no. 1, p. 53-57.

Gruszecka, G., and Kabat, R., 1978, Digital interpolation of isolines of deposit parameters, *in* Dalkowski, T., ed., Information Systems and Operation Research in Mining: Olesnica, Poland, p. 227-234.

Gurnell, A.M., 1981, Mapping potential evapotranspiration: the smooth interpolation of isolines with low density station network: Applied Geography, v. 1, p. 167-183.

Guth, P.L., Ressler, E.K., and Bacastow, T.S., 1987, Microcomputer program for manipulating large digital terrain models: Computers & Geosciences, v. 13, p. 209-213.

Haas, A.G., and Viallix, J.R., 1976, Krigeage applied to geophysics: the answer to the problem of estimates and contouring: Geophys. Prospecting, v. 24, p. 49-69.

Hakopian, H.A., 1982, Multivariate divided differences and multivariate interpolation of Lagrange and Hermite type: Jour. Approximation Theory, v. 34, p. 286-305.

Hall, C.A., 1968, On error bounds for spline interpolation: Jour. Approximation Theory, v. 1, p. 209-218.

Hall, J.K., 1979, An algorithm for integrating polynomials over any closed boundary, and its application to calculation of volume under polynomial trend surfaces, *in* Gill, D., and Merriam, D.F., eds., Geomathematical and petrophysical studies in sedimentology: Pergamon Press, Oxford, p. 211-218.

Hamilton, L.H., and Tasker, B.S., 1984, Practical aspects of drilling for coal and stratiform deposits on triangular grids: The Coal Jour., March, p. 67-73.

Hammer, P.T.C., Hildebrand, J.A., and Parker, R.L., 1991, Gravity inversion using seminorm minimization - density modeling of Jasper Seamount: Geophys., v. 56, no. 1, p. 68-79.

Harada, K., and Nakamae, E., 1982a, An isotropic four-point interpolation based on cubic splines: Computer Graphics and Image Processing, v. 20, p. 283-287.

Harada, K., and Nakamae, E., 1982b, Application of the Bezier curve to data interpolation: Computer-aided Design, v. 14, no. 1, p. 55-59.

Harbaugh, J.W., and Merriam, D.F., 1968, Computer applications in stratigraphic analysis: John Wiley & Sons, New York, 282 p.

Harbaugh, J.W., and Sackin, M.J., 1968, FORTRAN IV program for harmonic trend analysis using double Fourier sreies and regularly gridded data for the GE625 computer: Kansas Geol. Survey Computer Contrib., v. 29, p. 1-6.

Harder, R.L., and Desmarais, R.N., 1972, Interpolation using surface splines: Jour. Aircraft, v. 9, no. 2, p. 189-191.

Harding, J.E., 1923, How to calculate tonnage and grade of an orebody: Engineering Mining Jour., v. 116, no. 11, p. 445-448.

Hardy, R.L., 1971, Multiquadric equations of topography and other irregular surfaces: Jour. Geophysical Research, v. 76, no. 8, p. 1905-1915.

Hardy, R.L., 1977, Least squares prediction: Photogrammetric Engineering and Remote Sensing, v. 43, no. 4, p. 475-492.

References

Hardy, R.L., 1990, Theory and applications of the multiquadric-biharmonic method: Computers Math. With Applic., v. 19, no. 8/9, p. 163-208.

Hardy, R.L., and Nelson, S.A., 1986, A multiquadric-biharmonic representation and approximation of disturbing potential: Geophys. Res. Letters, v. 13, no. 1, p. 18-21.

Hartley, P.J., and Judd, C.J., 1980, Parametrization and shape of B-spline curves for CAD: Computer-aided Design, v. 12, no. 5, p. 235-238.

Hayes, J.G., 1986, Advances in algorithms for surface fitting: *in* Mohamed, J.L., and Walsh, J.E., eds., Numerical algorithms, Clarendon Press, Oxford, p. 314-326.

Hayes, W.B., and Koch, G.S., 1984, Constructing and analyzing area-of-influence polygons by computer: Computers & Geosciences, v. 10, no. 4, p. 411-430.

Herron, G., 1985, A characterization of certain C^1 discrete triangular interpolants: SIAM Jour. Numerical Anal., v. 22, no. 4, p. 811-819.

Hessing, R.C., Lee, H.K., Pierce, A., and Powers, E.N., 1972, Automatic contouring using bicubic functions: Geophysics, v. 37, no. 4, p. 669-674.

Hilbert, R., 1981, Construction and display of three-dimensional polygonal-histograms: Computer Graphics, v. 15, no. 2, p. 230-241.

Hobson, R.D., 1967, FORTRAN IV programs to determine surface roughness in topography for the CDC 3400 computer: Kansas Geol. Survey Computer Contrib., v. 14, p. 1-27.

Hobson, R.D., 1972, Surface roughness in topography: quantitative approach, *in* Chorley, R.J., ed., Spatial analysis in geomorphology: Harper & Row, New York, p. 221-245.

Hoffmann, C., and Hopcroft, J., 1986, Quadratic blending surfaces: Computer-aided Design, v. 18, no. 6, p. 301-306.

Holroyd, M.T., and Bhattacharyya, B.K., 1970, Automatic contouring of geophysical data using bicubic spline interpolation: Geol. Survey of Canada, Paper 70-55, Ottawa, 40 p.

Hoover, H.C., 1909, Principles of mining: McGraw-Hill, New York, 199 p.

Horton, R.E., 1923, Rainfall interpolation: Mon. Wea. Rev., v. 51, no. 6, p. 291-304.

Hu, C.L., and Schumaker, L.L., 1986, Complete spline smoothing: Numerische Mathematik, v. 49, p. 1-10.

Hughes, R.J., 1959, Volume estimates from contours: Econ. Geol., v. 54, p. 730-744.

IBM, 1965, Numerical surface techniques and contour map plotting: IBM Data Processing Applications, White Plains, New York.

Inoue, H., 1986, A least-squares smooth fitting for irregularly spaced data: finite-element approach using the cubic B-spline basis: Geophysics, v. 51, no. 11, p. 2051-2066.

Isted, T.C., and Mendelsohn, F., 1967, Ore reserve estimations by geometric methods: Roan Selection Trust Technical Services Ltd., Geologic Reseach Unit, Report GR8, 33 p.

James, W.R., 1966, FORTRAN IV program using double Fourier series for surface fitting of irregularly spaced data: Kansas Geol. Survey Computer Contrib., v. 5, p. 1-18.

Jancaitis, J.R., and Junkins, J.L., 1973, Modeling irregular surfaces: Photogrammetric Engineering, v. 39, p. 413-420.

Johnston, F.R., and Harrison, P.J., 1984, Discount weighted moving averages: Jour. Operational Res. Soc., v. 35, no. 7, p. 629-635.

Jones, N.L., Wright, S.G., and Maidment, D.R., 1990, Watershed delineation with triangle-based terrain models: Jour. Hydraulic Engineering, v. 116, no. 10, p. 1232-1251.

Jones, R.L., 1971, A generalized digital contouring program: NASA Langley Research Center, Hampton, Va, NASA TDD-6022.

Jones, T.A., Hamilton, D.E., and Johnson, C.R., 1986, Contouring geologic surfaces with the computer: Van Nostrand Reinhold, New York, 314 p.

Junkins, J.L., Miller, G.W., and Jancaitis, J.R., 1973, A weighting function approach to modeling of irregular surfaces: Jour. Geophys. Research, v. 78, no. 11, p. 1794-1803.

References

Jupp, D.L., 1976, B-splines for smoothing and differentiating data sequences: Math. Geol., v. 8, no. 3, p. 243-266.

Kalkani, E.C., 1977, Evaluation technique to determine relative accuracies of contour maps: Geophysics, v. 42, no. 4, p. 860-867.

Kane, V.E., Begovich, C.L., Butz, T.R., and Myers, D.E., 1982, Interpolation of regional geochemistry using optimal interpolation parameters: Computers & Geosciences, v. 8, no. 2, p. 117-135.

Kashiyama, K., and Kawahara, M., 1985, Interpolation method for preparation of input data of water depth in finite element analysis of shallow water flow: Engineering Computations, v. 2, no. 4, p. 266-270.

Kennedy, S., and Tobler, W.R., 1983, Geographic interpolation: Geographical Analysis, v. 15, no. 2, p. 151-156.

Keppel, E., 1975, Approximating complex surfaces by triangulation of contour lines: IBM Jour. Res. Develop., v. 19, p. 2-11.

Kimberley, M.M., 1986, Geochemistry and structure of stratiform deposits with a portable microcomputer: Ore Geology Reviews, v. 1, p. 7-42.

Klincsek, G.T., 1980, Minimal triangulations of polygonal domains: Annals of Discrete Mathematics, v. 9, p. 121-123.

Klucewicz, I.M., 1978, A piecewise C^1 interpolant to arbitrarily spaced data: Computer Graphics and Image Processing, v. 8, p. 92-112.

Kochanek, D.H.U., and Bartels, R.H., 1984, Interpolating splines with local tension, continuity, and bias control: Computer Graphics, v. 18, no. 3, p. 33-41.

Kok, R., and Begin, J., 1981, Evaluation of automatic contouring methods for drainage design: Trans. of the ASAE, v. 24, no. 1, p. 87-96,102.

Kok, A.L., Blais, J.A.R., and Rangayyan, R.M., 1987, Filtering of digitally correlated Gestalt elevation data: Photogrammetric Engineering and Remote Sensing, v. 53, no. 5, p. 535-538.

Kratky, V., 1978, Reflexive prediction and digital terrain modelling: Photogrammetric Engineering and Remote Sensing, v. 44, no. 5, p. 569-574.

Kratky, V., 1981, Spectral analysis of interpolation: Photogrammetria, v. 37, p. 61-72.

Kraus, K., and Mikhail, E.M., 1972, Linear least-squares interpolation: Photogrammetric Engineering, v. 38, p. 1016-1029.

Krige, D.G., 1964, Recent developments in South Africa in the application of trend surface and multiple regression techniques to gold ore evaluation: Colorado School of Mines Quart., v. 59, p. 795-809.

Kurtze, D.A., 1986, Failures of local approximation in finite-element methods: Intern. Jour. Numerical Methods in Engineering, v. 23, p. 1483-1494.

Lacombe, C., and Bedard, C., 1986, Interpolation function of a general triangular mid-edge finite element: Computers Math. With Applic., v. 12A, no. 3, p. 363-373.

Lam, N.S-N., 1983, Spatial interpolation methods: a review: The Am. Cartographer, v. 10, no. 2, p. 129-149.

Lancaster, P., and Salkauskas, K., 1981, Surfaces generated by moving least squares methods: Mathematics of Computation, v. 37, p. 141-158.

Langridge, D.J., 1984, Detection of discontinuities in the first derivatives of surfaces: Computer Vision, Graphics and Image Processing, v. 27, p. 291-308.

LaPorte, M., 1962, Elaboration rapide de cartes gravimetriques deduites de l'anomalie de Bouguer a l'aide d'une calculatrice electronique: Geophys. Prospecting, v. 10, p. 238-257.

Larkin, F.M., 1967, Some techniques for rational interpolation: The Computer Jour., v. 10, p. 178-187.

Laslett, G.M., McBratney, A.B., Pahl, P.J., and Hutchinson, M.F., 1987, Comparison of several spatial prediction methods for soil pH: Jour. Soil Science, v. 38, p. 325-341.

Lawson, C.L., 1977, Software for C^1 surface interpolation: *in* Rice, J.R., ed., Mathematical software III, Academic Press, New York.

Leberl, F., 1975, Photogrammetric interpolation: Photogrammetric Engineering and Remote Sensing, v. 41, p. 603-612.

Lee, G.S., 1982, Piecewise linear approximation of multivariate functions: Bell System Technol. Jour., v. 61, no. 7, p. 1463-1486.

Lee, P.J., 1981, The most predictable surface (MPS) mapping method in petroleum exploration: Bull. Can. Petroleum Geol., v. 29, no. 2, p. 224-240.

Lee, Y.C., 1985, Comparison of planimetric and height accuracy of digital maps: Surveying and Mapping, v. 45, no. 4, p. 333-340.

Lee, S.L., and Phillips, G.M., 1987, Interpolation on the triangle: Commun. Appl. Numerical Methods, v. 3, p. 271-276.

Lee, D.T., and Schachter, B.J., 1980, Two algorithms for constructing a Delaunay triangulation: Intern. Jour. Comput. Inform. Sci., v. 9, p. 219-242.

Lenard, M., 1985, Spline interpolation in two variables: Studia Scient. Math. Hungarica, v. 20, p. 145-154.

Leventhal, S.H., Klein, M.H., and Culham, W.E., 1985, Curvilinear coordinate systems for reservoir simulation: Soc. Petroleum Engineers Jour., v. 25, no. 6, p. 893-901.

Lewis, B.A., and Robinson, J.S., 1978, Triangulation of planar regions with applications: The Computer Jour., v. 21, no. 4, p. 324-332.

Liszka, T., 1984, An interpolation method for an irregular net of nodes: Intern. Jour. Numerical Methods in Engineering, v. 20, p. 1599-1612.

Little, F.F., 1983, Convex combination surfaces, *in* Barnhill, R.E., and Boehm, W., eds., Surfaces in computer aided geometric design: North-Holland, Amsterdam, p. 99-107.

Lodwick, G.D., and Whittle, J., 1970, A technique for automatic contouring field survey data: The Australian Computer Jour., v. 2, no. 3, p. 104-109.

Loh, R., 1981, Convex B-spline surfaces: Computer-aided Design, v. 13, no. 3, p. 145-149.

Loikkanen, M.J., 1985, A 4-node thin hybrid plate finite element: Engineering Computations, v. 2, p. 151-154.

Long, C., 1987, Special Bezier quartics in three dimensional curve design and interpolation: Computer-aided Design, v. 19, no. 2, p. 77-84.

Lord, E.A., and Wilson, C.B., 1984, The mathematical description of shape and form: John Wiley & Sons, New York, 260 p.

Lord, M., 1987, Curve and surface representation by iterative B-spline fit to a data point set: Engineering in Medicine, v. 16, no. 1, p. 29-35.

Lorenc, A.C., 1981, A global three-dimensional multivariate statistical interpolation scheme: Mon. Wea. Rev., v. 109, p. 701-720.

Lorentz, G.G., and Lorentz, R.A., 1983, Multivariate interpolation, *in* Graves-Morris, P.R., Saff, E.B., and Varga, R.S., eds., Rational approximation and interpolation: Springer-Verlag, Berlin, p. 136-144.

Lowden, B., 1985, A storage efficient method for producing smooth contour maps: Infor, v. 23, no. 4, p. 447-468.

MacCallum, K.J., and Zhang, J.-M., 1986, Curve-smoothing techniques using B-splines: The Computer Jour., v. 29, no. 6, p. 564-571.

MacDonald, G.M., and Waters, N.M., 1987, An evaluation of automated mapping algorithms for the analysis of Quaternary pollen data: Review of Palaeobotany and Palynology, v. 51, p. 289-307.

MacGillivray, R.B., Hawkins, D.M., and Berjak, M., 1969, The computer mapping and assessment of borehole and sampling data for stable minerals, particularly as applied to coal mining: Jour. South African Inst. Min. Metall., v. 69, p. 250-265.

Mackay, J.R., 1951, Some problems and techniques in isopleth mapping: Econ. Geography, v. 27, no. 1, p. 1-9.

References

Maffini, G., 1987, Raster versus vector data encoding and handling: a commentary: Photogrammetric Engineering and Remote Sensing, v. 53, no. 10, p. 1397-1398.

Magnus, E.R., Joyce, C.C., and Scott, W.D., 1983, A spiral procedure for selecting a triangular grid from random data: Jour. Appl. Math., and Physics (ZAMP), v. 34, p. 231-235.

Manacher, G.K., and Zobrist, A.L., 1979, Neither the Greedy nor the Delaunay triangulation of a planar point set approximates the Optimal triangulation: Information Processing Letters, v. 9, p. 31-34.

Mansfield, L., 1980, Interpolation to scattered data in the plane by locally defined C^1 functions, *in* Cheney, E.W., ed., Approximation theory III, Academic Press, New York, p. 623-628.

Mason, J.C., 1984, BASIC matrix methods: Butterworths, London, 160 p.

Maus, A., 1984, Delaunay triangulation and the convex hull of n points in expected linear time: BIT, v. 24, p. 151-163.

Maxwell, J.C., 1870, On contour-lines and measurement of heights: Philosophical Magazine, v. 40, p. 421-427.

Mazzega, P., and Houry, S., 1989, An experiment to invert *Seasat* altimetry for the Mediterranean and Black Sea mean surfaces: Geophys. Jour., v. 96, no. 2, p. 259-272.

McAllister, D.F., and Roulier, J.A., 1981, An algorithm for computing a shape-preserving osculatory quadratic spline: ACM Trans. on Mathematical Software, v. 7, no. 3, p. 331-347.

McCullagh, M.J., 1981, Creation of smooth contours over irregularly distributed data using local surface patches: Geographical Analysis, v. 13, p. 51-63.

McCullagh, M.J., 1983, Computer mapping for minerals exploration: First Australian Conf. on Computer Graphics, Sydney, p. 178-184.

McLain, D.H., 1974, Drawing contours from arbitrary data points: The Computer Jour., v. 17, no. 4, p. 318-324.

McLain, D.H., 1976, Two dimensional interpolation from random data: The Computer Jour., v. 19, no. 2, p. 178-181; errata p. 384.

McLaughlin, H.W., 1983, Shape-preserving planar interpolation: an algorithm: IEEE Trans. on Computer Graphics and Applic., v. 6, p. 58-67.

Medvedev, N.N., 1986, The algorithm for three-dimensional Voronoi polyhedra: Jour. Computational Physics, v. 67, p. 223-229.

Meinguet, J., 1979, Multivariate interpolation at arbitrary points made simple: Jour. Appl. Math. and Physics (ZAMP), v. 30, p. 292-304.

Meinguet, J., 1983, Surface spline interpolation: basic theory and computational aspects: Seminaire Mathematique, 2eme Semestre, v. III, p. 1-15.

Mendelsohn, F., 1980, Some aspects of ore reserve estimation: University of Witswatersrand, Johannesburg, Economic Geology Unit, Information Cir. 147, 42 p.

Merriam, D.F., and Harbaugh, J.W., 1963, Computer helps map oil structures: Oil and Gas Jour., v. 61, no. 11, p. 158-163.

Mettke, H., 1983, Convex cubic HERMITE-spline interpolation: Jour. Computational Appl. Math., v. 9, p. 205-211.

Micchelli, C.A., 1980, A constructive approach to Kergin interpolation in R^k: multivariate B-splines and Lagrange interpolation: Rocky Mountain Jour. Math., v. 10, no. 3, p. 485-497.

Miller, J.R., 1986, Sculptured surfaces in solid models: issues and alternative approaches: IEEE Trans. on Computer Graphics and Applic., v. 6, p. 37-48.

Mirante, A., and Weingarten, N., 1982, The radial sweep algorithm for constructing triangulated networks: IEEE Trans. on Computer Graphics and Applic., v. 2, p. 11-21.

Moffat, A.J., Catt, J.A., Webster, R., and Brown, E.H., 1986, A re-examination of the evidence for a Plio-Pleistocene marine transgression on the Chiltern Hills, I, structures and surfaces: Earth Surface Processes and Landforms, v. 11, p. 95-106.

Monaghan, J.J., 1985, Extrapolating B splines for interpolation: Jour. Computational Physics, v. 60, p. 253-262.
Monmonier, M.S., 1982, Computer-assisted cartography: Prentice-Hall, Englewood Cliffs, 214 p.
Monro, D.M., 1985, Real discrete fast Fourier transform, in Griffiths, P., and Hill, I.D., eds., Applied statistics algorithms, Ellis Horwood, Chichester, 307 p.
Moore, I.G., 1977, Automatic contouring of geological data, APCOM 77 15th Intern. Symp. on the Application of Computers and Operations Research in the Minerals Industry, Australasian Inst. Min. Metall., p. 209-220.
Morrison, J.L., 1974, Observed statistical trends in various interpolation algorithms useful for first stage interpolation: The Can. Cartographer, v. 11, no. 2, p. 142-159.
Nagy, G., 1980, What is a "good" data structure for 2-D points?, in Freeman, H., and Pieroni, G.G., eds., Map data processing: Academic Press, New York, p. 119-135.
Navlakha, J.K., 1984, An analytical technique for 3-dimensional interpolation: BIT, v. 24, p. 119-122.
Navon, I.M., 1986, A review of variational and optimization methods in meteorology, in Sasaki, Y.K., ed., Variational methods in geosciences: Elsevier, Amsterdam, 309 p.
Neuman, A.E., 1987, Representing topography with second-degree bivariate polynomial functions fitted by least squares: Jour. Geol. Education, v. 35, p. 271-274.
Newman, W.M., and Sproull, R.F., 1979, Principles of interactive computer graphics: McGraw-Hill, London, 541 p.
Newton, R., 1968, Deriving contour maps from geological data: Can. Jour. Earth Sciences, v. 5, p. 165-166.
Nielson, G.M., 1979, The side-vertex method for interpolation in triangles: Jour. Approximation Theory, v. 25, p. 318-336.
Nielson, G.M., 1983, A method for interpolating scattered data based upon a minimum norm network: Math. Computation, v. 40, p. 253-271.
Nielson, G.M., 1986, Rectangular v-splines: IEEE Trans. on Computer Graphics and Applic., v. 6, p. 35-40.
Nielson, G.M., and Franke, R., 1983, Surface construction based upon triangulations, in Barnhill, R.E., and Boehm, W., eds., Surfaces in computer aided geometric design: North-Holland, Amsterdam, p. 163-177.
Nielson, G.M., and Franke, R., 1984, A method for construction of surfaces under tension: Rocky Mountain Jour. Math., v. 14, no. 1, p. 203-221.
Nielson, G.M., and Mangeron, D.J., 1981, Bilinear interpolation in triangles based upon a Mangeron theorem: Rev. Real Acad. Cienc. Exact. Fis. Natur. Madrid, v. 75, p. 89-96.
North, F.K., 1985, Petroleum geology: Allen & Unwin, London, 607 p.
Ojakangas, D.R., and Basham, W.L., 1964, Simplified computer contouring of exploration data: Stanford Univ. Publ. Geol. Sciences, v. 9, p. 757-770.
Oldknow, A., 1987, Drawing, curves, surfaces and solids: some recent applications of mathematics from computer graphics: Bull. Inst. Math. Applic, v. 23, p. 134-139.
O'Neill, M.A., 1988, Faster than fast Fourier: Byte, v. 13, no. 4, p. 293-300.
Onukogu, I.B., and Esele, I.I., 1987, Response surface exploration with spline functions: biased regressors: Statistica, v. 47, no. 2, p. 277-298.
Osborn, R.T., 1967, An automated procedure for producing contour charts: U.S. Naval Oceanographic Office, IM No. 67-4, 54 p.
Oxley, A., 1985, Surface fitting by triangulation: The Computer Jour., v. 28, no. 3, p. 335-339.
Paik, C.H., and Fox, M.D., 1988, Fast Hartley transforms for image processing: IEEE Trans. on Medical Imaging, v. 7, no. 2, p. 149-153.
Pal, T.K., 1980, Hybrid surface patch: Computer-aided Design, v. 12, no. 6, p. 283-287.
Palmer, J.A.B., 1969, Automated mapping: 4th Austral. Computer Conf. Proc., Australian Computer Soc., Adelaide, p. 463-468.

References

Panofsky, H.A., 1949, Objective weather-map analysis: Jour. Meteorology, v. 6, p. 386-392.
Papo, H.B., and Gelbman, E., 1984, Digital terrain models for slopes and curvatures: Photogrammetric Engineering and Remote Sensing, v. 50, no. 6, p. 695-701.
Pelto, C.R., Elkins, T.A., and Boyd, H.A., 1968, Automatic contouring of irregularly spaced data: Geophysics, v. 33, no. 3, p. 424-430.
Percell, P., 1976, On cubic and quartic Clough-Tocher finite elements: SIAM Jour. Numerical Anal., v. 13, no. 1, p. 100-103.
Petersen, R.G., 1985, Design and analysis of experiments: Marcel Dekker, New York, 429 p.
Petrie, G., and Kennie, T.J.M., 1987, Terrain modelling in surveying and civil engineering: Computer-aided Design, v. 19, no. 4, p. 171-187.
Peucker, T.K., 1980, The impact of different mathematical approaches to contouring: Cartgraphica, v. 17, p. 73-95.
Philip, G.M., and Watson, D.F., 1982a, Optimum drilling patterns for establishing coal reserves: Australian Jour. Coal Min. Tech. Res., v. 2, p. 65-68.
Philip, G.M., and Watson, D.F., 1982b, A precise method for determining contoured surfaces: Jour. Australian Petroleum Exploration Assoc., v. 22, p. 202-212.
Philip, G.M., and Watson, D.F., 1984, Drilling patterns and ore reserve assessments: Bull. Proc. Australasian Inst. Min. Metall., v. 289, p. 205-211.
Philip, G.M., and Watson, D.F., 1985a, Theoretical aspects of grade-tonnage calculations: Intern. Jour. Mining Engineering, v. 3, p. 149-154.
Philip, G.M., and Watson, D.F., 1985b, Some limitations in the geostatistical evaluation of ore deposits: Intern. Jour. Mining Engineering, v. 3, p. 155-159.
Philip, G.M., and Watson, D.F., 1985c, Reply to contributed discussion by P.S.B. Stewart: Bull. Proc. Australasian Inst. Min. Metall., v. 290, p. 77.
Philip, G.M., and Watson, D.F., 1985d, A deterministic approach to computing ore reserves, I, Grade estimation: Earth Resources Foundation Occasional Publication 2, The University of Sydney, 24 p.
Philip, G.M., and Watson, D.F., 1986a, Interpolation methods in finite element analysis: Engineering Computations, v. 3, p. 175.
Philip, G.M., and Watson, D.F., 1986b, Automatic interpolation methods for mapping piezometric surfaces: Automatica, v. 22, no. 6, p. 753-756.
Philip, G.M., and Watson, D.F., 1986c, A method for assessing local variation among scattered measurements: Math. Geol., v. 18, no. 8, p. 759-764.
Philip, G.M., and Watson, D.F., 1987, Neighborhood discontinuities in bivariate interpolation of scattered observations: Math. Geol., v. 19, no. 1, p. 69-74.
Piazza, A., Menozzi, P., and Cavalli-Sforza, L., 1981, The making and testing of gene-frequency maps: Biometrics, v. 37, p. 635-659.
Piegl, L., 1984, A generalization of the Bernstein-Bezier method: Compter-aided Design, v. 16, no. 4, p. 209-215.
Piegl, L., 1986, Representation of rational Bezier curves and surfaces by recursive algorithms: Computer-aided Design, v. 18, no. 7, p. 361-366.
Piegl, L., 1988, Hermite-and Coons-like interpolants using rational Bezier approximation form with infinite control points: Computer-aided Design, v. 20, no. 1, p. 2-10.
Piegl, L., and Tiller, W., 1987, Curve and surface constructions using rational B-splines: Computer-aided Design, v. 19, no. 9, p. 485-498.
Plaisted, D.A., and Hong, J., 1987, A heuristic triangulation algorithm: Jour. Algorithms, v. 8, p. 405-437.
Posdamer, J.L., 1983, Spatial sorting for sampled surface geometries: Proc. Soc. Photo-optical Instrument Engineers, v. 361, p. 152-163.
Pouzet, J., 1980, Estimation of a surface with known discontinuities for automatic contouring purposes: Math. Geol., v. 12, no. 6, p. 559-575.

Powell, M.J.D., and Sabin, M.A., 1977, Piecewise quadratic approximations on triangles: ACM Trans. on Mathematical Software, v. 3, no. 4, p. 316-325.
Preparata, F.P., and Muller, D.E., 1979, Finding the intersection of n half-spaces in time O($n \log n$): Theoretical Computer Science, v. 8, p. 45-55.
Press, W.H., Flannery, B.P., Teukolsky, S.A., and Vetterling, W.T., 1986, Numerical recipes: Cambridge University Press, Cambridge, 818 p.
Preusser, A., 1984a, Bivariate interpolation uber dreieckselemten durch polynome 5. ordnung mit C^1-kontinuitat: Zeitschrift Vermessungswesen, v. 109, no. 6, p. 292-301.
Preusser, A., 1984b, TRICP: a contour plot program for triangular meshes: ACM Trans. on Mathematical Software, v. 10, no. 4, p. 473-475.
Radian Corporation, 1979, CPS-1 user's manual: Radian Corporation, Austin, Texas, v. 1, 187 p.
Rattray, M.Jr., 1962, Interpolation errors and oceanographic sampling: Deep Sea Research, v. 9, p. 25-37.
Rayner, J.N., 1967, Correlation between surfaces by spectral methods: Kansas Geol. Survey Computer Contrib., v. 12, p. 31-37.
Read, J., 1982, Fourier transforms with VisiCalc: Practical Computing, November, p. 117-125.
Reedman, J.H., 1979, Techniques in mineral exploration: Applied Science Publishers, London, 533 p.
Reiniger, R.F., and Ross, C.K., 1968, A method of interpolation with application to oceanographic data: Deep Sea Research, v. 15, p. 185-193.
Renka, R.J., and Cline, A.K., 1984, A triangle-based C^1 interpolation method: Rocky Mountain Jour. Math., v. 14, no. 1, p. 223-237.
Rentrop, P., 1980, An algorithm for the computation of the exponential spline: Numerische Mathematik, v. 35, p. 81-93.
Renz, W., 1982, Interactive smoothing of digitized point data: Computer-aided Design, v. 14, no. 5, p. 267-269.
Rhind, D.W., 1975, A skeletal overview of spatial interpolation techniques: Computer Applications, v. 2, no. 3/4, p. 293-309.
Rhind, D.W., 1972, One-sided constraints in automated contouring of drift thickness: Bull. Geol. Soc. Am., v. 83, p. 2525-2532.
Rhynsburger, D., 1973, Analytic delineation of Thiessen polygons: Geographical Analysis, v. 5, p. 133-144.
Riggs, D.S., Guarnieri, J.A., and Addelman, S., 1978, Fitting straight lines when both variables are subject to error: Life Sciences, v. 22, p. 1305-1360.
Rikitake, T., Sato, R., and Hagiwara, Y., 1987, Applic. mathematics for earth scientists: Reidel, Dordrecht, 435 p.
Ripley, B.D., 1981, Spatial statistics, John Wiley & Sons, New York, 252 p.
Ripley, B.D., 1984, Spatial statistics: developments 1980-3: Intern. Stat. Review, v. 52, no. 2, p. 141-150.
Robinson, A.H., 1971, The genealogy of the isopleth: Cartographic Jour., v. 8, p. 49-53.
Robinson, J.E., 1982, Computer applications in petroleum geology: Hutchinson Ross, 164 p.
Rock, N.M.S., 1988, Numerical geology, Lecture notes in earth sciences, v. 18: Spriner-Verlag, Berlin, 427 p.
Rogers, C.A., 1964, Packing and covering: Cambridge University Press.
Rogers, D.F., and Adams, J.A., 1976, Mathematical elements for computer graphics: McGraw-Hill, New York, 239 p.
Rom, M., and Bergman, S., 1986, A new technique for automatic contouring and contour representation from machine-readable spatial data: The Computer Jour., v. 29, no. 5, p. 467-471.

References

Roulier, J.A., 1980, Constrained interpolation: SIAM Jour. Sci. Stat. Comput., v. 1, no. 3, p. 333-344.

Roulier, J.A., 1987, A convexity-preserving grid refinement algorithm for interpolation of bivariate functions: IEEE Trans. on Computer Graphics and Applic., v. 7, p. 57-62.

Sabin, M.A., 1980, Contouring - a review of methods for scattered data, *in* Brodlie, K.W., ed., Mathematical methods in computer graphics and design: Academic Press, New York, p. 63-86.

Sabin, M.A., 1985, Contouring - the state of the art, *in* Earnshaw, R.E., ed., Fundamental algorithms for computer graphics: Springer-Verlag, p. 411-482.

Sablonniere, P., 1987, Error bounds for Hermite interpolation by quadratic splines on an alpha-triangulation: IMA Jour. Numerical Anal., v. 7, p. 495-508.

Sager, T.W., 1983, Estimating modes and isopleths: Commun. Stat.-Theor. Meth., v. 12, no. 5, p. 529-557.

Sampson, R.J., 1975a, The SURFACE II graphics system, *in* Davis, J.C., and McCullagh, M.J., ed., Display and analysis of spatial data: John Wiley & Sons, London, p. 244-266.

Sampson, R.J., 1975b, SURFACE II graphics system: Kansas Geol. Survey, Spatial Analysis Ser. 1, 240 p.

Sampson, R.J., and Davis, J.C., 1967, Three-dimensional response surface program in FORTRAN II for the IBM 1620 computer: Kansas Geol. Survey Computer Contrib., v. 10, p. 1-7.

Sandwell, D.T., 1987, Biharmonic spline interpolation of GEOS-3 and SEASAT altimeter data: Geophys. Res. Letters, v. 14, no. 2, p. 139-142.

Satterfield, S.G., and Rogers, D.F., 1985a, A procedure for generating contour lines from a B-spline surface: IEEE Trans. on Computer Graphics and Applic., v. 5, p. 71-75.

Satterfield, S.G., and Rogers, D.F., 1985b, A procedure for generating contour lines from a B-spline surface, *in* Kunii, T.L., ed., Frontiers in Computer Graphics: Springer-Verlag, Tokyo, p. 66-73.

Sawkar, D.G., Shevare, G.R., and Koruthu, S.P., 1987, Contour plotting for scattered data: Computers & Graphics, v. 11, no. 2, p. 101-104.

Schagen, I.P., 1979, Interpolation in two dimensions, - a new technique: Jour. Inst. Math. Applic., v. 23, p. 53-59.

Schagen, I.P., 1982, Automatic contouring from scattered data points: The Computer Jour., v. 25, no. 1, p. 7-11.

Schellmoser, K., 1962, Zur berechnung von vorraten nach der isolinienmethode: Zeitschrift angew. Geologie, v. 8, no. 1, p. 15-18.

Schmid, C.F., and MacCannell, E.H., 1955, Basic problems, techniques, and theory of isopleth mapping: Jour. Am. Stat. Assoc., March, p. 220-239.

Schmidt, J.W., and Hess, W., 1987, Positive interpolation with rational quadratic splines: Computing, v. 38, p. 261-267.

Schroeder, W.J. and Shephard, M.S., 1990, A combined octree/Delaunay method for fully automatic 3-D mesh generation, Intern, Jour. Numerical Methods in Engineering, v. 29, no. 1, p. 37-55.

Schumaker, L.L., 1976, Fitting surfaces to scattered data, *in* Lorentz, G.G., ed., Approximation theory II: Academic Press, New York, 203-268.

Schut, G.H., 1974, Two interpolation methods: Photogrammetric Engineering and Remote Sensing, v. 40, p. 1447-1453.

Schut, G.H., 1976, Review of interpolation methods for digital terrain models: The Canadian Surveyor, v. 30, no. 5, p. 389-412.

Scott, D.S., 1984, The complexity of interpolating given data in three space with a convex function of two variables: Jour. Approximation Theory, v. 42, p. 52-63.

Sederberg, T.W., and Anderson, D.C., 1985, Steiner surface patches: IEEE Trans. on Computer Graphics and Applic., May, p. 23-36.

References

Sederberg, T.W., Anderson, D.C., and Goldman, R.N., 1985, Implicitization, inversion and intersection of planar rational cubic curves: Computer Vision, Graphics and Image Processing, v. 31, p. 89-102.

Sederberg, T.W., and Goldman, R.N., 1986, Algebraic geometry for computer-aided geometric design: IEEE Trans. on Computer Graphics and Applic., v. 6, p. 52-59.

Segalman, D.J., Woyak, D.B., and Rowlands, R.E., 1979, Smooth spline-like finite-element differentiation of full-field experimental data over arbitrary geometry: Experimental Mechanics, v. 19, p. 429-437.

Seldner, D., and Westermann, T., 1988, Algorithms for interpolation and localization in irregular 2D meshes: Jour. Comput. Physics, v. 79, p. 1-11.

Shafer, S.A., Kanade, T., and Kender, J., 1983, Gradient space under orthography and perspective: Computer Vision, Graphics and Image Processing, v. 24, p. 182-199.

Shamos, M.I., and Hoey, D., 1975, Closest-point problems: IEEE 16th Annual Symposium Foundations Computer Science, IEEE, Los Angeles, p. 151-162.

Sharapov, I.P., 1962, Contributions to the theory of consistency of mineral deposits: Sci. Trans. of Coal Inst., Perm Publ., Perm.

Shepard, D., 1968, A two-dimensional interpolation function for irregularly-spaced data: Proc. 23rd National Conference ACM, ACM, p. 517-524.

Sibson, R., 1978, Locally equiangular triangulations: The Computer Jour., v. 21, no. 3, p. 243-245.

Sibson, R., 1980a, A vector identity for the Dirichlet tessellation: Math. Proc. Cambridge Phil. Soc., v. 87, p. 151-155.

Sibson, R., 1980b, The Dirichlet tessellation as an aid in data analysis: Scand. Jour. Statistics, v. 7, p. 14-20.

Sibson, R., 1981, A brief description of natural neighbour interpolation, *in* Barnett, V., ed., Interpreting multivariate data: John Wiley, p. 21-36.

Sibson, R., and Thomson, G.D., 1981, A seamed quadratic element for contouring: The Computer Jour., v. 24, no. 4, p. 378-382.

Simons, S.L.Jr., 1983, Make fast and simple plots on a microcomputer: Byte, v. 8, p. 487-492.

Simpson, S.M.Jr., 1954, Least squares polynomial fitting to gravitational data and density plotting by digital computers: Geophysics, v. 19, p. 255-269.

Sloan, S.W., and Houlsby, G.T., 1984, An implementation of Watson's algorithm for computing 2-dimensional Delaunay triangulations: Adv. Engineering Software, v. 6, no. 4, p. 192-197.

Smith, F.G., 1968, Three computer programs for contouring map data: Can. Jour. Earth Sciences, v. 5, p. 324-327.

Snyder, W.V., 1978, Algorithm 531 contour plotting: ACM Trans. on Mathematical Software, v. 4, no. 3, p. 290-294.

Sowerbutts, W.T.C., 1983, A surface-plotting program suitable for microcomputers: Computer-aided Design, v. 15, no. 6, p. 324-327.

Sowerbutts, W.T.C., and Mason, R.W.I., 1984, A microcomputer based system for small-scale geophysical surveys: Geophysics, v. 49, no. 2, p. 189-193.

Spillers, W.R., and Law, K.H., 1987, On the hidden line removal problem: Computers & Structures, v. 26, no. 4, p. 709-717.

Stead, S.E., 1984, Estimation of gradients from scattered data: Rocky Mountain Jour. Math., v. 14, no. 1, p. 265-279.

Stearns, F., 1968, A method for estimating the quantitative reliability of isoline maps: Ann. Assoc. Am. Geographers, v. 58, p. 590-600.

Stelzer, J.F., and Welzel, R., 1987, Plotting contours in a natural way: Intern. Jour. Numerical Methods in Engineering, v. 24, p. 1757-1759.

Steven, G.P., 1984, Use of minimum surface theory to aid finite element contour plotting: Intern. Jour. Numerical Methods in Engineering, v. 20, p. 1791-1796.

References

Stoer, J., 1982, Curve fitting with clothoidal splines: Jour. Research Nat. Bureau Stand., v. 87, no. 4, p. 317-346.
Suffern, K.G., 1984, Contouring functions of two variables: The Australian Computer Jour., v. 16, no. 3, p. 102-106.
Surkan, A.J., Denny, J.R., and Batcha, J., 1964, Computer contouring: new tool for evaluation and analysis of mines: Engineering Min. Jour., v. 165, no. 12, p. 72-76.
Sutcliffe, D.C., 1976, A remark on a contouring algorithm: The Computer Jour., v. 19, no. 4, p. 333-335.
Sutterlin, P.G., and Hastings, J.P., 1986, Trend-surface analysis revisited - a case history: Computers & Geosciences, v. 12, no. 4B, p. 537-562.
Swain, C.J., 1976, A FORTRAN IV program for interpolation of irregularly spaced data using the difference equations of minimum curvature: Computers & Geosciences, v. 1, p. 231-240.
Sweeney, M.A.J., and Bartels, R.H., 1986, Ray tracing free-form B-spline surfaces: IEEE Trans. on Computer Graphics and Applic., v. 6, p. 41-49.
Swindle, G., and Van Andel, T.H., 1969, Computer contouring of deep sea bathymetric data: Marine Geol., v. 7, p. 347-355.
Szabados, J., 1986, On the work of G. Freud in the theory of interpolation of functions: Jour. Approximation Theory, v. 46, p. 119-128.
Tanemura, M., Ogawa, T., and Ogita, N., 1983, A new algorithm for the three-dimensional Voronoi tessellation: Jour. Computational Physics, v. 51, p. 191-207.
Tartar, M.E., Freeman, W., and Hopkins, A., 1986, A FORTRAN implementation of univariate Fourier series density estimation: Commun. Stat.-Simulat., v. 15, p. 855-871.
Tasker, B.S., 1985, Polygonal blocks of influence in triangular grids, Bull. Proc. Australasian Inst. Min. Metall., v. 290, no. 3, p. 71-72.
Tempfli, K., and Makarovic, B., 1979, Transfer functions of interpolation methods: Geo-Processing, v. 1, p. 1-26.
Thiessen, A.H., 1911, Precipitation averages for large areas: Mon. Wea. Rev., July, p. 1082-1084.
Tiller, W., 1983, Rational B-splines for curve and surface representation: IEEE Trans. on Computer Graphics and Applic., v. 3, p. 61-69.
Timmer, H.G., 1980, Alternative representation for parametric cubic curves and surfaces: Computer-aided Design, v. 12, no. 1, p. 25-28.
Tobler, W.R., 1979a, Lattice tuning: Geographical Analysis, v. 11, no. 1, p. 36-44.
Tobler, W.R., 1979b, Smooth pycnophylactic interpolation for geographical regions: Jour. Am. Stat. Assoc., v. 74, p. 519-536.
Tobler, W.R., and Kennedy, S., 1985, Smooth multidimensional interpolation: Geographical Analysis, v. 17, no. 3, p. 251-257.
Todd, H.W., 1987, Computer stereograms of Oklahoma subsurface geology: Geobyte, v. 2, no. 3, p. 30-33.
Tornow, V., 1982, An exponential spline interpolation for unequally spaced points: Computer Physics Commun., v. 28, p. 61-67.
Turner, A.K., and Miles, R.D., 1967, Terrain analysis by computer: Proc. Indiana Acad. Sci., v. 77, p. 256-270.
Tzimbalario, J., 1980, Cardinal discrete splines: Applicable Analysis, v. 11, p. 85-101.
Ubhaya, V.A., 1987, An $O(n)$ algorithm for least squares quasi-convex approximation: Computers Math. With Applic., v. 14, no. 8, p. 583-590.
Unwin, D.J., 1974, An introduction to trend surface analysis: Concepts and Techniques in Modern Geography, no. 5, 40 p.
Unwin, D., 1981, Introductory spatial analysis, Methuen, London, 212 p.
Utreras, F., 1983, Natural spline functions, their associated eigenvalue problem: Numerische Mathematik, v. 42, p. 107-117.

References

Van Kuilenburg, J., De Gruijter, J.J., Marsman, B.A., and Bouma, J., 1982, Accuracy of spatial interpolation between point data on soil moisture supply capacity, compared with estimates from mapping units: Geoderma, v. 27, p. 311-325.

Villalobos, M., and Wahba, G., 1987, Inequality-constrained multivariate smoothing splines with application to the estimation of posterior probabilities: Jour. Am. Stat. Assoc., v. 82, no. 397, p. 239-248.

Virdee, T.S., and Kottegoda, N.T., 1984, A brief review of kriging and its application to optimal interpolation and observation well selection: Hydrological Sciences Jour., v. 29, p. 367-387.

Waggenspack, W.N., and Anderson, D.C., 1986, Converting standard bivariate polynomials to Bernstein form over arbitrary triangular regions: Computer-aided Design, v. 18, no. 10, p. 529-532.

Wahba, G., 1981, Numerical experiments with the thin plate histospline: Commun. Stat.-Theor. Meth., v. A10, no. 24, p. 2475-2514.

Wahba, G., 1986, Multivariate thin plate spline smoothing with positivity and other linear inequality constraints, *in* Wegman, E.J., and DePriest, D.J., eds., Satistical Image Processing and Graphics: Marcel Dekker, New York, p. 275-289.

Wahba, G., and Wendelberger, J., 1980, Some new mathematical methods for variational objective analysis using splines and cross validation: Mon. Wea. Rev., v. 108, p. 1122-1143.

Wallis, H., 1976, Map-making to 1900: Intern. Cartographic Assoc. 8th Intern. Conf. on Cartography, Moscow, 52 p.

Walters, R.F., 1969, Contouring by machine: a user's guide: Bull. Am. Assoc. Petroleum Geol., v. 53, no. 11, p. 2324-2340.

Wang, C.Y., 1981, Shape classification of the parametric cubic curve and parametric B-spline cubic curve: Computer-aided Design, v. 13, no. 4, p. 199-206.

Wang, C.Y., 1983, C^1 rational interpolation over an arbitrary triangle: Computer-aided Design, v. 15, no. 1, p. 33-36.

Watson, D.F., 1981, Computing the n-dimensional Delaunay tessellation with application to Voronoi polytopes: The Computer Jour., v. 24, no. 2, p. 167-172.

Watson, D.F., 1982, ACORD - Automatic contouring of raw data: Computers & Geosciences, v. 8, no. 1, p. 97-101.

Watson, D.F., 1983, Two images for three dimensions: Practical Computing, August, p. 104-107; Errata September, p. 8.

Watson, D.F., 1985, Natural neighbor sorting: The Australian Computer Jour., v. 17, no. 4, p. 189-193.

Watson, D.F., and Philip, G.M., 1984, Systematic triangulations: Computer Vision, Graphics and Image Processing, v. 26, p. 217-223.

Watson, D.F., and Philip, G.M., 1984, Triangle-based interpolation: Math. Geol., v. 16, no. 8, p. 779-795.

Watson, D.F., and Philip, G.M., 1985, A refinement of inverse distance weighted interpolation: Geo-Processing, v. 2, p. 315-327.

Watson, D.F., and Philip, G.M., 1986, Automatic mineral deposit assessment using triangular prisms: Computers & Geosciences, v. 12, no. 2, p. 221-224.

Watson, D.F., and Philip, G.M., 1986, A derivation of the Isted formula for average mineral grade of a triangular prism: Math. Geol., v. 18, no. 3, p. 329-333.

Watson, D.F., and Philip, G.M., 1987, Neighborhood-based interpolation: Geobyte, v. 2, no. 2, p. 12-16.

Watson, D.F., and Philip, G.M., 1988, Measures of variability for geological data: Math. Geol., v. 21, no. 2, p. 233-254.

Watson, G.S., 1972, Trend surface analysis and spatial correlation: Geol. Soc. Am. Special Paper, v. 146, p. 39-46.

References

Watson, G.S., 1984, Smoothing and interpolation by kriging and with splines: Math. Geol., v. 16, no. 6, p. 601-615.

Weatherill, N.P., 1990, The integrity of geometrical boundaries in the two-dimensional Delaunay triangulation: Commun. Appl. Numerical Methods, v. 6, p. 101-109.

Weaver, R.C., 1964, Relative merits of interpolation and approximating functions in the grade prediction problem, *in* Parks, G.A., ed., Computers in the mineral industries Part 1: Stanford Univ. Publ. in the Geological Sciences, v. 9, p. 171-185.

Wever, U., 1988, Non-negative exponential splines: Computer-aided Design, v. 20, no. 1, p. 11-16.

Whitehouse, D.J., and Phillips, M.J., 1985, Sampling in a two-dimensional plane: Jour. Physics A: Math. Gen., v. 18, p. 2465-2477.

Whitney, E.N., 1929, Areal rainfall estimates: Mon. Wea. Rev., November, p. 462-463.

Whitten, E.H.T., 1967, Fourier trend-surface analysis in the geometrical analysis of subsurface folds of the Michigan Basin: Kansas Geol. Survey Computer Contrib., v. 12, p. 10-11.

Whitten, E.H., and Koelling, M.E.V., 1973, Spline-surface interpolation, spatial filtering, and trend surfaces for geological mapped variables: Math. Geol., v. 5, no. 2, p. 111-126.

Wolfe, P., 1982, Checking the calculation of gradients, ACM Trans. on Mathematical Software, v. 8, no. 4, p. 337-343.

Wong, F.S., 1985, First-order, second-moment methods: Computers & Structures, v. 20, no. 4, p. 779-791.

Wren, A.E., 1975, Contouring and the contour map: a new perspective: Geophys. Prospecting, v. 23, p. 1-17.

Wright, J.K., 1942, Map makers are human comments on the subjective in maps, The Geographical Review, v. 32, no. 4, p. 527-544.

Wunsch, C., and Zlotnicki, V., 1984, The accuracy of altimetric surfaces: Geophys. Jour. Roy. Astr. Soc., v. 78, p. 795-808.

Yarnal, B., 1984, A procedure for the classification of synoptic weather maps from gridded atmospheric pressure surface map: Computers & Geosciences, v. 10, no. 4, p. 397-410.

Yates, S.R., 1987, Contur: a Fortran algorithm for two-dimensional high-quality contouring: Computers & Geosciences, v. 13, no. 1, p. 61-76.

Yeo, M.F., 1984, An interactive contour plotting program: Engineering Computations, v. 1, p. 273-279.

Yeske, L., Scarpace, F., and Green, T., 1975, Measurement of lake currents: Photogrammetric Engineering and Remote Sensing, v. 41, p. 637-646.

Yfantis, E.A., and Borgman, L.E., 1981, Fast Fourier transforms 2-3-5: Computers & Geosciences, v. 7, p. 99-108.

Ying, L.-A., 1982, Some 'special' interpolation formulae for triangular and quadrilateral elements: Intern. Jour. Numerical Methods in Engineering, v. 18, p. 959-966.

Yoeli, P., 1977, Computer executed interpolation of contours into arrays of randomly distributed height-points: The Cartographic Jour., v. 14, p. 103-108.

Young, G.S., Pielke, R.A., and Kessler, R.C., 1984, A comparison of the terrain height variance spectra of the Front Range with that of a hypothetical mountain: Jour. Atmospheric Sciences, v. 41, no. 7, p. 1249-1250.

Zlamal, M., 1973, A remark on the serendipity family: Intern. Jour. Numerical Methods in Engineering, v. 7, p. 98-100.

Zhou, J.-M. and others, 1990, Computing constrained triangulation and Delaunay triangulation: a new algorithm: IEEE Trans. on Magnetics, v. 26, no. 2, p. 694-697.

Zwart, P.B., 1973, Multivariate splines with nondegenerate partitions, SIAM Jour. Numerical Anal., v. 10, no. 4, p. 665-673.

Zyda, M.J., Jones, A.R., and Hogan, P.G., 1987, Surface construction from planar contours: Computers & Graphics, v. 11, no. 4, p. 393-408.

Appendix
Included Software

SUMMARY

Twelve BASIC programs are provided on a DOS diskette in an envelope inside the back cover. They implement eighteen of the interpolation methods described in this book, and produce displays on the screen or pen-and-paper plotter. These programs may be used both to give a hands-on feel for the various concepts of contouring, and also as practical contouring programs for regular use. Line numbers are used only once, but some components occur in two or more programs and always have the same line numbers.

The following list of program components indicates which programs execute the various major operations.

INITIALIZATION AND DATA INPUT

SCRN\CON2R.BAS,   PLOT\CON2R.BAS
SCRN\NEIGHBOR.BAS,   PLOT\NEIGHBOR.BAS
SCRN\TRIANGLE.BAS,   PLOT\TRIANGLE.BAS
SCRN\DISTANCE.BAS,   PLOT\DISTANCE.BAS
SCRN\FITFUNCT.BAS,   PLOT\FITFUNCT.BAS
SCRN\RECTANGL.BAS,   PLOT\RECTANGL.BAS

Appendix

GRADIENT ESTIMATION

SCRN\CON2R.BAS	PLOT\CON2R.BAS
SCRN\NEIGHBOR.BAS,	PLOT\NEIGHBOR.BAS
SCRN\TRIANGLE.BAS,	PLOT\TRIANGLE.BAS
SCRN\DISTANCE.BAS,	PLOT\DISTANCE.BAS
SCRN\RECTANGL.BAS,	PLOT\RECTANGL.BAS

Options

(1) Neighborhood-based Linear (most robust) (**SL**=0)
(2) Triangle Cross Products (simplest) (**SL**=1)
(3) Least Squares Planes (fastest) (**SL**=2)
(4) Minimum Curvature Splines (slowest) (**SL**=3)
(5) Least Squares Quadratic (unreliable) (**SL**=4)
(Note: SCRN\CON2R.BAS and PLOT\CON2R.BAS only have option 1)

SURFACE CONSTRUCTION

Neighborhood-Based Interpolation

SCRN\CON2R.BAS	PLOT\CON2R.BAS
SCRN\NEIGHBOR.BAS,	PLOT\NEIGHBOR.BAS

Options

(1) Linear (fast and conservative) (**CL**=0)
(2) Linear Blended with Gradients (recommended) (**CL**=1)

Triangle-Based Interpolation

SCRN\TRIANGLE.BAS,   PLOT\TRIANGLE.BAS

Options

(1) Linear (fast) (**CL**=0)
(2) Linear Blended with Gradients (smoother) (**CL**=1)

Distance-Based Interpolation

SCRN\DISTANCE.BAS,   PLOT\DISTANCE.BAS

Options

(1) Polygonal Flat Top (proximal polygons) (**CL=0, PW>=8**)
(2) Polygonal Slant Top (**CL=1, PW>=8**)
(3) Inverse Distance Weighted Observations (IDWO)(**CL=0, PW<8**)
(4) IDWO Blended with Gradients (IDWG) (**CL=1, PW<8**)

Fitted Function Interpolation

    SCRN\FITFUNCT.BAS, PLOT\FITFUNCT.BAS

Options

(1) Collocation, Multiquadic (**TB=5**)
(2) Collocation, Inverted Multiquadric (**TB=8**)
(3) Minimum Curvature Splines Interpolation (**TB=3**)
(4) Minimum Curvature Splines Approximation (**RO<>0**)
(5) Covariance Interpolation (for known covariance)
(6) Covariance Approximation (**RO<>0**)
(7) Polynomial Trend Approximation (**TB=4**)

Rectangle-Based Interpolation

    SCRN\RECTANGL.BAS, PLOT\RECTANGL.BAS

Options

(1) Bilinear (very fast) (**CL=0, TB=6**)
(2) Bilinear Blended with Gradients (**CL=1, TB=6**)
(3) Minimum Curvature Splines (**TB=7**)

DISPLAY CONSTRUCTION

Screen Output

    SCRN\NEIGHBOR.BAS, SCRN\CON2R.BAS
    SCRN\TRIANGLE.BAS,   SCRN\DISTANCE.BAS
    SCRN\FITFUNCT.BAS,   SCRN\RECTANGL.BAS

Appendix

Options

(1) Isolines (**CR=0, DP=0, SR=0**)
(2) Isometric isolines (**CR=0, DP=1, SR=0**)
(3) Isometric isoline stereogram (**CR=0, DP=1, SR=1**)
(4) Isometric orthogonal profiles (**CR=0, DP=2, SR=0**)
(5) Isometric orthogonal profiles stereogram (**DP=2, SR=1**)
(6) Color-filled isochors (**CR=1, DP=0**)
(7) Isometric color-filled isolines (**CR=1, DP=1**)

Plotter Output

PLOT\NEIGHBOR.BAS,   PLOT\CON2R.BAS
PLOT\TRIANGLE.BAS,   PLOT\DISTANCE.BAS
PLOT\FITFUNCT.BAS,   PLOT\RECTANGL.BAS

Options

(1) Isolines (**DP=0**)
(2) Isometric isolines (**DP=1**)
(3) Isometric isoline stereogram (**SR=1**)
(4) Isometric orthogonal profiles (**DP=2**)
(5) Isometric orthogonal profiles stereogram

THE TWELVE BASIC PROGRAMS

```
1000 REM SCRN\CON2R.BAS NEIGHBORHOOD-BASED INTERPOLATION
     Copyright (c) 1988 D.F.Watson
1010 DIM P1#(299,6), P2#(593,3), P5#(50), P8(80), P9(2,3),S1(2033), S2(3),S4(50),
     PO(4,2)
1020 DIM RI#(3,3), R3#(3,3), R4#(3,2), R5#(3,3), R7#(3), R8(2,2), S3(2,3), BX(6,3)
1030 DIM P3%(593,3),P4%(593),P6%(50),P7%(25,2),R2%(2,2),R6%(3,2),SP%(16)
1040 DATA 0,8,1,4,5,9,6,12,13,2,7,3,10,11,14,15:REM spectrum
1050 DATA 1,0,0,1,1,0,0,1,0,0,0,0,1,0,0,1,0,1,-1,-1,0,1,2,5,-1,0,2,3,-1,5,1E37,3,1
1090 P9(1,1)=1E37:P9(1,2)=1E37:P9(1,3)=1E37:P9(2,1)=1E37:P9(2,2)=1E37:P9(2,3)=
     1E37
1100 SR=0:SZ=480:GR=29:DP=0:CL=1:PT=1:CR=0:WD=640:HI=350:LW=5:AZ=25:
     TL=-15:RD=2:BI=1.5:BJ=7:IF DP=0 THEN SR=0
1140 OPEN "A:\HILL.DAT" FOR INPUT AS #1
1150 REM OPEN "A:\JDAVIS.DAT" FOR INPUT AS #1
1170 FOR I1%=1 TO 16:READ SP%(I1%):NEXT I1%
1180 INPUT#1,ND%,NC%:RM=13/ND%:FOR I1%=1 TO NC%:INPUT#1,P8(I1%):NEXT
     I1%:IF CR>0 AND NC%>31 THEN NC%=31
1185 FOR I1%=1 TO ND%:P4%(I1%)=I1%:NEXT I1%:FOR I1%=1 TO
     ND%:I2%=INT(RND*ND%+.5):IF I2%<1 THEN I2%=1
1187 IF I2%>ND% THEN I2%=ND%
1188 I3%=P4%(I1%):P4%(I1%)=P4%(I2%):P4%(I2%)=I3%:NEXT I1%
1190 INPUT#1,XS,YS,DT:I2%=1
1220 I1%=P4%(I2%)
```

```
1222 INPUT#1, P1#(I1%,1), P1#(I1%,2), P1#(I1%,3): P1#(I1%,1) = (P1#(I1%,1)-XS) / DT:
     P1#(I1%,2) = (P1#(I1%,2 )- YS) / DT: IF (P1#(I1%,1) - .5)^2+(P1#(I1%,2) - 0.5)^2 < RD
     THEN GOTO 1270
1230 ND%=ND%-1:IF I2%>ND% THEN GOTO 1360
1260 GOTO 1222
1270 IF P9(1,1)<P1#(I1%,1) THEN P9(1,1)=P1#(I1%,1)
1280 IF P9(2,1)>P1#(I1%,1) THEN P9(2,1)=P1#(I1%,1)
1290 IF P9(1,2)<P1#(I1%,2) THEN P9(1,2)=P1#(I1%,2)
1300 IF P9(2,2)>P1#(I1%,2) THEN P9(2,2)=P1#(I1%,2)
1310 IF P9(1,3)<P1#(I1%,3) THEN P9(1,3)=P1#(I1%,3)
1320 IF P9(2,3)>P1#(I1%,3) THEN P9(2,3)=P1#(I1%,3)
1350 I2%=I2%+1:IF I2%<=ND% THEN GOTO 1220
1360 IF ND%<4 THEN STOP
1370 XF=P9(1,1) - P9(2,1): YF=P9(1,2)- P9(2,2): VS=P9(1,3)-P9(2,3):R8(2,1)=AZ:
     R8(2,2)=TL
1380 FOR I1%=1 TO 6:READ BX(I1%,1),BX(I1%,2),BX(I1%,3):NEXT I1%
1390 BN%=1:IF DP<1 THEN GOTO 1450
1400 BN%=2:BX(2,1)=0:BX(2,3)=1:IF R8(2,1)>0 THEN GOTO 1420
1410 BX(2,1)=1: BX(3,1)=1: BX(4,1)=1: BX(5,1)=0: BX(6,1)=0
1420 IF DP<2 OR CR>0 THEN GOTO 1450
1430 NC1%=INT(SZ/10):  NC%=NC1%+1:  U0=.9999/NC1%: U1=.0000511-U0
1440 FOR I1%=1 TO NC%: U1=U1+U0: P8(I1%)=U1: NEXT I1%: GOTO 1460
1450 FOR I1%=1 TO NC%: P8(I1%)=(P8(I1%)-P9(2,3))/VS+.00001*(RND-.5): NEXT I1%
1460 FOR I1%= 1 TO ND%: P1#(I1%,1)= P1(I1%,1)+.0001*(RND-.5): P1#(I1%,2)=
     P1#(I1%,2)+.0001*(RND-.5) : P1#(I1%,3)= (P1#(I1%,3)-P9(2,3))/VS: NEXT I1%
1470 FOR I1%=1 TO 3: READ R1#(I1%, 1), R1#(I1%, 2), R1#(I1%,3), R6%(I1%,1),
     R6%(I1%,2)
1480 P1#(I1% + ND%,1) = R1#(I1%,1)*XF + P9(2,1): P1#(I1% + ND%,2) =
     R1#(I1%,2)*YF+P9(2,2): P1#(I1%+ND%,3) = 0: NEXT I1%
1490 IF DP<1 THEN GOTO 1560
1500 FOR I1%=1 TO 2: R8(I1%,1)=COS(R8(2,I1%)/57.29578): R8(I1%,2)=SIN(R8(2,I1%)/
     57.29578): NEXT I1%
1510 FOR I1%=2 TO 6
1520 YY=(BX(I1%,2)-.5)*R8(1,1)+(BX(I1%,1)-.5)*R8(1,2)+.5: BX(I1%,1)=(BX(I1%,1)-
     .5)*R8(1,1)-(BX(I1%,2)-.5)*R8(1,2)+.5
1530 BX3=(BX(I1%,3)-.5)*R8(2,1)+(YY-.5)*R8(2,2)+.5: BX(I1%,2)=(YY-.5)*R8(2,1)-
     (BX(I1%,3)-.5)*R8(2,2)+.5: BX(I1%,3)=BX3: NEXT I1%
1540 REM end data input
1560 IF CL<1 THEN GOTO 3110
1570 PRINT " AT " TIME$ " BEGIN GRADIENT ESTIMATION - PLEASE WAIT"
1580 FOR I0%=1 TO ND%
1590 FOR I1%=1 TO 3: P3%(1,I1%)=ND%+I1%: P3%(2,I1%)=ND%+I1%:
     P2#(1,I1%)=R1#(I1%,3): P2#(2,I1%)=R1#(I1%,3) : NEXT I1%
1600 FOR I1%=1 TO 593: P4%(I1%)=I1%: NEXT I1%: R2%(1,1)=1: R2%(1,2)=1:
     R2%(2,1)=3 : R2%(2,2)=4: K1%=I0%: K2%=I0%: K3%=ND%+4: K4=-.5
1610 FOR I1%=K1% TO K2%
1620 IF I1%=K3% THEN GOTO 2000
1630 K5=K4 1640 K6=1.5+K5: IF K3%>ND% THEN K5=-K5
1650 U0=0: K7%=0: L6%=K6-2
1660 FOR I2%=1 TO R2%(1,K6)
1670 L6%=L6%+2: IF P3%(L6%,1)=0 THEN GOTO 1670
1680 U1#=P2#(L6%,3)-(P1#(I1%,1)-P2#(L6%,1))^2: IF U1#<0 THEN GOTO 1840
1690 U1#=U1#-(P1#(I1%,2)-P2#(L6%,2))^2: IF U1#<0 THEN GOTO 1840
1700 R2%(2,K6)=R2%(2,K6)-2: P4%(R2%(2,K6))=L6%: U0=U0-1
1710 FOR I3%=1 TO 3
1720 L1%=1: IF L1%=I3% THEN L1%=L1%+1
1730 L2%=L1%+1: IF L2%=I3% THEN L2%=L2%+1
1740 IF K7%<1 THEN GOTO 1810
1750 K9%=K7%
1760 FOR I4%=1 TO K9%
1770 IF P3%(L6%,L1%)<>P7%(I4%,1) THEN GOTO 1800
1780 IF P3%(L6%,L2%)<>P7%(I4%,2) THEN GOTO 1800
1790 P7%(I4%,1)=P7%(K9%,1): P7%(I4%,2)=P7%(K9%,2): K7%=K7%-1: GOTO 1820
1800 NEXT I4%
1810 K7%=K7%+1: P7%(K7%,1)=P3%(L6%,L1%): P7%(K7%,2)=P3%(L6%,L2%)
```

Appendix

```
1820 NEXT I3%
1830 P3%(L6%,1) = 0
1840 NEXT I2%
1850 IF K7%<1 THEN GOTO 2000
1860 REM make new triangles
1870 FOR I2% = 1 TO K7%
1880 IF K5>0 THEN GOTO 1900
1890 IF P7%(I2%,1)<>I0% AND P7%(I2%,2)<>I0% THEN GOTO 1970
1900 FOR I3% = 1 TO 2
1910 R3#(I3%,1) = P1#(P7%(I2%,I3%),1) - P1#(I1%,1): R3#(I3%,2) =
     P1#(P7%(I2%,I3%),2) - P1#(I1%,2)
1920 R3#(I3%,3) = R3#(I3%,1)*(P1#(P7%(I2%,I3%),1)+P1#(I1%,1))/2+R3#(I3%,2)
     *(P1#(P7%(I2%,I3%),2)+P1#(I1%,2))/2
1930 NEXT I3%
1940 U1# = R3#(1,1)*R3#(2,2) - R3#(2,1)*R3#(1,2): P2#(P4#(R2%(2,K6)),1) =
     (R3#(1,3)*R3#(2,2) - R3#(2,3)*R3#(1,2))/U1#: P2#(P4#(R2%(2,K6)),2) =
     (R3#(1,1)*R3#(2,3) - R3#(2,1)*R3#(1,3))/U1#
1950 P2#(P4#(R2%(2,K6)),3) = (P1#(I1%,1) - P2#(P4#(R2%(2,K6)),1))^2+(P1#(I1%,2) -
     P2#(P4#(R2%(2,K6)),2))^2
1960 P3%(P4#(R2%(2,K6)),1) = P7%(I2%,1): P3%(P4#(R2%(2,K6)),2) = P7%(I2%,2):
     P3%(P4#(R2%(2,K6)),3) = I1%: R2%(2,K6) = R2%(2,K6)+2: U0 = U0+1
1970 NEXT I2%
1980 R2%(1,K6) = R2%(1,K6)+U0: IF K5>0 THEN GOTO 2000
1990 K5 = - K5: GOTO 1640
2000 NEXT I1%
2010 IF K2%>K1% GOTO 2040
2020 K1% = 1: K2% = ND%: K3% = I0%: GOTO 1610
2030 REM order triangles positively
2040 FOR I1% = 1 TO 2
2050 L8% = I1% - 2
2060 FOR I2% = 1 TO R2%(1,I1%)
2070 L8% = L8%+2: IF P3%(L8%,1) = 0 THEN GOTO 2070
2080 IF P3%(L8%,1)>ND% THEN GOTO 2140
2090 FOR I3% = 1 TO 2
2100 R3#(1,I3%) = P1#(P3%(L8%,2),I3%) - P1#(P3%(L8%,1),I3%): R3#(2,I3%) =
     P1#(P3%(L8%,3),I3%) - P1#(P3%(L8%,1),I3%)
2110 NEXT I3%
2120 IF R3#(1,1)*R3#(2,2) - R3#(1,2)*R3#(2,1)> = 0 THEN GOTO 2140
2130 K8 = P3%(L8%,3): P3%(L8%,3) = P3%(L8%,2): P3%(L8%,2) = K8
2140 NEXT I2%
2150 NEXT I1%
2170 REM calculate natural neighbor coordinates
2180 R5#(1,1) = P1#(I0%,1): R5#(1,2) = P1#(I0%,2): R5#(2,1) = R5#(1,1)+.0001 :
     R5#(2,2) = R5#(1,2): R5#(3,1) = R5#(1,1): R5#(3,2) = R5#(1,2)+.0001
2190 FOR I1% = 1 TO 3
2200 L4% = 0: L5% = 0: L6% = 0
2210 FOR I2% = 1 TO R2%(1,2)
2220 L6% = L6%+2: IF P3%(L6%,1) = 0 THEN GOTO 2220
2230 IF P3%(L6%,1)>ND% THEN GOTO 2410
2240 U1# = P2#(L6%,3) - (R5#(I1%,1) - P2#(L6%,1))^2: IF U1#<0 THEN GOTO 2410
2250 U1# = U1# - (R5#(I1%,2) - P2#(L6%,2))^2: IF U1#<0 THEN GOTO 2410
2260 FOR I3% = 1 TO 3
2270 FOR I4% = 1 TO 2
2280 R3#(I4%,1) = P1#(P3%(L6%,R6%(I3%,I4%)),1) - R5#(I1%,1): R3#(I4%,2) =
     P1#(P3%(L6%,R6%(I3%,I4%)),2) - R5#(I1%,2)
2290 R3#(I4%,3) = R3#(I4%,1)*(P1#(P3%(L6%,R6%(I3%,I4%)),1)+R5#(I1%,1))/
     2+R3#(I4%,2) *(P1#(P3%(L6%,R6%(I3%,I4%)),2)+R5#(I1%,2))/2: NEXT I4%
2300 U1# = R3#(1,1)*R3#(2,2) - R3#(1,2)*R3#(2,1)
2310 R4#(R6%(I3%,2),1) = (R3#(1,3)*R3#(2,2) - R3#(2,3)*R3#(1,2))/U1#:
     R4#(R6%(I3%,2),2) = (R3#(1,1)*R3#(2,3) - R3#(2,1)*R3#(1,3))/U1#: NEXT I3%: L3% =
     0
2320 FOR I3% = 1 TO 3: R3#(3,I3%) = ((R4#(R6%(I3%,1),1) -
     P2#(L6%,1))*(R4#(R6%(I3%,2),2) - P2#(L6%,2)) - (R4#(R6%(I3%,2),1) -
     P2#(L6%,1))*(R4#(R6%(I3%,1),2) - P2#(L6%,2)))/2: IF R3#(3,I3%)>0 THEN L3% =
     L3%+1
2330 NEXT I3%
```

```
2340 IF L3%>2 THEN L4% = 1
2350 FOR I3% = 1 TO 3: IF L5%<1 THEN GOTO 2390
2360 FOR I4% = 1 TO L5%: IF P3%(L6%,I3%)<>P6%(I4%) THEN GOTO 2380
2370 P5#(I4%) = P5#(I4%)+R3#(3,I3%): GOTO 2400
2380 NEXT I4%
2390 L5% = L5%+1: P6%(L5%) = P3%(L6%,I3%): P5#(L5%) = R3#(3,I3%)
2400 NEXT I3%
2410 NEXT I2%: IF L4%<1 THEN GOTO 2480
2420 U4 = 0
2430 FOR I2% = 1 TO L5%: U4 = U4+P5#(I2%): NEXT I2%: R7#(I1%) = 0
2440 FOR I2% = 1 TO L5%: P5#(I2%) = P5#(I2%)/U4: R7#(I1%) =
    R7#(I1%)+P5#(I2%)*P1#(P6%(I2%),3): NEXT I2%: NEXT I1%
2450 P1#(I0%,6) = P1#(I0%,3) - R7#(1): P1#(I0%,4) = (R7#(1) - R7#(2))/.0001:
    P1#(I0%,5) = (R7#(1) - R7#(3))/.0001 : GOTO 3080
2460 REM cross product gradients
2480 P1#(I0%,4) = 0: P1#(I0%,5) = 0: P1#(I0%,6) = 0: P7# = 0: L7% = - 1
2490 FOR I1% = 1 TO R2%(1,1)
2500 L7% = L7%+2: IF P3%(L7%,1) = 0 THEN GOTO 2500
2510 IF P3%(L7%,1)>ND% THEN GOTO 2570
2520 FOR I2% = 1 TO 2: FOR I3% = 1 TO 3: R3#(I2%,I3%) = P1#(P3%(L7%,1),I3%) -
    P1#(P3%(L7%,I2%+1),I3%): NEXT I3% : NEXT I2%
2530 R3#(3,1) = R3#(1,2)*R3#(2,3) - R3#(2,2)*R3#(1,3): R3#(3,2) = R3#(1,3)*R3#(2,1) -
    R3#(2,3)*R3#(1,1): R3#(3,3) = R3#(1,1)*R3#(2,2) - R3#(2,1)*R3#(1,2): U3 = 1
2540 IF R3#(3,3)<0 THEN U3 = - 1
2550 U2 = (R3#(3,1)^2+R3#(3,2)^2+R3#(3,3)^2)^.5: P7# = P7#+U2
2560 FOR I2% = 1 TO 3: P1#(I0%,I2%+3) = P1#(I0%,I2%+3)+R3#(3,I2%)*U3: NEXT I2%
2570 NEXT I1%
2580 U2 = P1#(I0%,4)^2+P1#(I0%,5)^2+P1#(I0%,6)^2: P7# = 1 - U2^.5/P7#
2590 P1#(I0%,4) = P1#(I0%,4)/P1#(I0%,6): P1#(I0%,5) = P1#(I0%,5)/P1#(I0%,6):
    P1#(I0%,6) = P7#
3080 NEXT I0%
3100 REM natural neighbor sort
3110 FOR I0% = 1 TO 3: P3%(1,I0%) = ND%+I0%: P2#(1,I0%) = R1#(I0%,3): NEXT I0%
3120 FOR I0% = 1 TO 593: P4%(I0%) = I0%: NEXT I0%: NT% = 1: L0% = 2
3130 FOR I0% = 1 TO ND%: L3% = 0
3140 FOR I1% = 1 TO NT%
3150 U1# = P2#(I1%,3) - (P1#(I0%,1) - P2#(I1%,1))^2: IF U1#<0 THEN GOTO 3300
3160 U1# = U1# - (P1#(I0%,2) - P2#(I1%,2))^2: IF U1#<0 THEN GOTO 3300
3170 L0% = L0% - 1: P4%(L0%) = I1%
3180 FOR I2% = 1 TO 3
3190 L1% = 1: IF L1% = I2% THEN L1% = L1%+1
3200 L2% = L1%+1: IF L2% = I2% THEN L2% = L2%+1
3210 IF L3%<1 THEN GOTO 3280
3220 L5% = L3%
3230 FOR I3% = 1 TO L5%
3240 IF P3%(I1%,L1%)<>P7%(I3%,1) THEN GOTO 3270
3250 IF P3%(I1%,L2%)<>P7%(I3%,2) THEN GOTO 3270
3260 P7%(I3%,1) = P7%(L5%,1): P7%(I3%,2) = P7%(L5%,2): L3% = L3% - 1: GOTO
    3290
3270 NEXT I3%
3280 L3% = L3%+1: P7%(L3%,1) = P3%(I1%,L1%): P7%(L3%,2) = P3%(I1%,L2%)
3290 NEXT I2%
3300 NEXT I1%
3310 FOR I1% = 1 TO L3%
3320 FOR I2% = 1 TO 2
3330 R3#(I2%,1) = P1#(P7%(I1%,I2%),1) - P1#(I0%,1): R3#(I2%,2) =
    P1#(P7%(I1%,I2%),2) - P1#(I0%,2): R3#(I2%,3) = R3#(I2%,1)*(P1#(P7%(I1%,I2%),1) +
    P1#(I0%,1))/2
3340 R3#(I2%,3) = R3#(I2%,3) + R3#(I2%,2)*(P1#(P7%(I1%,I2%),2) + P1#(I0%,2))/2
3350 NEXT I2%
3360 U1# = R3#(1,1)*R3#(2,2) - R3#(2,1)*R3#(1,2): P2#(P4%(L0%),1) =
    (R3#(1,3)*R3#(2,2) - R3#(2,3)*R3#(1,2))/U1#: P2#(P4%(L0%),2) = (R3#(1,1)*R3#(2,3)
    - R3#(2,1)*R3#(1,3))/U1#
3370 P2#(P4%(L0%),3) = (P1#(I0%,1) - P2#(P4%(L0%),1))^2 + (P1#(I0%,2) -
    P2#(P4%(L0%),2))^2
```

Appendix

```
3380 P3%(P4%(L0%),1) = P7%(I1%,1): P3%(P4%(L0%),2) = P7%(I1%,2):
     P3%(P4%(L0%),3) = I0%: L0% = L0%+1 3390 NEXT I1%
3400 NT% = NT%+2: NEXT I0%
3410 FOR I0% = 1 TO NT%: IF P3%(I0%,1)>ND% THEN GOTO 3460
3420 FOR I1% = 1 TO 2
3430 R3#(1,I1%) = P1#(P3%(I0%,2),I1%) - P1#(P3%(I0%,1),I1%): R3#(2,I1%) =
     P1#(P3%(I0%,3),I1%) - P1#(P3%(I0%,1),I1%): NEXT I1%
3440 IF R3#(1,1)*R3#(2,2) - R3#(1,2)*R3#(2,1)> = 0 THEN GOTO 3460
3450 L4% = P3%(I0%,3): P3%(I0%,3) = P3%(I0%,2): P3%(I0%,2) = L4%
3460 NEXT I0%
3820 REM make triangular grid
3830 PRINT " AT " TIME$ " BEGIN SURFACE INTERPOLATION  -  PLEASE WAIT":
     SOUND 123,5: SOUND 185,5: SOUND 208,5 : SOUND 165,5
3840 I0% = GR+1: IS = 1: I2% = 2*I0%+1: I3% = 1+INT(GR/3^.5): I4% = I3%*I2%+I0%:
     SX = 1/GR: SY = 1/(2*I3%): S2(3) = .000001
3850 WX# = S2(3): S2(1) = 1.00001: AA# = SX*SY*.5*DT*DT: SM# = 0: SA# = 0: WY# =
     1: U8 = 1
3860 FOR L1% = 1 TO I4%
3870 S1(L1%) = - 999: WXO# = WX#: WYO# = WY#
3880 REM .....calculate natural neighbor coordinates of WX#,WY# **********
3890 K4% = 0: K5% = 0
3900 FOR L2% = 1 TO NT%
3910 IF P3%(L2%,1)>ND% THEN GOTO 4280
3915 IF P2#(L2%,3)>RM THEN GOTO 4280
3920 U1# = P2#(L2%,3) - (WX# - P2#(L2%,1))^2: IF U1#<0 THEN GOTO 4280
3930 U1# = U1# - (WY# - P2#(L2%,2))^2: IF U1#<0 THEN GOTO 4280
3940 FOR L3% = 1 TO 3
3950 L4% = 1: IF L4% = L3% THEN L4% = L4%+1
3960 L5% = L4%+1: IF L5% = L3% THEN L5% = L5%+1
3970 IF ABS((P1#(P3%(L2%,L5%),1) - P1#(P3%(L2%,L4%),1))*(WY# -
     P1#(P3%(L2%,L4%),2)) - (WX# - P1#(P3%(L2%,L4%),1))*(P1#(P3%(L2%,L5%),2) -
     P1#(P3%(L2%,L4%),2)))>.000001 THEN GOTO 4030
4000 DEN# = ABS(P1#(P3%(L2%,L4%),2) -
     P1#(P3%(L2%,L5%),2))+ABS(P1#(P3%(L2%,L5%),1) - P1#(P3%(L2%,L4%),1))
4010 WX# = WX#+(P1#(P3%(L2%,L4%),2) - P1#(P3%(L2%,L5%),2))*.0001/DEN#: WY# =
     WY#+(P1#(P3%(L2%,L5%),1) - P1#(P3%(L2%,L4%),1))*.0001/DEN#
4030 NEXT L3%
4040 FOR L3% = 1 TO 3
4050 FOR L4% = 1 TO 2
4060 R3#(L4%,1) = P1#(P3%(L2%,R6%(L3%,L4%)),1) - WX#: R3#(L4%,2) =
     P1#(P3%(L2%,R6%(L3%,L4%)),2) - WY#
4070 R3#(L4%,3) = R3#(L4%,1)*(P1#(P3%(L2%,R6%(L3%,L4%)),1)+WX#)/2+R3#(L4%,2)
     *(P1#(P3%(L2%,R6%(L3%,L4%)),2)+WY#)/2
4080 NEXT L4%
4090 U1# = R3#(1,1)*R3#(2,2) - R3#(1,2)*R3#(2,1)
4100 R4#(R6%(L3%,2),1) = (R3#(1,3)*R3#(2,2) - R3#(2,3)*R3#(1,2))/U1#
4110 R4#(R6%(L3%,2),2) = (R3#(1,1)*R3#(2,3) - R3#(2,1)*R3#(1,3))/U1#
4120 NEXT L3%
4130 K3% = 0
4140 FOR L3% = 1 TO 3
4150 R3#(3,L3%) = ((R4#(R6%(L3%,1),1) - P2#(L2%,1))*(R4#(R6%(L3%,2),2) -
     P2#(L2%,2)) - (R4#(R6%(L3%,2),1) - P2#(L2%,1))*(R4#(R6%(L3%,1),2) -
     P2#(L2%,2)))/2
4160 IF R3#(3,L3%)>0 THEN K3% = K3%+1
4170 NEXT L3%
4180 IF K3%>2 THEN K4% = 1
4190 FOR L3% = 1 TO 3
4200 IF K5%<1 THEN GOTO 4260
4210 FOR L4% = 1 TO K5%
4220 IF P3%(L2%,L3%)<>P6%(L4%) THEN GOTO 4250
4230 P5#(L4%) = P5#(L4%)+R3#(3,L3%)
4240 GOTO 4270
4250 NEXT L4%
4260 K5% = K5%+1: P6%(K5%) = P3%(L2%,L3%): P5#(K5%) = R3#(3,L3%)
4270 NEXT L3%
4280 NEXT L2%
```

```
4300 IF K4%<1 THEN GOTO 4940
4310 REM ....end determining K5% local coordinates *********
4560 S1(L1%) = 0: U4 = 0
4570 FOR L2% = 1 TO K5%: U4 = U4+P5#(L2%): NEXT L2%
4580 FOR L2% = 1 TO K5%: P5#(L2%) = P5#(L2%)/U4: S1(L1%) =
     S1(L1%)+P5#(L2%)*P1#(P6%(L2%),3): NEXT L2%
4590 IF CL<1 THEN GOTO 4940
4620 FOR L2% = 1 TO K5%: S4(L2%) = 0
4630 IF P5#(L2%)<.00001 OR P5#(L2%)>1 THEN GOTO 4690
4640 IF ABS(P1#(P6%(L2%),6))<.00001 THEN GOTO 4690
4650 RS# = ABS(P1#(P6%(L2%),6)) + BI: RT# = RS#*BJ: RB# = 1/RT#: BD# =
     P5#(L2%)^RT#: BB# = BD#*2: IF BD#>.5 THEN BB# = (1 - BD#)*2
4660 BB# = BB#^RS#/2: IF BD#>.5 THEN BB# = 1 - BB#
4670 HP# = BB#^RB#
4680 S4(L2%) = ((P1#(P6%(L2%),4)*P1#(P6%(L2%),1) +
     P1#(P6%(L2%),5)*P1#(P6%(L2%),2) + P1#(P6%(L2%),3) - P1#(P6%(L2%),4)*WX# -
     P1#(P6%(L2%),5)*WY#) - S1(L1%))*HP#
4690 NEXT L2%
4700 FOR L2% = 1 TO K5%: S1(L1%) = S1(L1%) + S4(L2%): NEXT L2%
4940 IS = IS + SX*U8: IF IS>0 THEN GOTO 4960
4950 WXO# = WXO# + SX/2: IS = 1.001 - SX
4960 WX# = WXO# + SX*U8: WY# = WYO#: IF U8*WX#<S2(2 - U8) THEN GOTO 4980
4970 U8 = - U8: WX# = S2(2 + U8): WY# = WY# - SY: IS = U8
4980 NEXT L1%
5060 REM end interpolation to triangular grid and begin display construction
5080 SCREEN 9: COLOR 2,0: CLS: SOUND 208,5: SOUND 165,5: SOUND 185,5:
     SOUND 123,5: IF DP>0 THEN SZ = INT(SZ*2/3)
5090 IF SR>0 AND SZ>240 THEN SZ = 240
5100 ZS = INT(SZ*.71): SW = (WD - SZ)\2: ZW = (HI - LW)\2 + ZS\2: IF DP>0 THEN ZW
     = ZW - LW - 10
5110 IF SR>0 AND CR<1 AND DP>0 THEN SW = (WD - 2*SZ)\3 - 10
5120 SV = SW + SZ + 60: IF SR<1 THEN GOTO 5130
5125 FOR L1% = 2 TO 6: BX1 = (BX(L1%,1) - .5)*.99875 - (BX(L1%,3) - .5)*.05 + .5:
     BX(L1%,3) = (BX(L1%,3) - .5)*.99875 + (BX(L1%,1) - .5)*.05 + .5: BX(L1%,1) = BX1:
     NEXT L1%
5130 FOR L1% = BN% + 1 TO BN% + 4: X1 = BX(L1% - 1,1)*SZ + SW: Y1 = ZW -
     BX(L1% - 1,2)*ZS: X2 = SW + BX(L1%,1)*SZ: Y2 = ZW - BX(L1%,2)*ZS: LINE(X1,Y1)
     - (X2,Y2),14: NEXT L1%: IF SR<1 OR CR>0 THEN GOTO 5210
5140 FOR L1% = 2 TO 6: BX(L1%,1) = (BX(L1%,1) - .5)*.995 + (BX(L1%,3) - .5)*.1 + .5:
     NEXT L1%
5150 FOR L1% = BN% + 1 TO BN% + 4: X1 = BX(L1% - 1,1)*SZ + SV: Y1 = ZW -
     BX(L1% - 1,2)*ZS: X2 = SV + BX(L1%,1)*SZ: Y2 = ZW - BX(L1%,2)*ZS: LINE(X1,Y1) -
     (X2,Y2),14: NEXT L1%
5200 REM plot contours
5210 SW = SW + 1: SZ = SZ - 2: ZS = ZS - 2: ZW = ZW - 1: ZX = SX: ZY = SY: ZZ = .5:
     R3#(2,2) = 1 + 2*SY: R3#(2,1) = 0
5220 FOR L1% = 1 TO I3%
5230 R3#(1,2) = R3#(2,2) - 2*SY - 2*ZY: R3#(1,1) = R3#(2,1): N1% = (2*L1% - .5 +
     ZZ)*I0% + L1% + .5/ZZ: R3#(1,3) = S1(N1%) : R3#(2,2) = R3#(2,2) - 2*SY - ZY:
     R3#(2,1) = .5 - ZZ: N2% = (2*L1% - .5 + ZZ)*I0% + L1%
5240 R3#(2,3) = S1(N2%): R3#(3,2) = R3#(2,2): R3#(3,1) = R3#(2,1) + ZX/2: N3% = N2%
     - .5/ZZ: R3#(3,3) = S1(N3%): AR# = AA#/2
5250 GOSUB 5820
5260 R3#(1,2) = R3#(1,2) + 2*ZY: N2% = (2*L1% - .5/ZZ - .5 - ZZ)*I0% + L1%: R3#(1,3) =
     S1(N2%)
5270 GOSUB 5820
5280 AR# = AA#
5290 FOR L2% = 1 TO GR
5300 R3#(2,2) = R3#(1,2): R3#(2,1) = R3#(1,1) + ZX: N4% = N2% + .5/ZZ: R3#(2,3) =
     S1(N4%)
5310 GOSUB 5820
5320 R3#(1,2) = R3#(3,2): R3#(1,1) = R3#(3,1) + ZX
5330 IF L2% = GR THEN R3#(1,1) = R3#(1,1) - ZX/2
5340 N6% = N3% - .5/ZZ: R3#(1,3) = S1(N6%): IF L2% = GR THEN AR# = AA#/2
5350 GOSUB 5820
5360 R3#(2,2) = R3#(2,2) - 2*ZY: N5% = N1% + .5/ZZ: R3#(2,3) = S1(N5%)
```

231

Appendix

```
5370 GOSUB 5820
5380 R3#(1,2) = R3#(2,2): R3#(1,1) = R3#(2,1) - ZX: R3#(1,3) = S1(N1%): IF L2% = GR
     THEN AR# = AA#
5390 GOSUB 5820
5400 N1% = N5%: N2% = N4%: N3% = N6%: R3#(1,2) = R3#(1,2) + 2*ZY: R3#(1,1) =
     R3#(2,1): R3#(1,3) = S1(N4%) : R3#(3,1) = R3#(3,1) + ZX: R3#(3,3) = S1(N6%)
5410 NEXT L2%
5420 ZY = - ZY: ZX = - ZX: ZZ = - ZZ
5430 NEXT L1%
5440 REM end plotting contours
5450 REM plot data points
5470 IF PT<1 OR CR>0 THEN GOTO 5790
5480 FOR L1% = 1 TO ND%
5490 IF P1#(L1%,1)< - .0001 OR P1#(L1%,1)>1.0001 OR P1#(L1%,2)< - .0001 OR
     P1#(L1%,2)>1.0001 THEN GOTO 5610
5500 PP1# = P1#(L1%,1): PP2# = P1#(L1%,2): PP3# = P1#(L1%,3): IF DP<1 THEN
     GOTO 5526
5510 YY = (PP2# - .5)*R8(1,1) + (PP1# - .5)*R8(1,2) + .5: PP1# = (PP1# - .5)*R8(1,1) -
     (PP2# - .5)*R8(1,2) + .5
5520 ZZ3 = (PP3# - .5)*R8(2,1) + (YY - .5)*R8(2,2) + .5: PP2# = (YY - .5)*R8(2,1) - (PP3#
     - .5)*R8(2,2) + .5: PP3# = ZZ3: IF SR<1 THEN GOTO 5526
5525 PPP# = PP1#: PP1# = (PP1# - .5)*.99875 - (PP3# - .5)*.05 + .5
5526 S3(1,1) = PP1#*SZ + SW - 2: S3(1,2) = ZW - PP2#*ZS - 2
5527 FOR L2% = 1 TO 3: FOR L3% = 1 TO 3: PSET(S3(1,1) + L2%,S3(1,2) + L3%),4:
     NEXT L3%: NEXT L2%
5528 IF SR<1 THEN GOTO 5610
5529 PP1# = (PPP# - .5)*.99875 + (PP3# - .5)*.05 + .5
5530 S3(1,1) = PP1#*SZ + SV - 2: S3(1,2) = ZW - PP2#*ZS - 2
5531 FOR L2% = 1 TO 3: FOR L3% = 1 TO 3: PSET(S3(1,1) + L2%,S3(1,2) + L3%),4:
     NEXT L3%: NEXT L2%
5610 NEXT L1%
5790 SOUND 247,5: SOUND 153,5: SOUND 185,5: SOUND 123,10: TM$ = TIME$: IF
     SR>0 THEN GOTO 5802
5800 SM# = SM#/SA#: LOCATE 12,1: PRINT "Average": PRINT "height of": PRINT
     "surface": PRINT USING "#####.###";SM#
5802 LOCATE 19,1: PRINT " ALL": PRINT "DONE!": PRINT " At": PRINT TM$
5804 PRINT "Press": INPUT "ENTER ",PM$: END
5810 REM curvemaker
5820 ST# = R3#(1,3)
5830 IF ST#>R3#(2,3) THEN ST# = R3#(2,3)
5840 IF ST#>R3#(3,3) THEN ST# = R3#(3,3)
5850 IF ST#< - 999 THEN RETURN
5860 SM# = SM# + AR#*((R3#(1,3) + R3#(2,3) + R3#(3,3))/3*VS + P9(2,3)): SA# = SA# +
     AR#: IF CR<1 THEN GOTO 6210
5870 PO(1,1) = - 1E37: PO(2,1) = PO(1,1): PO(1,2) = 1E37: PO(2,2) = PO(1,2): S3(1,1) =
     R3#(1,3) : S3(1,2) = R3#(2,3): S3(1,3) = R3#(3,3): IF DP<1 THEN GOTO 5910
5880 FOR L3% = 1 TO 3: R1#(L3%,1) = R3#(L3%,1): R1#(L3%,2) = R3#(L3%,2):
     R1#(L3%,3) = R3#(L3%,3)
5890 YY = (R3#(L3%,2) - .5)*R8(1,1) + (R3#(L3%,1) - .5)*R8(1,2): R3#(L3%,1) =
     (R3#(L3%,1) - .5)*R8(1,1) - (R3#(L3%,2) - .5)*R8(1,2) + .5
5900 R33# = (R3#(L3%,3) - .5)*R8(2,1) + YY*R8(2,2) + .5: R3#(L3%,2) = YY*R8(2,1) -
     (R3#(L3%,3) - .5)*R8(2,2) + .5: R3#(L3%,3) = R33#: NEXT L3%
5910 FOR L3% = 1 TO 3
5920 IF PO(1,1)<R3#(L3%,1) THEN PO(1,1) = R3#(L3%,1)
5930 IF PO(2,1)<R3#(L3%,2) THEN PO(2,1) = R3#(L3%,2)
5940 IF PO(1,2)>R3#(L3%,1) THEN PO(1,2) = R3#(L3%,1)
5950 IF PO(2,2)>R3#(L3%,2) THEN PO(2,2) = R3#(L3%,2)
5960 NEXT L3%
5970 X1 = SW + INT(PO(1,2)*SZ): XH = X1 - (X1\2)*2 - .5: X2 = SW + INT(PO(1,1)*SZ):
     Y1 = ZW - INT(PO(2,2)*ZS): Y2 = ZW - INT(PO(2,1)*ZS): YG = Y2 - (Y2\2)*2 - .5
5980 ZJ# = ((R3#(2,1) - R3#(1,1))*(R3#(3,2) - R3#(1,2))) - ((R3#(3,1) - R3#(1,1))*(R3#(2,2)
     - R3#(1,2)))
5990 FOR L3% = Y2 TO Y1
6000 YG = - YG: YY = (ZW - L3%)/ZS: XG = XH
6010 FOR L4% = X1 TO X2
6020 XG = - XG: XX = (L4% - SW)/SZ: ZI = 1
```

```
6030 FOR L5% = 1 TO 3
6040 L8% = 1: IF L8% = L5% THEN L8% = L8% + 1
6050 L9% = L8% + 1: IF L9% = L5% THEN L9% = L9% + 1
6060 S3(2,L5%) = (((R3#(L8%,1) - XX)*(R3#(L9%,2) - YY)) - ((R3#(L9%,1) -
     XX)*(R3#(L8%,2) - YY)))/ZJ#*ZI: IF S3(2,L5%)<0 THEN GOTO 6150
6070 ZI = - ZI: NEXT L5%
6080 ZC = S3(1,1)*S3(2,1) + S3(1,2)*S3(2,2) + S3(1,3)*S3(2,3): IF ZC<P8(1) THEN
     GOTO 6150
6090 FOR L5% = 2 TO NC%
6100 IF P8(L5%)<ZC THEN GOTO 6140
6110 LC% = INT(L5%/2 + XG/YG*.25 + .75): IF LC%<1 THEN GOTO 6150
6120 IF LC%>16 THEN LC% = 1
6130 PSET(L4%,L3%),SP%(LC%): GOTO 6150
6140 NEXT L5%
6150 NEXT L4%: NEXT L3%
6160 IF DP<1 THEN GOTO 6200
6170 FOR L3% = 1 TO 3
6180 R3#(L3%,1) = R1#(L3%,1): R3#(L3%,2) = R1#(L3%,2): R3#(L3%,3) = R1#(L3%,3)
6190 NEXT L3%
6200 RETURN
6210 L8% = 0: IF DP<2 THEN GOTO 6310
6220 L8% = 1
6230 REM rotate triangle for orthogonal profiles
6240 FOR L3% = 1 TO 3
6250 Z# = R3#(L3%,3): R3#(L3%,3) = 1 - R3#(L3%,L8%): R3#(L3%,L8%) = Z#
6260 NEXT L3%
6270 ST# = R3#(1,3)
6280 IF ST#>R3#(2,3) THEN ST# = R3#(2,3)
6290 IF ST#>R3#(3,3) THEN ST# = R3#(3,3)
6300 IF ST# < = - 999 THEN RETURN
6310 TP# = R3#(1,3)
6320 IF TP#<R3#(2,3) THEN TP# = R3#(2,3)
6330 IF TP#<R3#(3,3) THEN TP# = R3#(3,3)
6340 IF TP# = ST# THEN RETURN
6350 FOR L0% = 1 TO NC%
6360 REM slice the triangle
6370 IF P8(L0%)>TP# OR P8(L0%)<ST# THEN GOTO 6610
6380 L5% = 1: L4% = 0
6390 L4% = L4% + 1
6400 L7% = 1: IF L7% = L4% THEN L7% = L7% + 1
6410 L3% = L7% + 1: IF L3% = L4% THEN L3% = L3% + 1
6420 IF R3#(L7%,3) = R3#(L3%,3) THEN GOTO 6460
6430 F = (P8(L0%) - R3#(L7%,3))/(R3#(L3%,3) - R3#(L7%,3)): IF F<0 OR F>1 THEN
     GOTO 6460
6440 S3(L5%,1) = R3#(L7%,1) + (R3#(L3%,1) - R3#(L7%,1))*F: S3(L5%,2) = R3#(L7%,2)
     + (R3#(L3%,2) - R3#(L7%,2))*F
6450 L5% = L5% + 1
6460 IF L5%<3 THEN GOTO 6390
6470 S3(1,3) = P8(L0%): S3(2,3) = P8(L0%): IF DP<2 THEN GOTO 6500
6480 REM reverse rotate the intersection trace
6490 FOR L3% = 1 TO 2: Z# = S3(L3%,3): S3(L3%,3) = S3(L3%,L8%): S3(L3%,L8%) = 1
     - Z#: NEXT L3%
6500 IF DP = 0 THEN GOTO 6552
6510 REM apply perspective
6520 FOR L3% = 1 TO 2
6530 YY = (S3(L3%,2) - .5)*R8(1,1) + (S3(L3%,1) - .5)*R8(1,2): S3(L3%,1) = (S3(L3%,1) -
     .5)*R8(1,1) - (S3(L3%,2) - .5)*R8(1,2) + .5
6540 S33 = (S3(L3%,3) - .5)*R8(2,1) + YY*R8(2,2) + .5: S3(L3%,2) = YY*R8(2,1) -
     (S3(L3%,3) - .5)*R8(2,2) + .5: S3(L3%,3) = S33
6550 NEXT L3%
6552 IF SR<1 THEN GOTO 6580
6554 S3(1,1) = (S3(1,1) - .5)*.99875 - (S3(1,3) - .5)*.05 + .5: S3(1,3) = (S3(1,3) -
     .5)*.99875 + (S3(1,1) - .5)*.05 + .5: S3(2,1) = (S3(2,1) - .5)*.99875 - (S3(2,3) - .5)*.05 +
     .5: S3(2,3) = (S3(2,3) - .5)*.99875 + (S3(2,1) - .5)*.05 + .5
6560 REM draw a line from (S3(1,1),S3(1,2)) to (S3(2,1),S3(2,2))
```

Appendix

```
6580 LINE((SW + S3(1,1)*SZ),(ZW - S3(1,2)*ZS)) - ((SW + S3(2,1)*SZ),(ZW -
     S3(2,2)*ZS)),14: IF SR<1 OR DP<1 THEN GOTO 6610
6590 S3(1,1) = (S3(1,1) - .5)*.995 + (S3(1,3) - .5)*.1 + .5: S3(2,1) = (S3(2,1) - .5)*.995 +
     (S3(2,3) - .5)*.1 + .5
6600 LINE((SV + S3(1,1)*SZ),(ZW - S3(1,2)*ZS)) - ((SV + S3(2,1)*SZ),(ZW -
     S3(2,2)*ZS)),14
6610 NEXT L0%
6620 IF L8%<1 THEN RETURN
6630 REM reverse rotate the triangle
6640 FOR L3% = 1 TO 3
6650 Z# = R3#(L3%,3): R3#(L3%,3) = R3#(L3%,L8%): R3#(L3%,L8%) = 1 - Z#
6660 NEXT L3%
6670 L8% = L8% + 1: IF L8%<3 THEN GOTO 6240 ELSE RETURN


1000 REM SCRN\NEIGHBOR.BAS NEIGHBORHOOD - BASED INTERPOLATION
     Copyright (c) 1988 D.F.Watson
1010 DIM P1#(55,6), P2#(237,3), P5#(25), P8#(80), P9(2,3), S1(7742), S2(3), S4(25),
     PO(4,2)
1020 DIM R1#(3,3), R3#(25,25), R4#(3,2), R5#(3,3), R7#(3), R8(2,2), S3(2,3), BX(6,3),
     R9#(25), B#(25)
1030 DIM P3%(237,3),P4%(237),P6%(25),P7%(25,2),R2%(2,2),R6%(3,2),SP%(16)
1040 DATA 0,8,1,4,5,9,6,12,13,2,7,3,10,11,14,15: REM spectrum
1050 DATA 1,0,0,1,1,0,0,1,0,0,0,0,1,0,0,1,0,1, - 1, - 1,0,1,2,5, - 1,0,2,3, - 1,5,1E37,3,1
1060 DEF FNCV(RT#) = RT#^2*LOG(RT#): REM minimum curvature
1090 P9(1,1) = - 1E37: P9(1,2) = - 1E37: P9(1,3) = - 1E37: P9(2,1) = 1E37: P9(2,2) =
     1E37: P9(2,3) = 1E37
1100 SL = 0: SR = 1: SZ = 480: GR = 19: DP = 2: CL = 1: PT = 0: CR = 0: TT = 1: WD =
     640: HI = 350: LW = 5: AZ = - 25: TL = - 15: RD = 2: BI = 1.5: BJ = 7: IF DP = 0
     THEN SR = 0
1140 OPEN "A:\HILL.DAT" FOR INPUT AS #1
1150 REM OPEN "A:\JDAVIS.DAT" FOR INPUT AS #1
1170 FOR I1% = 1 TO 16: READ SP%(I1%): NEXT I1%
1180 INPUT#1,ND%,NC%: FOR I1% = 1 TO NC%: INPUT#1,P8(I1%): NEXT I1%: IF
     CR>0 AND NC%>31 THEN NC% = 31
1185 FOR I1% = 1 TO ND%: P4%(I1%) = I1%: NEXT I1%: FOR I1% = 1 TO ND%: I2% =
     INT(RND*ND% + .5): IF I2%<1 THEN I2% = 1
1187 IF I2%>ND% THEN I2% = ND%
1188 I3% = P4%(I1%): P4%(I1%) = P4%(I2%): P4%(I2%) = I3%: NEXT I1%
1190 INPUT#1,XS,YS,DT: I2% = 1
1220 I1% = P4%(I2%)
1222 INPUT#1,P1#(I1%,1),P1#(I1%,2),P1#(I1%,3): P1#(I1%,1) = (P1#(I1%,1) - XS)/DT:
     P1#(I1%,2) = (P1#(I1%,2) - YS)/DT: IF (P1#(I1%,1) - .5)^2 + (P1#(I1%,2) - .5)^2<RD
     THEN GOTO 1270
1230 ND% = ND% - 1: IF I2%>ND% THEN GOTO 1360
1260 GOTO 1222
1270 IF P9(1,1)<P1#(I1%,1) THEN P9(1,1) = P1#(I1%,1)
1280 IF P9(2,1)>P1#(I1%,1) THEN P9(2,1) = P1#(I1%,1)
1290 IF P9(1,2)<P1#(I1%,2) THEN P9(1,2) = P1#(I1%,2)
1300 IF P9(2,2)>P1#(I1%,2) THEN P9(2,2) = P1#(I1%,2)
1310 IF P9(1,3)<P1#(I1%,3) THEN P9(1,3) = P1#(I1%,3)
1320 IF P9(2,3)>P1#(I1%,3) THEN P9(2,3) = P1#(I1%,3)
1350 I2% = I2% + 1: IF I2%< = ND% THEN GOTO 1220
1360 IF ND%<4 THEN STOP
1370 XF = P9(1,1) - P9(2,1): YF = P9(1,2) - P9(2,2): VS = P9(1,3) - P9(2,3): R8(2,1) = AZ:
     R8(2,2) = TL
1380 FOR I1% = 1 TO 6: READ BX(I1%,1),BX(I1%,2),BX(I1%,3): NEXT I1%
1390 BN% = 1: IF DP<1 THEN GOTO 1450
1400 BN% = 2: BX(2,1) = 0: BX(2,3) = 1: IF R8(2,1)>0 THEN GOTO 1420
1410 BX(2,1) = 1: BX(3,1) = 1: BX(4,1) = 1: BX(5,1) = 0: BX(6,1) = 0
1420 IF DP<2 OR CR>0 THEN GOTO 1450
1430 NC1% = INT(SZ/10): NC% = NC1% + 1: U0 = .9999/NC1%: U1 = .0000511 - U0
1440 FOR I1% = 1 TO NC%: U1 = U1 + U0: P8(I1%) = U1: NEXT I1%: GOTO 1460
1450 FOR I1% = 1 TO NC%: P8(I1%) = (P8(I1%) - P9(2,3))/VS + .00001*(RND - .5):
     NEXT I1%
```

BASIC Programs

```
1460 FOR I1% = 1 TO ND%: P1#(I1%,1) = P1#(I1%,1) + .0001*(RND - .5): P1#(I1%,2) =
     P1#(I1%,2) + .0001*(RND - .5) : P1#(I1%,3) = (P1#(I1%,3) - P9(2,3))/VS: NEXT I1%
1470 FOR I1% = 1 TO 3: READ R1#(I1%,1), R1#(I1%,2), R1#(I1%,3), R6%(I1%,1),
     R6%(I1%,2)
1480 P1#(I1% + ND%,1) = R1#(I1%,1)*XF + P9(2,1): P1#(I1% + ND%,2) = R1#(I1%,2)*YF
     + P9(2,2): P1#(I1% + ND%,3) = 0: NEXT I1%
1490 IF DP<1 THEN GOTO 1550
1500 FOR I1% = 1 TO 2: R8(I1%,1) = COS(R8(2,I1%)/57.29578): R8(I1%,2) =
     SIN(R8(2,I1%)/57.29578): NEXT I1%
1510 FOR I1% = 2 TO 6
1520 YY = (BX(I1%,2) - .5)*R8(1,1) + (BX(I1%,1) - .5)*R8(1,2) + .5: BX(I1%,1) =
     (BX(I1%,1) - .5)*R8(1,1) - (BX(I1%,2) - .5)*R8(1,2) + .5
1530 BX3 = (BX(I1%,3) - .5)*R8(2,1) + (YY - .5)*R8(2,2) + .5: BX(I1%,2) = (YY -
     .5)*R8(2,1) - (BX(I1%,3) - .5)*R8(2,2) + .5: BX(I1%,3) = BX3: NEXT I1%
1540 REM end data input, begin gradient estimation
1550 SOUND 208,5: SOUND 165,5: SOUND 185,5: SOUND 123,5
1560 IF CL<1 THEN GOTO 3110
1570 PRINT " AT " TIME$ " BEGIN GRADIENT ESTIMATION  -  PLEASE WAIT"
1580 FOR I0% = 1 TO ND%
1590 FOR I1% = 1 TO 3: P3%(1,I1%) = ND% + I1%: P3%(2,I1%) = ND% + I1%:
     P2#(1,I1%) = R1#(I1%,3): P2#(2,I1%) = R1#(I1%,3) : NEXT I1%
1600 FOR I1% = 1 TO 237: P4%(I1%) = I1%: NEXT I1%: R2%(1,1) = 1: R2%(1,2) = 1:
     R2%(2,1) = 3 : R2%(2,2) = 4: K1% = I0%: K2% = I0%: K3% = ND% + 4: K4 = - .5
1610 FOR I1% = K1% TO K2%
1620 IF I1% = K3% THEN GOTO 2000
1630 K5 = K4
1640 K6 = 1.5 + K5: IF K3%>ND% THEN K5 =  - K5
1650 U0 = 0: K7% = 0: L6% = K6 - 2
1660 FOR I2% = 1 TO R2%(1,K6)
1670 L6% = L6% + 2: IF P3%(L6%,1) = 0 THEN GOTO 1670
1680 U1# = P2#(L6%,3) - (P1#(I1%,1) - P2#(L6%,1))^2: IF U1#<0 THEN GOTO 1840
1690 U1# = U1# - (P1#(I1%,2) - P2#(L6%,2))^2: IF U1#<0 THEN GOTO 1840
1700 R2%(2,K6) = R2%(2,K6) - 2: P4%(R2%(2,K6)) = L6%: U0 = U0 - 1
1710 FOR I3% = 1 TO 3
1720 L1% = 1: IF L1% = I3% THEN L1% = L1% + 1
1730 L2% = L1% + 1: IF L2% = I3% THEN L2% = L2% + 1
1740 IF K7%<1 THEN GOTO 1810
1750 K9% = K7%
1760 FOR I4% = 1 TO K9%
1770 IF P3%(L6%,L1%)<>P7%(I4%,1) THEN GOTO 1800
1780 IF P3%(L6%,L2%)<>P7%(I4%,2) THEN GOTO 1800
1790 P7%(I4%,1) = P7%(K9%,1): P7%(I4%,2) = P7%(K9%,2): K7% = K7% - 1: GOTO
     1820
1800 NEXT I4%
1810 K7% = K7% + 1: P7%(K7%,1) = P3%(L6%,L1%): P7%(K7%,2) = P3%(L6%,L2%)
1820 NEXT I3%
1830 P3%(L6%,1) = 0
1840 NEXT I2%
1850 IF K7%<1 THEN GOTO 2000
1860 REM make new triangles
1870 FOR I2% = 1 TO K7%
1880 IF K5>0 THEN GOTO 1900
1890 IF P7%(I2%,1)<>I0% AND P7%(I2%,2)<>I0% THEN GOTO 1970
1900 FOR I3% = 1 TO 2
1910 R3#(I3%,1) = P1#(P7%(I2%,I3%),1) - P1#(I1%,1): R3#(I3%,2) =
     P1#(P7%(I2%,I3%),2) - P1#(I1%,2) 1920 R3#(I3%,3) =
     R3#(I3%,1)*(P1#(P7%(I2%,I3%),1) + P1#(I1%,1))/2 + R3#(I3%,2)
     *(P1#(P7%(I2%,I3%),2) + P1#(I1%,2))/2
1930 NEXT I3%
1940 U1# = R3#(1,1)*R3#(2,2) - R3#(2,1)*R3#(1,2): P2#(P4%(R2%(2,K6)),1) =
     (R3#(1,3)*R3#(2,2) - R3#(2,3)*R3#(1,2))/U1#: P2#(P4%(R2%(2,K6)),2) =
     (R3#(1,1)*R3#(2,3) - R3#(2,1)*R3#(1,3))/U1#
1950 P2#(P4%(R2%(2,K6)),3) = (P1#(I1%,1) - P2#(P4%(R2%(2,K6)),1))^2 + (P1#(I1%,2) -
     P2#(P4%(R2%(2,K6)),2))^2
1960 P3%(P4%(R2%(2,K6)),1) = P7%(I2%,1): P3%(P4%(R2%(2,K6)),2) = P7%(I2%,2) :
     P3%(P4%(R2%(2,K6)),3) = I1%: R2%(2,K6) = R2%(2,K6) + 2: U0 = U0 + 1
```

235

Appendix

```
1970 NEXT I2%
1980 R2%(1,K6) = R2%(1,K6) + U0: IF K5>0 THEN GOTO 2000
1990 K5 = - K5: GOTO 1640
2000 NEXT I1%
2010 IF K2%>K1% GOTO 2040
2020 K1% = 1: K2% = ND%: K3% = I0%: GOTO 1610
2030 REM order triangles positively
2040 FOR I1% = 1 TO 2
2050 L8% = I1% - 2
2060 FOR I2% = 1 TO R2%(1,I1%)
2070 L8% = L8% + 2: IF P3%(L8%,1) = 0 THEN GOTO 2070
2080 IF P3%(L8%,1)>ND% THEN GOTO 2140
2090 FOR I3% = 1 TO 2
2100 R3#(1,I3%) = P1#(P3%(L8%,2),I3%) - P1#(P3%(L8%,1),I3%): R3#(2,I3%) =
    P1#(P3%(L8%,3),I3%) - P1#(P3%(L8%,1),I3%)
2110 NEXT I3%
2120 IF R3#(1,1)*R3#(2,2) - R3#(1,2)*R3#(2,1)> = 0 THEN GOTO 2140
2130 K8 = P3%(L8%,3): P3%(L8%,3) = P3%(L8%,2): P3%(L8%,2) = K8
2140 NEXT I2%
2150 NEXT I1%
2160 IF SL>0 THEN GOTO 2470
2170 REM calculate natural neighbor coordinates
2180 R5#(1,1) = P1#(I0%,1): R5#(1,2) = P1#(I0%,2): R5#(2,1) = R5#(1,1) + .0001 :
    R5#(2,2) = R5#(1,2): R5#(3,1) = R5#(1,1): R5#(3,2) = R5#(1,2) + .0001
2190 FOR I1% = 1 TO 3
2200 L4% = 0: L5% = 0: L6% = 0
2210 FOR I2% = 1 TO R2%(1,2)
2220 L6% = L6% + 2: IF P3%(L6%,1) = 0 THEN GOTO 2220
2230 IF P3%(L6%,1)>ND% THEN GOTO 2410
2240 U1# = P2#(L6%,3) - (R5#(I1%,1) - P2#(L6%,1))^2: IF U1#<0 THEN GOTO 2410
2250 U1# = U1# - (R5#(I1%,2) - P2#(L6%,2))^2: IF U1#<0 THEN GOTO 2410
2260 FOR I3% = 1 TO 3
2270 FOR I4% = 1 TO 2
2280 R3#(I4%,1) = P1#(P3%(L6%,R6%(I3%,I4%)),1) - R5#(I1%,1): R3#(I4%,2) =
    P1#(P3%(L6%,R6%(I3%,I4%)),2) - R5#(I1%,2)
2290 R3#(I4%,3) = R3#(I4%,1)*(P1#(P3%(L6%,R6%(I3%,I4%)),1) + R5#(I1%,1))/2 +
    R3#(I4%,2) *(P1#(P3%(L6%,R6%(I3%,I4%)),2) + R5#(I1%,2))/2: NEXT I4%
2300 U1# = R3#(1,1)*R3#(2,2) - R3#(1,2)*R3#(2,1)
2310 R4#(R6%(I3%,2),1) = (R3#(1,3)*R3#(2,2) - R3#(2,3)*R3#(1,2))/U1#:
    R4#(R6%(I3%,2),2) = (R3#(1,1)*R3#(2,3) - R3#(2,1)*R3#(1,3))/U1#: NEXT I3%: L3%=0
2320 FOR I3% = 1 TO 3: R3#(3,I3%) = ((R4#(R6%(I3%,1),1) -
    P2#(L6%,1))*(R4#(R6%(I3%,2),2) - P2#(L6%,2)) - (R4#(R6%(I3%,2),1) -
    P2#(L6%,1))*(R4#(R6%(I3%,1),2) - P2#(L6%,2)))/2: IF R3#(3,I3%)>0 THEN L3% =
    L3% + 1
2330 NEXT I3%
2340 IF L3%>2 THEN L4% = 1
2350 FOR I3% = 1 TO 3: IF L5%<1 THEN GOTO 2390
2360 FOR I4% = 1 TO L5%: IF P3%(L6%,I3%)<>P6%(I4%) THEN GOTO 2380
2370 P5#(I4%) = P5#(I4%) + R3#(3,I3%): GOTO 2400
2380 NEXT I4%
2390 L5% = L5% + 1: P6%(L5%) = P3%(L6%,I3%): P5#(L5%) = R3#(3,I3%)
2400 NEXT I3%
2410 NEXT I2%: IF L4%<1 THEN GOTO 2470
2420 U4 = 0
2430 FOR I2% = 1 TO L5%: U4 = U4 + P5#(I2%): NEXT I2%: R7#(I1%) = 0
2440 FOR I2% = 1 TO L5%: P5#(I2%) = P5#(I2%)/U4: R7#(I1%) = R7#(I1%) +
    P5#(I2%)*P1#(P6%(I2%),3): NEXT I2%: NEXT I1%
2450 P1#(I0%,6) = P1#(I0%,3) - R7#(1): P1#(I0%,4) = (R7#(1) - R7#(2))/.0001:
    P1#(I0%,5) = (R7#(1) - R7#(3))/.0001 : GOTO 3080
2460 REM cross product gradients
2470 IF SL>1 THEN GOTO 2610
2480 P1#(I0%,4) = 0: P1#(I0%,5) = 0: P1#(I0%,6) = 0: P7# = 0: L7% = - 1
2490 FOR I1% = 1 TO R2%(1,1) 2500 L7% = L7% + 2: IF P3%(L7%,1) = 0 THEN GOTO
    2500
2510 IF P3%(L7%,1)>ND% THEN GOTO 2570
```

```
2520 FOR I2% = 1 TO 2: FOR I3% = 1 TO 3: R3#(I2%,I3%) = P1#(P3%(L7%,1),I3%) -
     P1#(P3%(L7%,I2% + 1),I3%): NEXT I3% : NEXT I2%
2530 R3#(3,1) = R3#(1,2)*R3#(2,3) - R3#(2,2)*R3#(1,3): R3#(3,2) = R3#(1,3)*R3#(2,1) -
     R3#(2,3)*R3#(1,1): R3#(3,3) = R3#(1,1)*R3#(2,2) - R3#(2,1)*R3#(1,2): U3 = 1
2540 IF R3#(3,3)<0 THEN U3 = - 1
2550 U2 = (R3#(3,1)^2 + R3#(3,2)^2 + R3#(3,3)^2)^.5: P7# = P7# + U2
2560 FOR I2% = 1 TO 3: P1#(I0%,I2% + 3) = P1#(I0%,I2% + 3) + R3#(3,I2%)*U3: NEXT
     I2%
2570 NEXT I1%
2580 U2 = P1#(I0%,4)^2 + P1#(I0%,5)^2 + P1#(I0%,6)^2: P7# = 1 - U2^.5/P7#
2590 P1#(I0%,4) = P1#(I0%,4)/P1#(I0%,6): P1#(I0%,5) = P1#(I0%,5)/P1#(I0%,6):
     P1#(I0%,6) = P7#
2600 GOTO 3080
2610 IF SL>2 THEN GOTO 2810 : REM least squares gradients
2620 P1#(I0%,4) = 0: P1#(I0%,5) = 0: P1#(I0%,6) = .2: L7% = - 1: P6%(1) = I0%: L5% =1
2630 R4#(1,1) = P1#(I0%,3): R4#(2,1) = P1#(I0%,1)*P1#(I0%,3): R4#(3,1) =
     P1#(I0%,2)*P1#(I0%,3)
2640 R3#(2,1) = P1#(I0%,1): R3#(2,2) = P1#(I0%,1)^2: R3#(3,1) = P1#(I0%,2) : R3#(3,2)
     = P1#(I0%,1)*P1#(I0%,2): R3#(3,3) = P1#(I0%,2)^2
2650 FOR I1% = 1 TO R2%(1,1) 2660 L7% = L7% + 2
2670 IF P3%(L7%,1) = 0 THEN GOTO 2660
2680 IF P3%(L7%,1)>ND THEN GOTO 2760
2690 FOR I2% = 1 TO 3
2700 FOR I3% = 1 TO L5%
2710 IF P3%(L7%,I2%) = P6%(I3%) THEN GOTO 2750
2720 NEXT I3%
2730 L5% = L5% + 1: P6%(L5%) = P3%(L7%,I2%): R4#(1,1) = R4#(1,1) +
     P1#(P6%(L5%),3) : R4#(2,1) = R4#(2,1) + P1#(P6%(L5%),1)*P1#(P6%(L5%),3):
     R4#(3,1) = R4#(3,1) + P1#(P6%(L5%),2)*P1#(P6%(L5%),3)
2740 R3#(2,1) = R3#(2,1) + P1#(P6%(L5%),1): R3#(2,2) = R3#(2,2) +
     P1#(P6%(L5%),1)^2: R3#(3,1) = R3#(3,1) + P1#(P6%(L5%),2): R3#(3,2) = R3#(3,2) +
     P1#(P6%(L5%),1)*P1#(P6%(L5%),2): R3#(3,3) = R3#(3,3) + P1#(P6%(L5%),2)^2
2750 NEXT I2%
2760 NEXT I1%
2770 U1# = L5%*(R3#(2,2)*R3#(3,3) - R3#(3,2)^2) - R3#(2,1)*(R3#(2,1)*R3#(3,3) -
     R3#(3,1)*R3#(3,2)) + R3#(3,1)*(R3#(2,1)*R3#(3,2) - R3#(3,1)*R3#(2,2))
2780 P1#(I0%,4) = - (L5%*(R4#(2,1)*R3#(3,3) - R3#(3,2)*R4#(3,1)) -
     R4#(1,1)*(R3#(2,1)*R3#(3,3) - R3#(3,1)*R3#(3,2)) + R3#(3,1)*(R3#(2,1)*R4#(3,1) -
     R3#(3,1)*R4#(2,1)))/U1#
2790 P1#(I0%,5) = - (L5%*(R3#(2,2)*R4#(3,1) - R3#(3,2)*R4#(2,1)) -
     R3#(2,1)*(R3#(2,1)*R4#(3,1) - R3#(3,1)*R4#(2,1)) + R4#(1,1)*(R3#(2,1)*R3#(3,2) -
     R3#(3,1)*R3#(2,2)))/U1#
2800 GOTO 3080 : REM spline gradients
2810 IF SL>3 THEN GOTO 3035
2815 P1#(I0%,4) = 0: P1#(I0%,5) = 0: P1#(I0%,6) = .2: L7% = - 1: L5% = 1: P6%(1) =
     I0%
2820 R5#(1,1) = P1#(I0%,1): R5#(1,2) = P1#(I0%,2): R5#(2,1) = R5#(1,1) + .0001 :
     R5#(2,2) = R5#(1,2): R5#(3,1) = R5#(1,1): R5#(3,2) = R5#(1,2) + .0001
2830 FOR I1% = 1 TO R2%(1,1)
2840 L7% = L7% + 2: IF P3%(L7%,1) = 0 THEN GOTO 2840
2850 IF P3%(L7%,1)>ND% THEN GOTO 2890
2860 FOR I2% = 1 TO 3: FOR I3% = 1 TO L5%: IF P3%(L7%,I2%) = P6%(I3%) THEN
     GOTO 2880
2870 NEXT I3%: L5% = L5% + 1: P6%(L5%) = P3%(L7%,I2%) 2880 NEXT I2% 2890
     NEXT I1%
2900 LL% = L5% + 3
2910 FOR I1% = 1 TO L5%
2920 R3#(LL%,1) = 1: R3#(I1%,I1% + 3) = 1: R3#(I1%,2) = P1#(P6%(I1%),1): R3#(L5% +
     1,I1% + 3) = R3#(I1%,2)
2930 R3#(I1%,3) = P1#(P6%(I1%),2): R3#(L5% + 2,I1% + 3) = R3#(I1%,3): R9#(I1%) =
     P1#(P6%(I1%),3): NEXT I1%
2940 FOR I1% = 1 TO 3: FOR I2% = 1 TO 3: R3#(I1% + L5%,I2%) = 0: NEXT I2%:
     R9#(I1% + L5%) = 0: NEXT I1%
2950 FOR I1% = 1 TO L5%: R3#(I1%,I1% + 3) = 0: IJ = I1% + 1: IF IJ>L5% THEN GOTO
     2970
```

Appendix

```
2960 FOR I2% = IJ TO L5%: RT# = ((R3#(I1%,2) - R3#(I2%,2))^2 + (R3#(I1%,3) -
     R3#(I2%,3))^2)^.5: R3#(I2%,I1% + 3) = FNCV(RT#): R3#(I1%,I2% + 3) = R3#(I2%,I1%
     + 3): NEXT I2%
2970 NEXT I1%: GOSUB 6690
2980 TR# = 0: TS# = 0
2990 FOR I1% = 4 TO LL%
3000 RT# = ((P1#(P6%(I1% - 3),1) - R5#(2,1))^2 + (P1#(P6%(I1% - 3),2) -
     R5#(2,2))^2)^.5: TR# = TR# + B#(I1%)*FNCV(RT#)
3010 RS# = ((P1#(P6%(I1% - 3),1) - R5#(3,1))^2 + (P1#(P6%(I1% - 3),2) -
     R5#(3,2))^2)^.5: TS# = TS# + B#(I1%)*FNCV(RS#) : NEXT I1%
3020 R5#(2,3) = B#(1) + B#(2)*R5#(2,1) + B#(3)*R5#(2,2) + TR#: P1#(I0%,4) =
     (P1#(I0%,3) - R5#(2,3))/.0001
3030 R5#(3,3) = B#(1) + B#(2)*R5#(3,1) + B#(3)*R5#(3,2) + TS#: P1#(I0%,5) =
     (P1#(I0%,3) - R5#(3,3))/.0001: GOTO 3080
3033 REM hyperboloid gradients
3035 R3#(5,1) = 1: FOR I1% = 1 TO 4: R9#(I1%) = 0: FOR I2% = 1 TO 4: R3#(I1%,I2%) =
     0: NEXT I2%: NEXT I1%: LL% = 4
3036 P1#(I0%,4) = 0: P1#(I0%,5) = 0: P1#(I0%,6) = .2: L7% = - 1: L5% = 0: FOR I1% = 1
     TO R2%(1,1)
3037 L7% = L7% + 2: IF P3%(L7%,1) = 0 THEN GOTO 3037
3038 IF P3%(L7%,1)>ND% THEN GOTO 3070
3039 FOR I2% = 1 TO 3: IF L5%<1 THEN GOTO 3044
3041 FOR I3% = 1 TO L5%: IF P3%(L7%,I2%) = P6%(I3%) THEN GOTO 3060
3042 NEXT I3%
3044 L5% = L5% + 1: P6%(L5%) = P3%(L7%,I2%)
3050 R3#(5,2) = P1#(P6%(L5%),1): R3#(5,3) = P1#(P6%(L5%),2): R3#(5,4) =
     P1#(P6%(L5%),1)*P1#(P6%(L5%),2)
3055 FOR I3% = 1 TO 4: FOR I4% = 1 TO 4: R3#(I3%,I4%) = R3#(I3%,I4%) +
     R3#(5,I3%)*R3#(5,I4%): NEXT I4% : R9#(I3%) = R9#(I3%) +
     R3#(5,I3%)*P1#(P6%(L5%),3): NEXT I3%
3060 NEXT I2%
3070 NEXT I1%: GOSUB 6690: P1#(I0%,4) =  - B#(2) - B#(4)*P1#(I0%,2): P1#(I0%,5) =  -
     B#(3) - B#(4)*P1#(I0%,1): P1#(I0%,6) = .2
3080 NEXT I0%
3100 REM natural neighbor sort
3110 FOR I0% = 1 TO 3: P3%(1,I0%) = ND% + I0%: P2#(1,I0%) = R1#(I0%,3): NEXT I0%
3120 FOR I0% = 1 TO 237: P4%(I0%) = I0%: NEXT I0%: NT% = 1: L0% = 2
3130 FOR I0% = 1 TO ND%: L3% = 0
3140 FOR I1% = 1 TO NT%
3150 U1# = P2#(I1%,3) - (P1#(I0%,1) - P2#(I1%,1))^2: IF U1#<0 THEN GOTO 3300
3160 U1# = U1# - (P1#(I0%,2) - P2#(I1%,2))^2: IF U1#<0 THEN GOTO 3300
3170 L0% = L0% - 1: P4%(L0%) = I1%
3180 FOR I2% = 1 TO 3
3190 L1% = 1: IF L1% = I2% THEN L1% = L1% + 1
3200 L2% = L1% + 1: IF L2% = I2% THEN L2% = L2% + 1
3210 IF L3%<1 THEN GOTO 3280
3220 L5% = L3%
3230 FOR I3% = 1 TO L5%
3240 IF P3%(I1%,L1%)<>P7%(I3%,1) THEN GOTO 3270
3250 IF P3%(I1%,L2%)<>P7%(I3%,2) THEN GOTO 3270
3260 P7%(I3%,1) = P7%(L5%,1): P7%(I3%,2) = P7%(L5%,2): L3% = L3% - 1: GOTO
     3290
3270 NEXT I3%
3280 L3% = L3% + 1: P7%(L3%,1) = P3%(I1%,L1%): P7%(L3%,2) = P3%(I1%,L2%)
3290 NEXT I2%
3300 NEXT I1%
3310 FOR I1% = 1 TO L3%
3320 FOR I2% = 1 TO 2
3330 R3#(I2%,1) = P1#(P7%(I1%,I2%),1) - P1#(I0%,1): R3#(I2%,2) =
     P1#(P7%(I1%,I2%),2) - P1#(I0%,2): R3#(I2%,3) = R3#(I2%,1)*(P1#(P7%(I1%,I2%),1) +
     P1#(I0%,1))/2
3340 R3#(I2%,3) = R3#(I2%,3) + R3#(I2%,2)*(P1#(P7%(I1%,I2%),2) + P1#(I0%,2))/2
3350 NEXT I2%
3360 U1# = R3#(1,1)*R3#(2,2) - R3#(2,1)*R3#(1,2): P2#(P4%(L0%),1) =
     (R3#(1,3)*R3#(2,2) - R3#(2,3)*R3#(1,2))/U1#: P2#(P4%(L0%),2) = (R3#(1,1)*R3#(2,3)
     - R3#(2,1)*R3#(1,3))/U1#
```

```
3370 P2#(P4%(L0%),3) = (P1#(I0%,1) - P2#(P4%(L0%),1))^2 + (P1#(I0%,2) -
     P2#(P4%(L0%),2))^2
3380 P3%(P4%(L0%),1) = P7%(I1%,1): P3%(P4%(L0%),2) = P7%(I1%,2):
     P3%(P4%(L0%),3) = I0%: L0% = L0% + 1
3390 NEXT I1%
3400 NT% = NT% + 2: NEXT I0%
3410 FOR I0% = 1 TO NT%: IF P3%(I0%,1)>ND% THEN GOTO 3460
3420 FOR I1% = 1 TO 2
3430 R3#(1,I1%) = P1#(P3%(I0%,2),I1%) - P1#(P3%(I0%,1),I1%): R3#(2,I1%) =
     P1#(P3%(I0%,3),I1%) - P1#(P3%(I0%,1),I1%): NEXT I1%
3440 IF R3#(1,1)*R3#(2,2) - R3#(1,2)*R3#(2,1)> = 0 THEN GOTO 3460
3450 L4% = P3%(I0%,3): P3%(I0%,3) = P3%(I0%,2): P3%(I0%,2) = L4%
3460 NEXT I0%
3820 REM make triangular grid
3830 PRINT " AT " TIME$ " BEGIN SURFACE INTERPOLATION  -  PLEASE WAIT":
     SOUND 123,5: SOUND 185,5: SOUND 208,5 : SOUND 165,5
3840 I0% = GR + 1: IS = 1: I2% = 2*I0% + 1: I3% = 1 + INT(GR/3^.5): I4% = I3%*I2% +
     I0%: SX = 1/GR: SY = 1/(2*I3%): S2(3) = .000001
3850 WX# = S2(3): S2(1) = 1.00001: AA# = SX*SY*.5*DT*DT: SM# = 0: SA# = 0: WY# =
     1: U8 = 1
3860 FOR L1% = 1 TO I4%
3870 S1(L1%) = - 999: WXO# = WX#: WYO# = WY#
3880 REM .....calculate natural neighbor coordinates of WX#,WY# **********
3890 K4% = 0: K5% = 0
3900 FOR L2% = 1 TO NT%
3910 IF P3%(L2%,1)>ND% THEN GOTO 4280
3920 U1# = P2#(L2%,3) - (WX# - P2#(L2%,1))^2: IF U1#<0 THEN GOTO 4280
3930 U1# = U1# - (WY# - P2#(L2%,2))^2: IF U1#<0 THEN GOTO 4280
3940 FOR L3% = 1 TO 3
3950 L4% = 1: IF L4% = L3% THEN L4% = L4% + 1
3960 L5% = L4% + 1: IF L5% = L3% THEN L5% = L5% + 1
3970 IF ABS((P1#(P3%(L2%,L5%),1) - P1#(P3%(L2%,L4%),1))*(WY# -
     P1#(P3%(L2%,L4%),2)) - (WX# - P1#(P3%(L2%,L4%),1))*(P1#(P3%(L2%,L5%),2) -
     P1#(P3%(L2%,L4%),2)))>.000001 THEN GOTO 4030
4000 DEN# = ABS(P1#(P3%(L2%,L4%),2) - P1#(P3%(L2%,L5%),2)) +
     ABS(P1#(P3%(L2%,L5%),1) - P1#(P3%(L2%,L4%),1))
4010 WX# = WX# + (P1#(P3%(L2%,L4%),2) - P1#(P3%(L2%,L5%),2))*.0001/DEN#: WY#
     = WY# + (P1#(P3%(L2%,L5%),1) - P1#(P3%(L2%,L4%),1))*.0001/DEN#
4030 NEXT L3%
4040 FOR L3% = 1 TO 3
4050 FOR L4% = 1 TO 2
4060 R3#(L4%,1) = P1#(P3%(L2%,R6%(L3%,L4%)),1) - WX#: R3#(L4%,2) =
     P1#(P3%(L2%,R6%(L3%,L4%)),2) - WY#
4070 R3#(L4%,3) = R3#(L4%,1)*(P1#(P3%(L2%,R6%(L3%,L4%)),1) + WX#)/2 +
     R3#(L4%,2) *(P1#(P3%(L2%,R6%(L3%,L4%)),2) + WY#)/2
4080 NEXT L4%
4090 U1# = R3#(1,1)*R3#(2,2) - R3#(1,2)*R3#(2,1)
4100 R4#(R6%(L3%,2),1) = (R3#(1,3)*R3#(2,2) - R3#(2,3)*R3#(1,2))/U1#
4110 R4#(R6%(L3%,2),2) = (R3#(1,1)*R3#(2,3) - R3#(2,1)*R3#(1,3))/U1#
4120 NEXT L3%
4130 K3% = 0
4140 FOR L3% = 1 TO 3
4150 R3#(3,L3%) = ((R4#(R6%(L3%,1),1) - P2#(L2%,1))*(R4#(R6%(L3%,2),2) -
     P2#(L2%,2)) - (R4#(R6%(L3%,2),1) - P2#(L2%,1))*(R4#(R6%(L3%,1),2) -
     P2#(L2%,2)))/2
4160 IF R3#(3,L3%)>0 THEN K3% = K3% + 1
4170 NEXT L3%
4180 IF K3%>2 THEN K4% = 1
4190 FOR L3% = 1 TO 3
4200 IF K5%<1 THEN GOTO 4260
4210 FOR L4% = 1 TO K5%
4220 IF P3%(L2%,L3%)<>P6%(L4%) THEN GOTO 4250
4230 P5#(L4%) = P5#(L4%) + R3#(3,L3%)
4240 GOTO 4270
4250 NEXT L4%
4260 K5% = K5% + 1: P6%(K5%) = P3%(L2%,L3%): P5#(K5%) = R3#(3,L3%)
```

Appendix

```
4270 NEXT L3%
4280 NEXT L2%
4300 IF K4%<1 THEN GOTO 4940 ELSE GOTO 4560
4310 REM ....end determining K5% local coordinates **********
4560 S1(L1%) = 0: U4 = 0
4570 FOR L2% = 1 TO K5%: U4 = U4 + P5#(L2%): NEXT L2%
4580 FOR L2% = 1 TO K5%: P5#(L2%) = P5#(L2%)/U4: S1(L1%) = S1(L1%) +
     P5#(L2%)*P1#(P6%(L2%),3): NEXT L2%
4590 IF CL<1 THEN GOTO 4940
4620 FOR L2% = 1 TO K5%: S4(L2%) = 0
4630 IF P5#(L2%)<.00001 OR P5#(L2%)>1 THEN GOTO 4690
4640 IF ABS(P1#(P6%(L2%),6))<.00001 THEN GOTO 4690
4650 RS# = ABS(P1#(P6%(L2%),6)) + BI: RT# = RS#*BJ: RB# = 1/RT#: BD# =
     P5#(L2%)^RT#: BB# = BD#*2: IF BD#>.5 THEN BB# = (1 - BD#)*2
4660 BB# = BB#^RS#/2: IF BD#>.5 THEN BB# = 1 - BB#
4670 HP# = BB#*RB#
4680 S4(L2%) = ((P1#(P6%(L2%),4)*P1#(P6%(L2%),1) +
     P1#(P6%(L2%),5)*P1#(P6%(L2%),2) + P1#(P6%(L2%),3) - P1#(P6%(L2%),4)*WX# -
     P1#(P6%(L2%),5)*WY#) - S1(L1%))*HP#
4690 NEXT L2%
4700 FOR L2% = 1 TO K5%: S1(L1%) = S1(L1%) + S4(L2%): NEXT L2%: GOTO 4940
4940 IS = IS + SX*U8: IF IS>0 THEN GOTO 4960
4950 WXO# = WXO# + SX/2: IS = 1.001 - SX
4960 WX# = WXO# + SX*U8: WY# = WYO#: IF U8*WX#<S2(2 - U8) THEN GOTO 4980
4970 U8 =  - U8: WX# = S2(2 + U8): WY# = WY# - SY: IS = U8
4980 NEXT L1%
5060 REM end interpolation to triangular grid and begin display construction
5080 SCREEN 9: COLOR 2,0: CLS: SOUND 208,5: SOUND 165,5: SOUND 185,5:
     SOUND 123,5: IF DP>0 THEN SZ = INT(SZ*2/3)
5090 IF SR>0 AND SZ>240 THEN SZ = 240
5100 ZS = INT(SZ*.71): SW = (WD - SZ)\2: ZW = (HI - LW)\2 + ZS\2: IF DP>0 THEN ZW
     = ZW - LW - 10
5110 IF SR>0 AND CR<1 AND DP>0 THEN SW = (WD - 2*SZ)\3 - 10
5120 SV = SW + SZ + 60: IF SR<1 THEN GOTO 5130
5125 FOR L1% = 2 TO 6: BX(L1%,1) = (BX(L1%,1) - .5)*.99875 - (BX(L1%,3) - .5)*.05 +
     .5: BX(L1%,3) = (BX(L1%,3) - .5)*.99875 + (BX(L1%,1) - .5)*.05 + .5: NEXT L1%
5130 FOR L1% = BN% + 1 TO BN% + 4: X1 = SW + BX(L1% - 1,1)*SZ: Y1 = ZW -
     BX(L1% - 1,2)*ZS: X2 = SW + BX(L1%,1)*SZ: Y2 = ZW - BX(L1%,2)*ZS: LINE(X1,Y1)
     - (X2,Y2),14: NEXT L1%: IF SR<1 OR CR>0 THEN GOTO 5210
5140 FOR L1% = 2 TO 6: BX(L1%,1) = (BX(L1%,1) - .5)*.995 + (BX(L1%,3) - .5)*.1 + .5:
     NEXT L1%
5150 FOR L1% = BN% + 1 TO BN% + 4: X1 = SV + BX(L1% - 1,1)*SZ: Y1 = ZW -
     BX(L1% - 1,2)*ZS: X2 = SV + BX(L1%,1)*SZ: Y2 = ZW - BX(L1%,2)*ZS: LINE(X1,Y1) -
     (X2,Y2),14: NEXT L1%
5200 REM plot contours
5210 SW = SW + 1: SZ = SZ - 2: ZS = ZS - 2: ZW = ZW - 1: ZX = SX: ZY = SY: ZZ = .5:
     R3#(2,2) = 1 + 2*SY: R3#(2,1) = 0
5220 FOR L1% = 1 TO I3%
5230 R3#(1,2) = R3#(2,2) - 2*SY - 2*ZY: R3#(1,1) = R3#(2,1): N1% = (2*L1% - .5 +
     ZZ)*I0% + L1% + .5/ZZ: R3#(1,3) = S1(N1%): R3#(2,2) = R3#(2,2) - 2*SY - ZY:
     R3#(2,1) = .5 - ZZ: N2% = (2*L1% - .5 + ZZ)*I0% + L1%
5240 R3#(2,3) = S1(N2%): R3#(3,2) = R3#(2,2): R3#(3,1) = R3#(2,1) + ZX/2: N3% = N2%
     - .5/ZZ: R3#(3,3) = S1(N3%): AR# = AA#/2
5250 GOSUB 5820
5260 R3#(1,2) = R3#(1,2) + 2*ZY: N2% = (2*L1% - .5/ZZ - .5 - ZZ)*I0% + L1%: R3#(1,3) =
     S1(N2%)
5270 GOSUB 5820
5280 AR# = AA#
5290 FOR L2% = 1 TO GR
5300 R3#(2,2) = R3#(1,2): R3#(2,1) = R3#(1,1) + ZX: N4% = N2% + .5/ZZ: R3#(2,3) =
     S1(N4%)
5310 GOSUB 5820
5320 R3#(1,2) = R3#(3,2): R3#(1,1) = R3#(3,1) + ZX
5330 IF L2% = GR THEN R3#(1,1) = R3#(1,1) - ZX/2
5340 N6% = N3% - .5/ZZ: R3#(1,3) = S1(N6%): IF L2% = GR THEN AR# = AA#/2
5350 GOSUB 5820
```

```
5360 R3#(2,2) = R3#(2,2) - 2*ZY: N5% = N1% + .5/ZZ: R3#(2,3) = S1(N5%)
5370 GOSUB 5820
5380 R3#(1,2) = R3#(2,2): R3#(1,1) = R3#(2,1) - ZX: R3#(1,3) = S1(N1%): IF L2% = GR
     THEN AR# = AA#
5390 GOSUB 5820
5400 N1% = N5%: N2% = N4%: N3% = N6%: R3#(1,2) = R3#(1,2) + 2*ZY: R3#(1,1) =
     R3#(2,1): R3#(1,3) = S1(N4%): R3#(3,1) = R3#(3,1) + ZX: R3#(3,3) = S1(N6%)
5410 NEXT L2%
5420 ZY = - ZY: ZX = - ZX: ZZ = - ZZ
5430 NEXT L1%
5440 REM end plotting contours
5450 REM plot data points
5460 IF PT<1 THEN GOTO 5620
5470 IF CR>0 THEN GOTO 5790
5480 FOR L1% = 1 TO ND%
5490 IF P1#(L1%,1)< - .0001 OR P1#(L1%,1)>1.0001 OR P1#(L1%,2)< - .0001 OR
     P1#(L1%,2)>1.0001 THEN GOTO 5610
5500 PP1# = P1#(L1%,1): PP2# = P1#(L1%,2): PP3# = P1#(L1%,3): IF DP<1 THEN
     GOTO 5526
5510 YY = (PP2# - .5)*R8(1,1) + (PP1# - .5)*R8(1,2) + .5: PP1# = (PP1# - .5)*R8(1,1) -
     (PP2# - .5)*R8(1,2) + .5
5520 ZZ3 = (PP3# - .5)*R8(2,1) + (YY - .5)*R8(2,2) + .5: PP2# = (YY - .5)*R8(2,1) - (PP3#
     - .5)*R8(2,2) + .5: PP3# = ZZ3: IF SR<1 THEN GOTO 5526
5525 PPP# = PP1#: PP1# = (PP1# - .5)*.99875 - (PP3# - .5)*.05 + .5
5526 S3(1,1) = PP1#*SZ + SW - 2: S3(1,2) = ZW - PP2#*ZS - 2
5527 FOR L2% = 1 TO 3: FOR L3% = 1 TO 3: PSET(S3(1,1) + L2%,S3(1,2) + L3%),4:
     NEXT L3%: NEXT L2%
5528 IF SR<1 THEN GOTO 5610
5529 PP1# = (PPP# - .5)*.99875 + (PP3# - .5)*.05 + .5
5530 S3(1,1) = PP1#*SZ + SV - 2: S3(1,2) = ZW - PP2#*ZS - 2
5531 FOR L2% = 1 TO 3: FOR L3% = 1 TO 3: PSET(S3(1,1) + L2%,S3(1,2) + L3%),4:
     NEXT L3%: NEXT L2%
5610 NEXT L1%
5620 IF TT<1 THEN GOTO 5790
5630 REM draw triangles
5640 FOR L1% = 1 TO NT%
5650 IF P3%(L1%,1)>ND% THEN GOTO 5750
5651 FOR L2% = 1 TO 3: IF P1#(P3%(L1%,L2%),1)< - .01 OR
     P1#(P3%(L1%,L2%),1)>1.01 OR P1#(P3%(L1%,L2%),2)< - .01 OR
     P1#(P3%(L1%,L2%),2)>1.01 THEN GOTO 5750
5652 NEXT L2%
5654 PP1# = P1#(P3%(L1%,3),1): PP2# = P1#(P3%(L1%,3),2): PP3# =
     P1#(P3%(L1%,3),3): IF DP<1 THEN GOTO 5700
5656 YY = (PP2# - .5)*R8(1,1) + (PP1# - .5)*R8(1,2) + .5: PP1# = (PP1# - .5)*R8(1,1) -
     (PP2# - .5)*R8(1,2) + .5
5657 ZZ3 = (PP3# - .5)*R8(2,1) + (YY - .5)*R8(2,2) + .5: PP2# = (YY - .5)*R8(2,1) - (PP3#
     - .5)*R8(2,2) + .5: PP3# = ZZ3: IF SR<1 THEN GOTO 5700
5659 PPP# = PP1#: PP1# = (PP1# - .5)*.99875 - (PP3# - .5)*.05 + .5
5700 X1 = PP1#*SZ + SW: Y1 = ZW - PP2#*ZS: IF SR<1 THEN GOTO 5710
5709 PP1# = (PPP# - .5)*.99875 + (PP3# - .5)*.05 + .5: X3 = PP1#*SZ + SV: Y3 = ZW -
     PP2#*ZS
5710 FOR L2% = 1 TO 3: PP1# = P1#(P3%(L1%,L2%),1): PP2# =
     P1#(P3%(L1%,L2%),2): PP3# = P1#(P3%(L1%,L2%),3): IF DP<1 THEN GOTO 5715
5712 YY = (PP2# - .5)*R8(1,1) + (PP1# - .5)*R8(1,2) + .5: PP1# = (PP1# - .5)*R8(1,1) -
     (PP2# - .5)*R8(1,2) + .5
5713 ZZ3 = (PP3# - .5)*R8(2,1) + (YY - .5)*R8(2,2) + .5: PP2# = (YY - .5)*R8(2,1) - (PP3#
     - .5)*R8(2,2) + .5: PP3# = ZZ3: IF SR<1 THEN GOTO 5715
5714 PPP# = PP1#: PP1# = (PP1# - .5)*.99875 - (PP3# - .5)*.05 + .5
5715 X2 = PP1#*SZ + SW: Y2 = ZW - PP2#*ZS: IF SR<1 THEN GOTO 5718
5717 PP1# = (PPP# - .5)*.99875 + (PP3# - .5)*.05 + .5: X4 = PP1#*SZ + SV: Y4 = ZW -
     PP2#*ZS
5718 LINE(X1,Y1) - (X2,Y2),9: X1 = X2: Y1 = Y2: IF SR<1 THEN GOTO 5749
5719 LINE(X3,Y3) - (X4,Y4),9: X3 = X4: Y3 = Y4
5749 NEXT L2%
5750 NEXT L1%
```

Appendix

```
5790 SOUND 247,5: SOUND 153,5: SOUND 185,5: SOUND 123,10: TM$ = TIME$: IF
     SR>0 THEN GOTO 5802
5800 SM# = SM#/SA#: LOCATE 12,1: PRINT "Average": PRINT "height of": PRINT
     "surface": PRINT USING "#####.###";SM#
5802 LOCATE 19,1: PRINT " ALL": PRINT "DONE!": PRINT " At": PRINT TM$ 5804
     PRINT "Press": INPUT "ENTER ",PM$: END
5810 REM curvemaker
5820 ST# = R3#(1,3)
5830 IF ST#>R3#(2,3) THEN ST# = R3#(2,3)
5840 IF ST#>R3#(3,3) THEN ST# = R3#(3,3)
5850 IF ST#< = - 999 THEN RETURN
5860 SM# = SM# + AR#*((R3#(1,3) + R3#(2,3) + R3#(3,3))/3*VS + P9(2,3)): SA# = SA# +
     AR#: IF CR<1 THEN GOTO 6210
5870 PO(1,1) = - 1E37: PO(2,1) = PO(1,1): PO(1,2) = 1E37: PO(2,2) = PO(1,2): S3(1,1) =
     R3#(1,3) : S3(1,2) = R3#(2,3): S3(1,3) = R3#(3,3): IF DP<1 THEN GOTO 5910
5880 FOR L3% = 1 TO 3: R1#(L3%,1) = R3#(L3%,1): R1#(L3%,2) = R3#(L3%,2):
     R1#(L3%,3) = R3#(L3%,3)
5890 YY = (R3#(L3%,2) - .5)*R8(1,1) + (R3#(L3%,1) - .5)*R8(1,2): R3#(L3%,1) =
     (R3#(L3%,1) - .5)*R8(1,1) - (R3#(L3%,2) - .5)*R8(1,2) + .5
5900 R33# = (R3#(L3%,3) - .5)*R8(2,1) + YY*R8(2,2) + .5: R3#(L3%,2) = YY*R8(2,1) -
     (R3#(L3%,3) - .5)*R8(2,2) + .5: R3#(L3%,3) = R33#: NEXT L3%
5910 FOR L3% = 1 TO 3
5920 IF PO(1,1)<R3#(L3%,1) THEN PO(1,1) = R3#(L3%,1)
5930 IF PO(2,1)<R3#(L3%,2) THEN PO(2,1) = R3#(L3%,2)
5940 IF PO(1,2)>R3#(L3%,1) THEN PO(1,2) = R3#(L3%,1)
5950 IF PO(2,2)>R3#(L3%,2) THEN PO(2,2) = R3#(L3%,2)
5960 NEXT L3%
5970 X1 = SW + INT(PO(1,2)*SZ): XH = X1 - (X1\2)*2 - .5: X2 = SW + INT(PO(1,1)*SZ):
     Y1 = ZW - INT(PO(2,2)*ZS): Y2 = ZW - INT(PO(2,1)*ZS): YG = Y2 - (Y2\2)*2 - .5
5980 ZJ# = ((R3#(2,1) - R3#(1,1))*(R3#(3,2) - R3#(1,2))) - ((R3#(3,1) - R3#(1,1))*(R3#(2,2)
     - R3#(1,2)))
5990 FOR L3% = Y2 TO Y1
6000 YG = - YG: YY = (ZW - L3%)/ZS: XG = XH
6010 FOR L4% = X1 TO X2
6020 XG = - XG: XX = (L4% - SW)/SZ: ZI = 1
6030 FOR L5% = 1 TO 3
6040 L8% = 1: IF L8% = L5% THEN L8% = L8% + 1
6050 L9% = L8% + 1: IF L9% = L5% THEN L9% = L9% + 1
6060 S3(2,L5%) = (((R3#(L8%,1) - XX)*(R3#(L9%,2) - YY) - ((R3#(L9%,1) -
     XX)*(R3#(L8%,2) - YY))/ZJ#*ZI: IF S3(2,L5%)<0 THEN GOTO 6150
6070 ZI = - ZI: NEXT L5%
6080 ZC = S3(1,1)*S3(2,1) + S3(1,2)*S3(2,2) + S3(1,3)*S3(2,3): IF ZC<P8(1) THEN
     GOTO 6150
6090 FOR L5% = 2 TO NC%
6100 IF P8(L5%)<ZC THEN GOTO 6140
6110 LC% = INT(L5%/2 + XG/YG*.25 + .75): IF LC%<1 THEN GOTO 6150
6120 IF LC%>16 THEN LC% = 1
6130 PSET(L4%,L3%),SP%(LC%): GOTO 6150
6140 NEXT L5%
6150 NEXT L4%: NEXT L3%
6160 IF DP<1 THEN GOTO 6200
6170 FOR L3% = 1 TO 3
6180 R3#(L3%,1) = R1#(L3%,1): R3#(L3%,2) = R1#(L3%,2): R3#(L3%,3) = R1#(L3%,3)
6190 NEXT L3%
6200 RETURN
6210 L8% = 0: IF DP<2 THEN GOTO 6310
6220 L8% = 1
6230 REM rotate triangle for orthogonal profiles
6240 FOR L3% = 1 TO 3
6250 Z# = R3#(L3%,3): R3#(L3%,3) = 1 - R3#(L3%,L8%): R3#(L3%,L8%) = Z#
6260 NEXT L3%
6270 ST# = R3#(1,3)
6280 IF ST#>R3#(2,3) THEN ST# = R3#(2,3)
6290 IF ST#>R3#(3,3) THEN ST# = R3#(3,3)
6300 IF ST# < = - 999 THEN RETURN
6310 TP# = R3#(1,3)
```

```
6320 IF TP#<R3#(2,3) THEN TP# = R3#(2,3)
6330 IF TP#<R3#(3,3) THEN TP# = R3#(3,3)
6340 IF TP# = ST# THEN RETURN
6350 FOR L0% = 1 TO NC%
6360 REM slice the triangle
6370 IF P8(L0%)>TP# OR P8(L0%)<ST# THEN GOTO 6610
6380 L5% = 1: L4% = 0 6390 L4% = L4% + 1
6400 L7% = 1: IF L7% = L4% THEN L7% = L7% + 1
6410 L3% = L7% + 1: IF L3% = L4% THEN L3% = L3% + 1
6420 IF R3#(L7%,3) = R3#(L3%,3) THEN GOTO 6460
6430 F = (P8(L0%) - R3#(L7%,3))/(R3#(L3%,3) - R3#(L7%,3)): IF F<0 OR F>1 THEN
     GOTO 6460
6440 S3(L5%,1) = R3#(L7%,1) + (R3#(L3%,1) - R3#(L7%,1))*F: S3(L5%,2) = R3#(L7%,2)
     + (R3#(L3%,2) - R3#(L7%,2))*F
6450 L5% = L5% + 1
6460 IF L5%<3 THEN GOTO 6390
6470 S3(1,3) = P8(L0%): S3(2,3) = P8(L0%): IF DP<2 THEN GOTO 6500
6480 REM reverse rotate the intersection trace
6490 FOR L3% = 1 TO 2: Z# = S3(L3%,3): S3(L3%,3) = S3(L3%,L8%): S3(L3%,L8%) = 1
     - Z#: NEXT L3%
6500 IF DP = 0 THEN GOTO 6552
6510 REM apply perspective
6520 FOR L3% = 1 TO 2
6530 YY = (S3(L3%,2) - .5)*R8(1,1) + (S3(L3%,1) - .5)*R8(1,2): S3(L3%,1) = (S3(L3%,1) -
     .5)*R8(1,1) - (S3(L3%,2) - .5)*R8(1,2) + .5
6540 S33 = (S3(L3%,3) - .5)*R8(2,1) + YY*R8(2,2) + .5: S3(L3%,2) = YY*R8(2,1) -
     (S3(L3%,3) - .5)*R8(2,2) + .5: S3(L3%,3) = S33
6550 NEXT L3%
6552 IF SR<1 THEN GOTO 6580
6554 S3(1,1) = (S3(1,1) - .5)*.99875 - (S3(1,3) - .5)*.05 + .5: S3(1,3) = (S3(1,3) -
     .5)*.99875 + (S3(1,1) - .5)*.05 + .5: S3(2,1) = (S3(2,1) - .5)*.99875 - (S3(2,3) - .5)*.05 +
     .5: S3(2,3) = (S3(2,3) - .5)*.99875 + (S3(2,1) - .5)*.05 + .5
6560 REM draw a line from (S3(1,1),S3(1,2)) to (S3(2,1),S3(2,2)) 6580 LINE((SW +
     S3(1,1)*SZ),(ZW - S3(1,2)*ZS)) - ((SW + S3(2,1)*SZ),(ZW - S3(2,2)*ZS)),14: IF SR<1
     OR DP<1 THEN GOTO 6610
6590 S3(1,1) = (S3(1,1) - .5)*.995 + (S3(1,3) - .5)*.1 + .5: S3(2,1) = (S3(2,1) - .5)*.995 +
     (S3(2,3) - .5)*.1 + .5
6600 LINE((SV + S3(1,1)*SZ),(ZW - S3(1,2)*ZS)) - ((SV + S3(2,1)*SZ),(ZW -
     S3(2,2)*ZS)),14
6610 NEXT L0%
6620 IF L8%<1 THEN RETURN
6630 REM reverse rotate the triangle
6640 FOR L3% = 1 TO 3
6650 Z# = R3#(L3%,3): R3#(L3%,3) = R3#(L3%,L8%): R3#(L3%,L8%) = 1 - Z#
6660 NEXT L3%
6670 L8% = L8% + 1: IF L8%<3 THEN GOTO 6240 ELSE RETURN
6680 REM solve R3#(LL%,LL%)xB#(LL%) = R9#(LL%)
6690 FOR J2% = 1 TO LL%: P4%(J2%) = J2%: B#(J2%) = 0
6700 FOR J3% = 1 TO LL%: IF ABS(R3#(J2%,J3%))>= B#(J2%) THEN B#(J2%) =
     ABS(R3#(J2%,J3%))
6710 NEXT J3%: NEXT J2%: LM1% = LL% - 1: LP1% = LL% + 1
6720 FOR J2% = 1 TO LM1%: X3# = 0
6730 FOR J3% = J2% TO LL%: X4# = ABS(R3#(P4%(J3%),J2%))/B#(P4%(J3%)): IF
     X4#< = X3# THEN GOTO 6750
6740 X3# = X4#: K3% = J3%
6750 NEXT J3%: IF K3% = J2% GOTO 6770
6760 K2% = P4%(J2%): P4%(J2%) = P4%(K3%): P4%(K3%) = K2%
6770 X3# = R3#(P4%(J2%),J2%): K4% = J2% + 1
6780 FOR J3% = K4% TO LL%: X4# = - R3#(P4%(J3%),J2%)/X3#: R3#(P4%(J3%),J2%)
     = - X4#
6790 FOR J4% = K4% TO LL%: R3#(P4%(J3%),J4%) = R3#(P4%(J3%),J4%) +
     X4#*R3#(P4%(J2%),J4%): NEXT J4%: NEXT J3% : NEXT J2%
6800 B#(1) = R9#(P4%(1))
6810 FOR J1% = 2 TO LL%
6820 I8% = J1% - 1: X3# = 0
6830 FOR J2% = 1 TO I8%: X3# = X3# + R3#(P4%(J1%),J2%)*B#(J2%):NEXT J2%
```

Appendix

```
6840 B#(J1%) = R9#(P4%(J1%)) - X3#: NEXT J1%
6850 B#(LL%) = B#(LL%)/R3#(P4%(LL%),LL%)
6860 FOR J1% = 2 TO LL%: I8% = LP1% - J1%: I9 = I8% + 1: X3# = 0
6870 FOR J2% = I9 TO LL%: X3# = X3# + R3#(P4%(I8%),J2%)*B#(J2%): NEXT J2%
6880 B#(I8%) = (B#(I8%) - X3#)/R3#(P4%(I8%),I8%): NEXT J1%: RETURN


1000 REM SCRN\TRIANGLE.BAS TRIANGLE - BASED INTERPOLATION
     Copyright (c) 1988 D.F.Watson
1010 DIM P1#(55,6), P2#(237,3), P5#(55), P8(80), P9(2,3), S1(3038), S2(3), S4(55),
     PO(4,2)
1020 DIM R1#(3,3), R3#(55,55), R4#(3,2), R5#(3,3), R7#(3), R8(2,2), S3(2,3), BX(6,3),
     R9#(55), B#(55)
1030 DIM P3%(237,3), P4%(237), P6%(55), P7%(25,2), R2%(2,2), R6%(3,2), SP%(16)
1040 DATA 0,8,1,4,5,9,6,12,13,2,7,3,10,11,14,15: REM spectrum
1050 DATA 1,0,0,1,1,0,0,1,0,0,0,0,1,0,0,1,0,1, - 1, - 1,0,1,2,5, - 1,0,2,3, - 1,5,1E37,3,1
1060 DEF FNCV(RT#) = RT#^2*LOG(RT#): REM minimum curvature: REM DEF
     FNCV(RT#) = RT#^2*(LOG(RT#) - 1): REM biharmonic spline: REM DEF FNCV(RT#)
     = RT#: REM dummy function
1070 REM DEF FNCV(RT#) = NG + SI*(1 -  EXP( - RT#/RG)): NG = .25: SI = 1: RG = 1/3:
     REM exponential semivariogram
1080 REM DEF FNCV(RT#) = NG + SI*(1.5*RT#/RG - (RT#/RG)^3/2): NG = .25: SI = 1:
     RG = 1: REM spherical semivariogram
1090 P9(1,1) =  - 1E37: P9(1,2) =  - 1E37: P9(1,3) =  - 1E37: P9(2,1) = 1E37: P9(2,2) =
     1E37: P9(2,3) = 1E37
1100 SL = 0: SR = 0: SZ = 480: GR = 19: DP = 1: CL = 1: PT = 1: CR = 1: TT = 0: WD =
     640: HI = 350: LW = 5: AZ =  - 25: TL =  - 15: RD = 2: BI = 1.5: BJ = 7: IF DP = 0
     THEN SR = 0
1140 OPEN "A: \HILL.DAT" FOR INPUT AS #1
1150 REM OPEN "A: \JDAVIS.DAT" FOR INPUT AS #1
1170 FOR I1% = 1 TO 16: READ SP%(I1%): NEXT I1%
1180 INPUT#1,ND%,NC%: FOR I1% = 1 TO NC%: INPUT#1,P8(I1%): NEXT I1%: IF
     CR>0 AND NC%>31 THEN NC% = 31
1185 FOR I1% = 1 TO ND%: P4%(I1%) = I1%: NEXT I1%: FOR I1% = 1 TO ND%: I2% =
     INT(RND*ND% + .5): IF I2%<1 THEN I2% = 1
1187 IF I2%>ND% THEN I2% = ND%
1188 I3% = P4%(I1%): P4%(I1%) = P4%(I2%): P4%(I2%) = I3%: NEXT I1%
1190 INPUT#1,XS,YS,DT: I2% = 1
1220 I1% = P4%(I2%)
1222 INPUT#1,P1#(I1%,1),P1#(I1%,2),P1#(I1%,3): P1#(I1%,1) = (P1#(I1%,1) - XS)/DT:
     P1#(I1%,2) = (P1#(I1%,2) - YS)/DT: IF (P1#(I1%,1) - .5)^2 + (P1#(I1%,1) - .5)^2<RD
     THEN GOTO 1270
1230 ND% = ND% - 1: IF I2%>ND% THEN GOTO 1360
1260 GOTO 1222
1270 IF P9(1,1)<P1#(I1%,1) THEN P9(1,1) = P1#(I1%,1)
1280 IF P9(2,1)>P1#(I1%,1) THEN P9(2,1) = P1#(I1%,1)
1290 IF P9(1,2)<P1#(I1%,2) THEN P9(1,2) = P1#(I1%,2)
1300 IF P9(2,2)>P1#(I1%,2) THEN P9(2,2) = P1#(I1%,2)
1310 IF P9(1,3)<P1#(I1%,3) THEN P9(1,3) = P1#(I1%,3)
1320 IF P9(2,3)>P1#(I1%,3) THEN P9(2,3) = P1#(I1%,3)
1350 I2% = I2% + 1: IF I2%< = ND% THEN GOTO 1220
1360 IF ND%<4 THEN STOP
1370 XF = P9(1,1) - P9(2,1): YF = P9(1,2) - P9(2,2): VS = P9(1,3) - P9(2,3): R8(2,1) = AZ:
     R8(2,2) = TL
1380 FOR I1% = 1 TO 6: READ BX(I1%,1),BX(I1%,2),BX(I1%,3): NEXT I1%
1390 BN% = 1: IF DP<1 THEN GOTO 1450
1400 BN% = 2: BX(2,1) = 0: BX(2,3) = 1: IF R8(2,1)>0 THEN GOTO 1420
1410 BX(2,1) = 1: BX(3,1) = 1: BX(4,1) = 1: BX(5,1) = 0: BX(6,1) = 0
1420 IF DP<2 OR CR>0 THEN GOTO 1450
1430 NC1% = INT(SZ/10): NC% = NC1% + 1: U0 = .9999/NC1%: U1 = .0000511 - U0
1440 FOR I1% = 1 TO NC%: U1 = U1 + U0: P8(I1%) = U1: NEXT I1%: GOTO 1460
1450 FOR I1% = 1 TO NC%: P8(I1%) = (P8(I1%) - P9(2,3))/VS + .00001*(RND - .5):
     NEXT I1%
1460 FOR I1% = 1 TO ND%: P1#(I1%,1) = P1#(I1%,1) + .0001*(RND - .5): P1#(I1%,2) =
     P1#(I1%,2) + .0001*(RND - .5) : P1#(I1%,3) = (P1#(I1%,3) - P9(2,3))/VS: NEXT I1%
```

```
1470 FOR I1% = 1 TO 3: READ R1#(I1%,1), R1#(I1%,2), R1#(I1%,3), R6%(I1%,1),
     R6%(I1%,2)
1480 P1#(I1% + ND%,1) = R1#(I1%,1)*XF + P9(2,1): P1#(I1% + ND%,2) = R1#(I1%,2)*YF
     + P9(2,2): P1#(I1% + ND%,3) = 0: NEXT I1%
1490 IF DP<1 THEN GOTO 1550
1500 FOR I1% = 1 TO 2: R8(I1%,1) = COS(R8(2,I1%)/57.29578): R8(I1%,2) =
     SIN(R8(2,I1%)/57.29578): NEXT I1%
1510 FOR I1% = 2 TO 6
1520 YY = (BX(I1%,2) - .5)*R8(1,1) + (BX(I1%,1) - .5)*R8(1,2) + .5: BX(I1%,1) =
     (BX(I1%,1) - .5)*R8(1,1) - (BX(I1%,2) - .5)*R8(1,2) + .5
1530 BX3 = (BX(I1%,3) - .5)*R8(2,1) + (YY - .5)*R8(2,2) + .5: BX(I1%,2) = (YY -
     .5)*R8(2,1) - (BX(I1%,3) - .5)*R8(2,2) + .5: BX(I1%,3) = BX3: NEXT I1%
1540 REM end data input, begin gradient estimation
1550 SOUND 208,5: SOUND 165,5: SOUND 185,5: SOUND 123,5
1560 IF CL<1 THEN GOTO 3110
1570 PRINT " AT " TIME$ " BEGIN GRADIENT ESTIMATION  -  PLEASE WAIT"
1580 FOR I0% = 1 TO ND%
1590 FOR I1% = 1 TO 3: P3%(1,I1%) = ND% + I1%: P3%(2,I1%) = ND% + I1%:
     P2#(1,I1%) = R1#(I1%,3): P2#(2,I1%) = R1#(I1%,3) : NEXT I1%
1600 FOR I1% = 1 TO 237: P4%(I1%) = I1%: NEXT I1%: R2%(1,1) = 1: R2%(1,2) = 1:
     R2%(2,1) = 3 : R2%(2,2) = 4: K1% = I0%: K2% = I0%: K3% = ND% + 4: K4 =  - .5
1610 FOR I1% = K1% TO K2%
1620 IF I1% = K3% THEN GOTO 2000
1630 K5 = K4
1640 K6 = 1.5 + K5: IF K3%>ND% THEN K5 =  - K5
1650 U0 = 0: K7% = 0: L6% = K6 - 2
1660 FOR I2% = 1 TO R2%(1,K6) 1670 L6% = L6% + 2: IF P3%(L6%,1) = 0 THEN GOTO
     1670
1680 U1# = P2#(L6%,3) - (P1#(I1%,1) - P2#(L6%,1))^2: IF U1#<0 THEN GOTO 1840
1690 U1# = U1# - (P1#(I1%,2) - P2#(L6%,2))^2: IF U1#<0 THEN GOTO 1840
1700 R2%(2,K6) = R2%(2,K6) - 2: P4%(R2%(2,K6)) = L6%: U0 = U0 - 1
1710 FOR I3% = 1 TO 3
1720 L1% = 1: IF L1% = I3% THEN L1% = L1% + 1
1730 L2% = L1% + 1: IF L2% = I3% THEN L2% = L2% + 1
1740 IF K7%<1 THEN GOTO 1810
1750 K9% = K7%
1760 FOR I4% = 1 TO K9%
1770 IF P3%(L6%,L1%)<>P7%(I4%,1) THEN GOTO 1800
1780 IF P3%(L6%,L2%)<>P7%(I4%,2) THEN GOTO 1800
1790 P7%(I4%,1) = P7%(K9%,1): P7%(I4%,2) = P7%(K9%,2): K7% = K7% - 1: GOTO
     1820
1800 NEXT I4%
1810 K7% = K7% + 1: P7%(K7%,1) = P3%(L6%,L1%): P7%(K7%,2) = P3%(L6%,L2%)
1820 NEXT I3%
1830 P3%(L6%,1) = 0
1840 NEXT I2%
1850 IF K7%<1 THEN GOTO 2000
1860 REM make new triangles
1870 FOR I2% = 1 TO K7%
1880 IF K5>0 THEN GOTO 1900
1890 IF P7%(I2%,1)<>I0% AND P7%(I2%,2)<>I0% THEN GOTO 1970
1900 FOR I3% = 1 TO 2
1910 R3#(I3%,1) = P1#(P7%(I2%,I3%),1) - P1#(I1%,1): R3#(I3%,2) =
     P1#(P7%(I2%,I3%),2) - P1#(I1%,2)
1920 R3#(I3%,3) = R3#(I3%,1)*(P1#(P7%(I2%,I3%),1) + P1#(I1%,1))/2 + R3#(I3%,2)
     *(P1#(P7%(I2%,I3%),2) + P1#(I1%,2))/2
1930 NEXT I3%
1940 U1# = R3#(1,1)*R3#(2,2) - R3#(2,1)*R3#(1,2): P2#(P4%(R2%(2,K6)),1) =
     (R3#(1,3)*R3#(2,2) - R3#(2,3)*R3#(1,2))/U1#: P2#(P4%(R2%(2,K6)),2) =
     (R3#(1,1)*R3#(2,3) - R3#(2,1)*R3#(1,3))/U1#
1950 P2#(P4%(R2%(2,K6)),3) = (P1#(I1%,1) - P2#(P4%(R2%(2,K6)),1))^2 + (P1#(I1%,2) -
     P2#(P4%(R2%(2,K6)),2))^2
1960 P3%(P4%(R2%(2,K6)),1) = P7%(I2%,1): P3%(P4%(R2%(2,K6)),2) = P7%(I2%,2) :
     P3%(P4%(R2%(2,K6)),3) = I1%: R2%(2,K6) = R2%(2,K6) + 2: U0 = U0 + 1
1970 NEXT I2%
1980 R2%(1,K6) = R2%(1,K6) + U0: IF K5>0 THEN GOTO 2000
```

Appendix

```
1990 K5 = - K5: GOTO 1640
2000 NEXT I1%
2010 IF K2%>K1% GOTO 2040
2020 K1% = 1: K2% = ND%: K3% = I0%: GOTO 1610
2030 REM order triangles positively
2040 FOR I1% = 1 TO 2
2050 L8% = I1% - 2
2060 FOR I2% = 1 TO R2%(1,I1%)
2070 L8% = L8% + 2: IF P3%(L8%,1) = 0 THEN GOTO 2070
2080 IF P3%(L8%,1)>ND% THEN GOTO 2140
2090 FOR I3% = 1 TO 2
2100 R3#(1,I3%) = P1#(P3%(L8%,2),I3%) - P1#(P3%(L8%,1),I3%): R3#(2,I3%) =
    P1#(P3%(L8%,3),I3%) - P1#(P3%(L8%,1),I3%)
2110 NEXT I3%
2120 IF R3#(1,1)*R3#(2,2) - R3#(1,2)*R3#(2,1)> = 0 THEN GOTO 2140
2130 K8 = P3%(L8%,3): P3%(L8%,3) = P3%(L8%,2): P3%(L8%,2) = K8
2140 NEXT I2%
2150 NEXT I1%
2160 IF SL>0 THEN GOTO 2470
2170 REM calculate natural neighbor coordinates
2180 R5#(1,1) = P1#(I0%,1): R5#(1,2) = P1#(I0%,2): R5#(2,1) = R5#(1,1) + .0001 :
    R5#(2,2) = R5#(1,2): R5#(3,1) = R5#(1,1): R5#(3,2) = R5#(1,2) + .0001
2190 FOR I1% = 1 TO 3
2200 L4% = 0: L5% = 0: L6% = 0
2210 FOR I2% = 1 TO R2%(1,2) 2220 L6% = L6% + 2: IF P3%(L6%,1) = 0 THEN GOTO
    2220
2230 IF P3%(L6%,1)>ND% THEN GOTO 2410
2240 U1# = P2#(L6%,3) - (R5#(I1%,1) - P2#(L6%,1))^2: IF U1#<0 THEN GOTO 2410
2250 U1# = U1# - (R5#(I1%,2) - P2#(L6%,2))^2: IF U1#<0 THEN GOTO 2410
2260 FOR I3% = 1 TO 3
2270 FOR I4% = 1 TO 2
2280 R3#(I4%,1) = P1#(P3%(L6%,R6%(I3%,I4%)),1) - R5#(I1%,1): R3#(I4%,2) =
    P1#(P3%(L6%,R6%(I3%,I4%)),2) - R5#(I1%,2)
2290 R3#(I4%,3) = R3#(I4%,1)*(P1#(P3%(L6%,R6%(I3%,I4%)),1) + R5#(I1%,1))/2 +
    R3#(I4%,2) *(P1#(P3%(L6%,R6%(I3%,I4%)),2) + R5#(I1%,2))/2: NEXT I4%
2300 U1# = R3#(1,1)*R3#(2,2) - R3#(1,2)*R3#(2,1)
2310 R4#(R6%(I3%,2),1) = (R3#(1,3)*R3#(2,2) - R3#(2,3)*R3#(1,2))/U1#:
    R4#(R6%(I3%,2),2) = (R3#(1,1)*R3#(2,3) - R3#(2,1)*R3#(1,3))/U1#: NEXT I3%: L3%=0
2320 FOR I3% = 1 TO 3: R3#(3,I3%) = ((R4#(R6%(I3%,1),1) -
    P2#(L6%,1))*(R4#(R6%(I3%,2),2) - P2#(L6%,2)) - (R4#(R6%(I3%,2),1) -
    P2#(L6%,1))*(R4#(R6%(I3%,1),2) - P2#(L6%,2)))/2: IF R3#(3,I3%)>0 THEN L3% =
    L3% + 1
2330 NEXT I3%
2340 IF L3%>2 THEN L4% = 1
2350 FOR I3% = 1 TO 3: IF L5%<1 THEN GOTO 2390
2360 FOR I4% = 1 TO L5%: IF P3%(L6%,I3%)<>P6%(I4%) THEN GOTO 2380 2370
    P5#(I4%) = P5#(I4%) + R3#(3,I3%): GOTO 2400
2380 NEXT I4%
2390 L5% = L5% + 1: P6%(L5%) = P3%(L6%,I3%): P5#(L5%) = R3#(3,I3%)
2400 NEXT I3%
2410 NEXT I2%: IF L4%<1 THEN GOTO 2470
2420 U4 = 0
2430 FOR I2% = 1 TO L5%: U4 = U4 + P5#(I2%): NEXT I2%: R7#(I1%) = 0
2440 FOR I2% = 1 TO L5%: P5#(I2%) = P5#(I2%)/U4: R7#(I1%) = R7#(I1%) +
    P5#(I2%)*P1#(P6%(I2%),3): NEXT I2%: NEXT I1%
2450 P1#(I0%,6) = P1#(I0%,3) - R7#(1): P1#(I0%,4) = (R7#(1) - R7#(2))/.0001:
    P1#(I0%,5) = (R7#(1) - R7#(3))/.0001 : GOTO 3080
2460 REM cross product gradients
2470 IF SL>1 THEN GOTO 2610
2480 P1#(I0%,4) = 0: P1#(I0%,5) = 0: P1#(I0%,6) = 0: P7# = 0: L7% = - 1
2490 FOR I1% = 1 TO R2%(1,1)
2500 L7% = L7% + 2: IF P3%(L7%,1) = 0 THEN GOTO 2500
2510 IF P3%(L7%,1)>ND% THEN GOTO 2570
2520 FOR I2% = 1 TO 2: FOR I3% = 1 TO 3: R3#(I2%,I3%) = P1#(P3%(L7%,1),I3%) -
    P1#(P3%(L7%,I2% + 1),I3%): NEXT I3% : NEXT I2%
```

```
2530 R3#(3,1) = R3#(1,2)*R3#(2,3) - R3#(2,2)*R3#(1,3): R3#(3,2) = R3#(1,3)*R3#(2,1) -
     R3#(2,3)*R3#(1,1): R3#(3,3) = R3#(1,1)*R3#(2,2) - R3#(2,1)*R3#(1,2): U3 = 1
2540 IF R3#(3,3)<0 THEN U3 = - 1
2550 U2 = (R3#(3,1)^2 + R3#(3,2)^2 + R3#(3,3)^2)^.5: P7# = P7# + U2
2560 FOR I2% = 1 TO 3: P1#(I0%,I2% + 3) = P1#(I0%,I2% + 3) + R3#(3,I2%)*U3: NEXT
     I2%
2570 NEXT I1%
2580 U2 = P1#(I0%,4)^2 + P1#(I0%,5)^2 + P1#(I0%,6)^2: P7# = 1 - U2^.5/P7#
2590 P1#(I0%,4) = P1#(I0%,4)/P1#(I0%,6): P1#(I0%,5) = P1#(I0%,5)/P1#(I0%,6):
     P1#(I0%,6) = P7#
2600 GOTO 3080
2610 IF SL>2 THEN GOTO 2810 : REM least squares gradients
2620 P1#(I0%,4) = 0: P1#(I0%,5) = 0: P1#(I0%,6) = .2: L7% = - 1: P6%(1) = I0%: L5% =1
2630 R4#(1,1) = P1#(I0%,3): R4#(2,1) = P1#(I0%,1)*P1#(I0%,3): R4#(3,1) =
     P1#(I0%,2)*P1#(I0%,3)
2640 R3#(2,1) = P1#(I0%,1): R3#(2,2) = P1#(I0%,1)^2: R3#(3,1) = P1#(I0%,2) : R3#(3,2)
     = P1#(I0%,1)*P1#(I0%,2): R3#(3,3) = P1#(I0%,2)^2
2650 FOR I1% = 1 TO R2%(1,1)
2660 L7% = L7% + 2
2670 IF P3%(L7%,1) = 0 THEN GOTO 2660
2680 IF P3%(L7%,1)>ND% THEN GOTO 2760
2690 FOR I2% = 1 TO 3
2700 FOR I3% = 1 TO L5%
2710 IF P3%(L7%,I2%) = P6%(I3%) THEN GOTO 2750
2720 NEXT I3%
2730 L5% = L5% + 1: P6%(L5%) = P3%(L7%,I2%): R4#(1,1) = R4#(1,1) +
     P1#(P6%(L5%),3) : R4#(2,1) = R4#(2,1) + P1#(P6%(L5%),1)*P1#(P6%(L5%),3):
     R4#(3,1) = R4#(3,1) + P1#(P6%(L5%),2)*P1#(P6%(L5%),3)
2740 R3#(2,1) = R3#(2,1) + P1#(P6%(L5%),1): R3#(2,2) = R3#(2,2) +
     P1#(P6%(L5%),1)^2: R3#(3,1) = R3#(3,1)  + P1#(P6%(L5%),2): R3#(3,2) = R3#(3,2) +
     P1#(P6%(L5%),1)*P1#(P6%(L5%),2): R3#(3,3) = R3#(3,3) + P1#(P6%(L5%),2)^2
2750 NEXT I2%
2760 NEXT I1%
2770 U1# = L5%*(R3#(2,2)*R3#(3,3) - R3#(3,2)^2) - R3#(2,1)*(R3#(2,1)*R3#(3,3) -
     R3#(3,1)*R3#(3,2)) + R3#(3,1)*(R3#(2,1)*R3#(3,2) - R3#(3,1)*R3#(2,2)) 2780
     P1#(I0%,4) = - (L5%*(R4#(2,1)*R3#(3,3) - R3#(3,2)*R4#(3,1)) -
     R4#(1,1)*(R3#(2,1)*R3#(3,3) - R3#(3,1)*R3#(3,2)) + R3#(3,1)*(R3#(2,1)*R4#(3,1) -
     R3#(3,1)*R4#(2,1)))/U1#
2790 P1#(I0%,5) = - (L5%*(R3#(2,2)*R4#(3,1) - R3#(3,2)*R4#(2,1)) -
     R3#(2,1)*(R3#(2,1)*R4#(3,1) - R3#(3,1)*R4#(2,1)) + R4#(1,1)*(R3#(3,1)*R3#(3,2) -
     R3#(3,1)*R3#(2,2)))/U1#
2800 GOTO 3080 : REM spline gradients
2810 IF SL>3 THEN GOTO 3035
2815 P1#(I0%,4) = 0: P1#(I0%,5) = 0: P1#(I0%,6) = .2: L7% = - 1: L5% = 1: P6%(1) =
     I0%
2820 R5#(1,1) = P1#(I0%,1): R5#(1,2) = P1#(I0%,2): R5#(2,1) = R5#(1,1) + .0001 :
     R5#(2,2) = R5#(1,2): R5#(3,1) = R5#(1,1): R5#(3,2) = R5#(1,2) + .0001
2830 FOR I1% = 1 TO R2%(1,1)
2840 L7% = L7% + 2: IF P3%(L7%,1) = 0 THEN GOTO 2840
2850 IF P3%(L7%,1)>ND% THEN GOTO 2890
2860 FOR I2% = 1 TO 3: FOR I3% = 1 TO L5%: IF P3%(L7%,I2%) = P6%(I3%) THEN
     GOTO 2880
2870 NEXT I3%: L5% = L5% + 1: P6%(L5%) = P3%(L7%,I2%) 2880 NEXT I2% 2890
     NEXT I1%
2900 LL% = L5% + 3
2910 FOR I1% = 1 TO L5%
2920 R3#(I1%,1) = 1: R3#(LL%,I1% + 3) = 1: R3#(I1%,2) = P1#(P6%(I1%),1): R3#(L5% +
     1,I1% + 3) = R3#(I1%,2)
2930 R3#(I1%,3) = P1#(P6%(I1%),2): R3#(L5% + 2,I1% + 3) = R3#(I1%,3): R9#(I1%) =
     P1#(P6%(I1%),3): NEXT I1%
2940 FOR I1% = 1 TO 3: FOR I2% = 1 TO 3: R3#(I1% + L5%,I2%) = 0: NEXT I2%:
     R9#(I1% + L5%) = 0: NEXT I1%
2950 FOR I1% = 1 TO L5%: R3#(I1%,I1% + 3) = 0: IJ = I1% + 1: IF IJ>L5% THEN GOTO
     2970
```

Appendix

```
2960 FOR I2% = IJ TO L5%: RT# = ((R3#(I1%,2) - R3#(I2%,2))^2 + (R3#(I1%,3) -
     R3#(I2%,3))^2)^.5: R3#(I2%,I1% + 3) = FNCV(RT#): R3#(I1%,I2% + 3) = R3#(I2%,I1%
     + 3): NEXT I2%
2970 NEXT I1%: GOSUB 6690
2980 TR# = 0: TS# = 0
2990 FOR I1% = 4 TO LL%
3000 RT# = ((P1#(P6%(I1% - 3),1) - R5#(2,1))^2 + (P1#(P6%(I1% - 3),2) -
     R5#(2,2))^2)^.5: TR# = TR# + B#(I1%)*FNCV(RT#)
3010 RS# = ((P1#(P6%(I1% - 3),1) - R5#(3,1))^2 + (P1#(P6%(I1% - 3),2) -
     R5#(3,2))^2)^.5: TS# = TS# + B#(I1%)*FNCV(RS#) : NEXT I1%
3020 R5#(2,3) = B#(1) + B#(2)*R5#(2,1) + B#(3)*R5#(2,2) + TR#: P1#(I0%,4) =
     (P1#(I0%,3) - R5#(2,3))/.0001
3030 R5#(3,3) = B#(1) + B#(2)*R5#(3,1) + B#(3)*R5#(3,2) + TS#: P1#(I0%,5) =
     (P1#(I0%,3) - R5#(3,3))/.0001: GOTO 3080
3033 REM hyperboloid gradients
3035 R3#(5,1) = 1: FOR I1% = 1 TO 4: R9%(I1%) = 0: FOR I2% = 1 TO 4: R3#(I1%,I2%) =
     0: NEXT I2%: NEXT I1%: LL% = 4
3036 P1#(I0%,4) = 0: P1#(I0%,5) = 0: P1#(I0%,6) = .2: L7% = - 1: L5% = 0: FOR I1% = 1
     TO R2%(1,1)
3037 L7% = L7% + 2: IF P3%(L7%,1) = 0 THEN GOTO 3037
3038 IF P3%(L7%,1)>ND% THEN GOTO 3070
3039 FOR I2% = 1 TO 3: IF L5%<1 THEN GOTO 3044
3041 FOR I3% = 1 TO L5%: IF P3%(L7%,I2%) = P6%(I3%) THEN GOTO 3060
3042 NEXT I3%
3044 L5% = L5% + 1: P6%(L5%) = P3%(L7%,I2%)
3050 R3#(5,2) = P1#(P6%(L5%),1): R3#(5,3) = P1#(P6%(L5%),2): R3#(5,4) =
     P1#(P6%(L5%),1)*P1#(P6%(L5%),2)
3055 FOR I3% = 1 TO 4: FOR I4% = 1 TO 4: R3#(I3%,I4%) = R3#(I3%,I4%) +
     R3#(5,I3%)*R3#(5,I4%): NEXT I4% : R9%(I3%) = R9%(I3%) +
     R3#(5,I3%)*P1#(P6%(L5%),3): NEXT I3%
3060 NEXT I2%
3070 NEXT I1%: GOSUB 6690: P1#(I0%,4) =  - B#(2) - B#(4)*P1#(I0%,2): P1#(I0%,5) =  -
     B#(3) - B#(4)*P1#(I0%,1): P1#(I0%,6) = .2
3080 NEXT I0%
3100 REM natural neighbor sort
3110 FOR I0% = 1 TO 3: P3%(1,I0%) = ND% + I0%: P2%(1,I0%) = R1#(I0%,3): NEXT I0%
3120 FOR I0% = 1 TO 237: P4%(I0%) = I0%: NEXT I0%: NT% = 1: L0% = 2
3130 FOR I0% = 1 TO ND%: L3% = 0
3140 FOR I1% = 1 TO NT%
3150 U1# = P2#(I1%,3) - (P1#(I0%,1) - P2#(I1%,1))^2: IF U1#<0 THEN GOTO 3300
3160 U1# = U1# - (P1#(I0%,2) - P2#(I1%,2))^2: IF U1#<0 THEN GOTO 3300
3170 L0% = L0% - 1: P4%(L0%) = I1%
3180 FOR I2% = 1 TO 3
3190 L1% = 1: IF L1% = I2% THEN L1% = L1% + 1
3200 L2% = L1% + 1: IF L2% = I2% THEN L2% = L2% + 1
3210 IF L3%<1 THEN GOTO 3280
3220 L5% = L3%
3230 FOR I3% = 1 TO L5%
3240 IF P3%(I1%,L1%)<>P7%(I3%,1) THEN GOTO 3270
3250 IF P3%(I1%,L2%)<>P7%(I3%,2) THEN GOTO 3270
3260 P7%(I3%,1) = P7%(L5%,1): P7%(I3%,2) = P7%(L5%,2): L3% = L3% - 1: GOTO
     3290
3270 NEXT I3%
3280 L3% = L3% + 1: P7%(L3%,1) = P3%(I1%,L1%): P7%(L3%,2) = P3%(I1%,L2%) 3290
     NEXT I2%
3300 NEXT I1%
3310 FOR I1% = 1 TO L3%
3320 FOR I2% = 1 TO 2
3330 R3#(I2%,1) = P1#(P7%(I1%,I2%),1) - P1#(I0%,1): R3#(I2%,2) =
     P1#(P7%(I1%,I2%),2) - P1#(I0%,2): R3#(I2%,3) = R3#(I2%,1)*(P1#(P7%(I1%,I2%),1) +
     P1#(I0%,1))/2
3340 R3#(I2%,3) = R3#(I2%,3) + R3#(I2%,2)*(P1#(P7%(I1%,I2%),2) + P1#(I0%,2))/2
3350 NEXT I2%
3360 U1# = R3#(1,1)*R3#(2,2) - R3#(2,1)*R3#(1,2): P2#(P4%(L0%),1) =
     (R3#(1,3)*R3#(2,2) - R3#(2,3)*R3#(1,2))/U1#: P2#(P4%(L0%),2) = (R3#(1,1)*R3#(2,3)
     - R3#(2,1)*R3#(1,3))/U1#
```

```
3370 P2#(P4%(L0%),3) = (P1#(I0%,1) - P2#(P4%(L0%),1))^2 + (P1#(I0%,2) -
     P2#(P4%(L0%),2))^2
3380 P3%(P4%(L0%),1) = P7%(I1%,1): P3%(P4%(L0%),2) = P7%(I1%,2):
     P3%(P4%(L0%),3) = I0%: L0% = L0% + 1
3390 NEXT I1%
3400 NT% = NT% + 2: NEXT I0%
3410 FOR I0% = 1 TO NT%: IF P3%(I0%,1)>ND% THEN GOTO 3460
3420 FOR I1% = 1 TO 2
3430 R3#(1,I1%) = P1#(P3%(I0%,2),I1%) - P1#(P3%(I0%,1),I1%): R3#(2,I1%) =
     P1#(P3%(I0%,3),I1%) - P1#(P3%(I0%,1),I1%): NEXT I1%
3440 IF R3#(1,1)*R3#(2,2) - R3#(1,2)*R3#(2,1)> = 0 THEN GOTO 3460
3450 L4% = P3%(I0%,3): P3%(I0%,3) = P3%(I0%,2): P3%(I0%,2) = L4% 3460 NEXT I0%
3820 REM make triangular grid
3830 PRINT " AT " TIME$ " BEGIN SURFACE INTERPOLATION - PLEASE WAIT":
     SOUND 123,5: SOUND 185,5: SOUND 208,5 : SOUND 165,5
3840 I0% = GR + 1: IS = 1: I2% = 2*I0% + 1: I3% = 1 + INT(GR/3^.5): I4% = I3%*I2% +
     I0%: SX = 1/GR: SY = 1/(2*I3%): S2(3) = .000001
3850 WX# = S2(3): S2(1) = 1.00001: AA# = SX*SY*.5*DT*DT: SM# = 0: SA# = 0: WY# =
     1: U8 = 1
3860 FOR L1% = 1 TO I4%
3870 S1(L1%) = - 999: WXO# = WX#: WYO# = WY#
4320 REM .....calculate barycentric coordinates of WX#,WY# **********
4340 K5% = 0
4350 FOR L2% = 1 TO NT%
4360 IF P3%(L2%,1)>ND% THEN GOTO 4460
4370 U1# = P2#(L2%,3) - (WX# - P2#(L2%,1))^2: IF U1#<0 THEN GOTO 4460
4380 U1# = U1# - (WY# - P2#(L2%,2))^2: IF U1#<0 THEN GOTO 4460
4390 K3% = 0
4400 FOR L3% = 1 TO 3
4410 P5#(L3%) = ((P1#(P3%(L2%,R6%(L3%,2)),1) -
     P1#(P3%(L2%,R6%(L3%,1)),1))*(WY# - P1#(P3%(L2%,R6%(L3%,1)),2)) - (WX# -
     P1#(P3%(L2%,R6%(L3%,1)),1))*(P1#(P3%(L2%,R6%(L3%,2)),2) -
     P1#(P3%(L2%,R6%(L3%,1)),2)))
4420 P6%(L3%) = P3%(L2%,(6 - R6%(L3%,1) - R6%(L3%,2))): IF P5#(L3%)>0 THEN
     K3% = K3% + 1
4430 NEXT L3%
4440 IF K3%<3 THEN GOTO 4460
4450 K5% = 3: GOTO 4560
4460 NEXT L2%
4470 IF K5%<1 GOTO 4940
4560 S1(L1%) = 0: U4 = 0
4570 FOR L2% = 1 TO K5%: U4 = U4 + P5#(L2%): NEXT L2%
4580 FOR L2% = 1 TO K5%: P5#(L2%) = P5#(L2%)/U4: S1(L1%) = S1(L1%) +
     P5#(L2%)*P1#(P6%(L2%),3): NEXT L2%
4590 IF CL<1 THEN GOTO 4940
4620 FOR L2% = 1 TO K5%: S4(L2%) = 0
4630 IF P5#(L2%)<.00001 OR P5#(L2%)>1 THEN GOTO 4690
4640 IF ABS(P1#(P6%(L2%),6))<.00001 THEN GOTO 4690
4650 RS# = ABS(P1#(P6%(L2%),6)) + BI: RT# = RS#*BJ: RB# = 1/RT#: BD# =
     P5#(L2%)^RT#: BB# = BD#*2: IF BD#>.5 THEN BB# = (1 - BD#)*2
4660 BB# = BB#^RS#/2: IF BD#>.5 THEN BB# = 1 - BB#
4670 HP# = BB#^RB#
4680 S4(L2%) = ((P1#(P6%(L2%),4)*P1#(P6%(L2%),1) +
     P1#(P6%(L2%),5)*P1#(P6%(L2%),2) + P1#(P6%(L2%),3) - P1#(P6%(L2%),4)*WX# -
     P1#(P6%(L2%),5)*WY#) - S1(L1%))*HP#
4690 NEXT L2%
4700 FOR L2% = 1 TO K5%: S1(L1%) = S1(L1%) + S4(L2%): NEXT L2%
4940 IS = IS + SX*U8: IF IS>0 THEN GOTO 4960
4950 WXO# = WXO# + SX/2: IS = 1.001 - SX
4960 WX# = WXO# + SX*U8: WY# = WYO#: IF U8*WX#<S2(2 - U8) THEN GOTO 4980
4970 U8 = - U8: WX# = S2(2 + U8): WY# = WY# - SY: IS = U8
4980 NEXT L1%
5060 REM end interpolation to triangular grid and begin display construction
5080 SCREEN 9: COLOR 2,0: CLS: SOUND 208,5: SOUND 165,5: SOUND 185,5:
     SOUND 123,5: IF DP>0 THEN SZ = INT(SZ*2/3)
5090 IF SR>0 AND SZ>240 THEN SZ = 240
```

Appendix

```
5100 ZS = INT(SZ*.71): SW = (WD - SZ)\2: ZW = (HI - LW)\2 + ZS\2: IF DP>0 THEN ZW
     = ZW - LW - 10
5110 IF SR>0 AND CR<1 AND DP>0 THEN SW = (WD - 2*SZ)\3 - 10
5120 SV = SW + SZ + 60: IF SR<1 THEN GOTO 5130
5125 FOR L1% = 2 TO 6: BX(L1%,1) = (BX(L1%,1) - .5)*.99875 - (BX(L1%,3) - .5)*.05 +
     .5: BX(L1%,3) = (BX(L1%,3) - .5)*.99875 + (BX(L1%,1) - .5)*.05 + .5: NEXT L1%
5130 FOR L1% = BN% + 1 TO BN% + 4: X1 = SW + BX(L1% - 1,1)*SZ: Y1 = ZW -
     BX(L1% - 1,2)*ZS: X2 = SW + BX(L1%,1)*SZ: Y2 = ZW - BX(L1%,2)*ZS: LINE(X1,Y1)
     - (X2,Y2),14: NEXT L1%: IF SR<1 OR CR>0 THEN GOTO 5210
5140 FOR L1% = 1 TO 6: BX(L1%,1) = (BX(L1%,1) - .5)*.995 + (BX(L1%,3) - .5)*.1 + .5:
     NEXT L1%
5150 FOR L1% = BN% + 1 TO BN% + 4: X1 = SV + BX(L1% - 1,1)*SZ: Y1 = ZW -
     BX(L1% - 1,2)*ZS: X2 = SV + BX(L1%,1)*SZ: Y2 = ZW - BX(L1%,2)*ZS: LINE(X1,Y1) -
     (X2,Y2),14: NEXT L1%
5200 REM plot contours
5210 SW = SW + 1: SZ = SZ - 2: ZS = ZS - 2: ZW = ZW - 1: ZX = SX: ZY = SY: ZZ = .5:
     R3#(2,2) = 1 + 2*SY: R3#(2,1) = 0
5220 FOR L1% = 1 TO I3%
5230 R3#(1,2) = R3#(2,2) - 2*SY - 2*ZY: R3#(1,1) = R3#(2,1): N1% = (2*L1% - .5 +
     ZZ)*I0% + L1% + .5/ZZ: R3#(1,3) = S1(N1%): R3#(2,2) = R3#(2,2) - 2*SY - ZY:
     R3#(2,1) = .5 - ZZ: N2% = (2*L1% - .5 + ZZ)*I0% + L1%
5240 R3#(2,3) = S1(N2%): R3#(3,2) = R3#(2,2): R3#(3,1) = R3#(2,1) + ZX/2: N3% = N2%
     - .5/ZZ: R3#(3,3) = S1(N3%): AR# = AA#/2
5250 GOSUB 5820
5260 R3#(1,2) = R3#(1,2) + 2*ZY: N2% = (2*L1% - .5/ZZ - .5 - ZZ)*I0% + L1%: R3#(1,3) =
     S1(N2%)
5270 GOSUB 5820
5280 AR# = AA#
5290 FOR L2% = 1 TO GR
5300 R3#(2,2) = R3#(1,2): R3#(2,1) = R3#(1,1) + ZX: N4% = N2% + .5/ZZ: R3#(2,3) =
     S1(N4%)
5310 GOSUB 5820
5320 R3#(1,2) = R3#(3,2): R3#(1,1) = R3#(3,1) + ZX
5330 IF L2% = GR THEN R3#(1,1) = R3#(1,1) - ZX/2
5340 N6% = N3% - .5/ZZ: R3#(1,3) = S1(N6%): IF L2% = GR THEN AR# = AA#/2
5350 GOSUB 5820
5360 R3#(2,2) = R3#(2,2) - 2*ZY: N5% = N1% + .5/ZZ: R3#(2,3) = S1(N5%)
5370 GOSUB 5820
5380 R3#(1,2) = R3#(2,2): R3#(1,1) = R3#(2,1) - ZX: R3#(1,3) = S1(N1%): IF L2% = GR
     THEN AR# = AA#
5390 GOSUB 5820
5400 N1% = N5%: N2% = N4%: N3% = N6%: R3#(1,2) = R3#(1,2) + 2*ZY: R3#(1,1) =
     R3#(2,1) : R3#(1,3) = S1(N4%): R3#(3,1) = R3#(3,1) + ZX: R3#(3,3) = S1(N6%)
5410 NEXT L2%
5420 ZY = -ZY: ZX = -ZX: ZZ = -ZZ
5430 NEXT L1%
5440 REM end plotting contours
5450 REM plot data points
5460 IF PT<1 THEN GOTO 5620
5470 IF CR>0 THEN GOTO 5790
5480 FOR L1% = 1 TO ND%
5490 IF P1#(L1%,1)< - .0001 OR P1#(L1%,1)>1.0001 OR P1#(L1%,2)< - .0001 OR
     P1#(L1%,2)>1.0001 THEN GOTO 5610
5500 PP1# = P1#(L1%,1): PP2# = P1#(L1%,2): PP3# = P1#(L1%,3): IF DP<1 THEN
     GOTO 5526
5510 YY = (PP2# - .5)*R8(1,1) + (PP1# - .5)*R8(1,2) + .5: PP1# = (PP1# - .5)*R8(1,1) -
     (PP2# - .5)*R8(1,2) + .5
5520 ZZ3 = (PP3# - .5)*R8(2,1) + (YY - .5)*R8(2,2) + .5: PP2# = (YY - .5)*R8(2,1) - (PP3#
     - .5)*R8(2,2) + .5: PP3# = ZZ3: IF SR<1 THEN GOTO 5526
5525 PPP# = PP1#: PP1# = (PP1# - .5)*.99875 - (PP3# - .5)*.05 + .5
5526 S3(1,1) = PP1#*SZ + SW - 2: S3(1,2) = ZW - PP2#*ZS - 2
5527 FOR L2% = 1 TO 3: FOR L3% = 1 TO 3: PSET(S3(1,1) + L2%,S3(1,2) + L3%),4:
     NEXT L3%: NEXT L2%
5528 IF SR<1 THEN GOTO 5610
5529 PP1# = (PPP# - .5)*.99875 + (PP3# - .5)*.05 + .5
5530 S3(1,1) = PP1#*SZ + SV - 2: S3(1,2) = ZW - PP2#*ZS - 2
```

```
5531 FOR L2% = 1 TO 3: FOR L3% = 1 TO 3: PSET(S3(1,1) + L2%,S3(1,2) + L3%),4:
     NEXT L3%: NEXT L2%
5610 NEXT L1%
5620 IF TT<1 THEN GOTO 5790
5630 REM draw triangles
5640 FOR L1% = 1 TO NT%
5650 IF P3%(L1%,1)>ND% THEN GOTO 5750
5651 FOR L2% = 1 TO 3: IF P1#(P3%(L1%,L2%),1)< - .01 OR
     P1#(P3%(L1%,L2%),1)>1.01 OR P1#(P3%(L1%,L2%),2)< - .01 OR
     P1#(P3%(L1%,L2%),2)>1.01 THEN GOTO 5750
5652 NEXT L2%
5654 PP1# = P1#(P3%(L1%,3),1): PP2# = P1#(P3%(L1%,3),2): PP3# =
     P1#(P3%(L1%,3),3): IF DP<1 THEN GOTO 5700
5656 YY = (PP2# - .5)*R8(1,1) + (PP1# - .5)*R8(1,2) + .5: PP1# = (PP1# - .5)*R8(1,1) -
     (PP2# - .5)*R8(1,2) + .5
5657 ZZ3 = (PP3# - .5)*R8(2,1) + (YY - .5)*R8(2,2) + .5: PP2# = (YY - .5)*R8(2,1) - (PP3#
     - .5)*R8(2,2) + .5: PP3# = ZZ3: IF SR<1 THEN GOTO 5700
5659 PPP# = PP1#: PP1# = (PP1# - .5)*.99875 - (PP3# - .5)*.05 + .5
5700 X1 = PP1#*SZ + SW: Y1 = ZW - PP2#*ZS: IF SR<1 THEN GOTO 5710
5709 PP1# = (PPP# - .5)*.99875 + (PP3# - .5)*.05 + .5: X3 = PP1#*SZ + SV: Y3 = ZW -
     PP2#*ZS
5710 FOR L2% = 1 TO 3: PP1# = P1#(P3%(L1%,L2%),1): PP2# =
     P1#(P3%(L1%,L2%),2): PP3# = P1#(P3%(L1%,L2%),3): IF DP<1 THEN GOTO 5715
5712 YY = (PP2# - .5)*R8(1,1) + (PP1# - .5)*R8(1,2) + .5: PP1# = (PP1# - .5)*R8(1,1) -
     (PP2# - .5)*R8(1,2) + .5
5713 ZZ3 = (PP3# - .5)*R8(2,1) + (YY - .5)*R8(2,2) + .5: PP2# = (YY - .5)*R8(2,1) - (PP3#
     - .5)*R8(2,2) + .5: PP3# = ZZ3: IF SR<1 THEN GOTO 5715
5714 PPP# = PP1#: PP1# = (PP1# - .5)*.99875 - (PP3# - .5)*.05 + .5
5715 X2 = PP1#*SZ + SW: Y2 = ZW - PP2#*ZS: IF SR<1 THEN GOTO 5718
5717 PP1# = (PPP# - .5)*.99875 + (PP3# - .5)*.05 + .5: X4 = PP1#*SZ + SV: Y4 = ZW -
     PP2#*ZS
5718 LINE(X1,Y1) - (X2,Y2),9: X1 = X2: Y1 = Y2: IF SR<1 THEN GOTO 5749
5719 LINE(X3,Y3) - (X4,Y4),9: X3 = X4: Y3 = Y4
5749 NEXT L2%
5750 NEXT L1%
5790 SOUND 247,5: SOUND 153,5: SOUND 185,5: SOUND 123,10: TM$ = TIME$: IF
     SR>0 THEN GOTO 5802
5800 SM# = SM#/SA#: LOCATE 12,1: PRINT "Average": PRINT "height of": PRINT
     "surface": PRINT USING "#####.###"; SM#
5802 LOCATE 19,1: PRINT " ALL": PRINT "DONE!": PRINT " At": PRINT TM$
5804 PRINT "Press": INPUT "ENTER ",PM$: END
5810 REM curvemaker
5820 ST# = R3#(1,3)
5830 IF ST#>R3#(2,3) THEN ST# = R3#(2,3) 5840 IF ST#>R3#(3,3) THEN ST# =
     R3#(3,3)
5850 IF ST#< = - 999 THEN RETURN
5860 SM# = SM# + AR#*((R3#(1,3) + R3#(2,3) + R3#(3,3))/3*VS + P9(2,3)): SA# = SA# +
     AR#: IF CR<1 THEN GOTO 6210
5870 PO(1,1) = - 1E37: PO(2,1) = PO(1,1): PO(1,2) = 1E37: PO(2,2) = PO(1,2): S3(1,1) =
     R3#(1,3) : S3(1,2) = R3#(2,3): S3(1,3) = R3#(3,3): IF DP<1 THEN GOTO 5910
5880 FOR L3% = 1 TO 3: R1#(L3%,1) = R3#(L3%,1): R1#(L3%,2) = R3#(L3%,2):
     R1#(L3%,3) = R3#(L3%,3)
5890 YY = (R3#(L3%,2) - .5)*R8(1,1) + (R3#(L3%,1) - .5)*R8(1,2): R3#(L3%,1) =
     (R3#(L3%,1) - .5)*R8(1,1) - (R3#(L3%,2) - .5)*R8(1,2) + .5
5900 R33# = (R3#(L3%,3) - .5)*R8(2,1) + YY*R8(2,2) + .5: R3#(L3%,2) = YY*R8(2,1) -
     (R3#(L3%,3) - .5)*R8(2,2) + .5: R3#(L3%,3) = R33#: NEXT L3%
5910 FOR L3% = 1 TO 3
5920 IF PO(1,1)<R3#(L3%,1) THEN PO(1,1) = R3#(L3%,1)
5930 IF PO(2,1)<R3#(L3%,2) THEN PO(2,1) = R3#(L3%,2)
5940 IF PO(1,2)>R3#(L3%,1) THEN PO(1,2) = R3#(L3%,1)
5950 IF PO(2,2)>R3#(L3%,2) THEN PO(2,2) = R3#(L3%,2)
5960 NEXT L3%
5970 X1 = SW + INT(PO(1,2)*SZ): XH = X1 - (X1\2)*2 - .5: X2 = SW + INT(PO(1,1)*SZ):
     Y1 = ZW - INT(PO(2,2)*ZS): Y2 = ZW - INT(PO(2,1)*ZS): YG = Y2 - (Y2\2)*2 - .5
5980 ZJ# = ((R3#(2,1) - R3#(1,1))*(R3#(3,2) - R3#(1,2))) - ((R3#(3,1) - R3#(1,1))*(R3#(2,2)
     - R3#(1,2)))
```

251

Appendix

```
5990 FOR L3% = Y2 TO Y1
6000 YG = - YG: YY = (ZW - L3%)/ZS: XG = XH
6010 FOR L4% = X1 TO X2
6020 XG = - XG: XX = (L4% - SW)/SZ: ZI = 1
6030 FOR L5% = 1 TO 3
6040 L8% = 1: IF L8% = L5% THEN L8% = L8% + 1
6050 L9% = L8% + 1: IF L9% = L5% THEN L9% = L9% + 1
6060 S3(2,L5%) = (((R3#(L8%,1) - XX)*(R3#(L9%,2) - YY)) - ((R3#(L9%,1) -
     XX)*(R3#(L8%,2) - YY)))/ZJ#*ZI: IF S3(2,L5%)<0 THEN GOTO 6150
6070 ZI = - ZI: NEXT L5%
6080 ZC = S3(1,1)*S3(2,1) + S3(1,2)*S3(2,2) + S3(1,3)*S3(2,3): IF ZC<P8(1) THEN
     GOTO 6150
6090 FOR L5% = 2 TO NC%
6100 IF P8(L5%)<ZC THEN GOTO 6140
6110 LC% = INT(L5%/2 + XG/YG*.25 + .75): IF LC%<1 THEN GOTO 6150
6120 IF LC%>16 THEN LC% = 1
6130 PSET(L4%,L3%),SP%(LC%): GOTO 6150
6140 NEXT L5%
6150 NEXT L4%: NEXT L3%
6160 IF DP<1 THEN GOTO 6200
6170 FOR L3% = 1 TO 3
6180 R3#(L3%,1) = R1#(L3%,1): R3#(L3%,2) = R1#(L3%,2): R3#(L3%,3) = R1#(L3%,3)
6190 NEXT L3%
6200 RETURN
6210 L8% = 0: IF DP<2 THEN GOTO 6310
6220 L8% = 1
6230 REM rotate triangle for orthogonal profiles
6240 FOR L3% = 1 TO 3
6250 Z# = R3#(L3%,3): R3#(L3%,3) = 1 - R3#(L3%,L8%): R3#(L3%,L8%) = Z#
6260 NEXT L3%
6270 ST# = R3#(1,3)
6280 IF ST#>R3#(2,3) THEN ST# = R3#(2,3)
6290 IF ST#>R3#(3,3) THEN ST# = R3#(3,3)
6300 IF ST# < = - 999 THEN RETURN
6310 TP# = R3#(1,3)
6320 IF TP#<R3#(2,3) THEN TP# = R3#(2,3)
6330 IF TP#<R3#(3,3) THEN TP# = R3#(3,3)
6340 IF TP# = ST# THEN RETURN
6350 FOR L0% = 1 TO NC%
6360 REM slice the triangle
6370 IF P8(L0%)>TP# OR P8(L0%)<ST# THEN GOTO 6610
6380 L5% = 1: L4% = 0
6390 L4% = L4% + 1
6400 L7% = 1: IF L7% = L4% THEN L7% = L7% + 1
6410 L3% = L7% + 1: IF L3% = L4% THEN L3% = L3% + 1
6420 IF R3#(L7%,3) = R3#(L3%,3) THEN GOTO 6460
6430 F = (P8(L0%) - R3#(L7%,3))/(R3#(L3%,3) - R3#(L7%,3)): IF F<0 OR F>1 THEN
     GOTO 6460
6440 S3(L5%,1) = R3#(L7%,1) + (R3#(L3%,1) - R3#(L7%,1))*F: S3(L5%,2) = R3#(L7%,2)
     + (R3#(L3%,2) - R3#(L7%,2))*F
6450 L5% = L5% + 1
6460 IF L5%<3 THEN GOTO 6390
6470 S3(1,3) = P8(L0%): S3(2,3) = P8(L0%): IF DP<2 THEN GOTO 6500
6480 REM reverse rotate the intersection trace
6490 FOR L3% = 1 TO 2: Z# = S3(L3%,3): S3(L3%,3) = S3(L3%,L8%): S3(L3%,L8%) = 1
     - Z#: NEXT L3%
6500 IF DP = 0 THEN GOTO 6552
6510 REM apply perspective
6520 FOR L3% = 1 TO 2
6530 YY = (S3(L3%,2) - .5)*R8(1,1) + (S3(L3%,1) - .5)*R8(1,2): S3(L3%,1) = (S3(L3%,1) -
     .5)*R8(1,1) - (S3(L3%,2) - .5)*R8(1,2) + .5
6540 S33 = (S3(L3%,3) - .5)*R8(2,1) + YY*R8(2,2) + .5: S3(L3%,2) = YY*R8(2,1) -
     (S3(L3%,3) - .5)*R8(2,2) + .5: S3(L3%,3) = S33
6550 NEXT L3%
6552 IF SR<1 THEN GOTO 6580
```

```
6554 S3(1,1) = (S3(1,1) - .5)*.99875 - (S3(1,3) - .5)*.05 + .5: S3(1,3) = (S3(1,3) -
     .5)*.99875 + (S3(1,1) - .5)*.05 + .5: S3(2,1) = (S3(2,1) - .5)*.99875 - (S3(2,3) - .5)*.05 +
     .5: S3(2,3) = (S3(2,3) - .5)*.99875 + (S3(2,1) - .5)*.05 + .5
6560 REM draw a line from (S3(1,1),S3(1,2)) to (S3(2,1),S3(2,2))
6580 LINE((SW + S3(1,1)*SZ),(ZW - S3(1,2)*ZS)) - ((SW + S3(2,1)*SZ),(ZW -
     S3(2,2)*ZS)),14: IF SR<1 OR DP<1 THEN GOTO 6610
6590 S3(1,1) = (S3(1,1) - .5)*.995 + (S3(1,3) - .5)*.1 + .5: S3(2,1) = (S3(2,1) - .5)*.995 +
     (S3(2,3) - .5)*.1 + .5
6600 LINE((SV + S3(1,1)*SZ),(ZW - S3(1,2)*ZS)) - ((SV + S3(2,1)*SZ),(ZW -
     S3(2,2)*ZS)),14
6610 NEXT L0%
6620 IF L8%<1 THEN RETURN
6630 REM reverse rotate the triangle
6640 FOR L3% = 1 TO 3
6650 Z# = R3#(L3%,3): R3#(L3%,3) = R3#(L3%,L8%): R3#(L3%,L8%) = 1 - Z#
6660 NEXT L3%
6670 L8% = L8% + 1: IF L8%<3 THEN GOTO
6240 ELSE RETURN
6680 REM solve R3#(LL%,LL%)xB#(LL%) = R9#(LL%)
6690 FOR J2% = 1 TO LL%: P4%(J2%) = J2%: B#(J2%) = 0
6700 FOR J3% = 1 TO LL%: IF ABS(R3#(J2%,J3%))> = B#(J2%) THEN B#(J2%) =
     ABS(R3#(J2%,J3%))
6710 NEXT J3%: NEXT J2%: LM1% = LL% - 1: LP1% = LL% + 1
6720 FOR J2% = 1 TO LM1%: X3# = 0
6730 FOR J3% = J2% TO LL%: X4# = ABS(R3#(P4%(J3%),J2%))/B#(P4%(J3%)): IF
     X4#< = X3# THEN GOTO 6750
6740 X3# = X4#: K3% = J3%
6750 NEXT J3%: IF K3% = J2% GOTO 6770
6760 K2% = P4%(J2%): P4%(J2%) = P4%(K3%): P4%(K3%) = K2%
6770 X3# = R3#(P4%(J2%),J2%): K4 = J2% + 1
6780 FOR J3% = K4 TO LL%: X4# =  - R3#(P4%(J3%),J2%)/X3#: R3#(P4%(J3%),J2%) =
      - X4#
6790 FOR J4% = K4 TO LL%: R3#(P4%(J3%),J4%) = R3#(P4%(J3%),J4%) +
     X4#*R3#(P4%(J2%),J4%): NEXT J4%: NEXT J3% : NEXT J2%
6800 B#(1) = R9#(P4%(1))
6810 FOR J1% = 2 TO LL%
6820 I8% = J1% - 1: X3# = 0
6830 FOR J2% = 1 TO I8%: X3# = X3# + R3#(P4%(J1%),J2%)*B#(J2%): NEXT J2%
6840 B#(J1%) = R9#(P4%(J1%)) - X3#: NEXT J1%
6850 B#(LL%) = B#(LL%)/R3#(P4%(LL%),LL%)
6860 FOR J1% = 2 TO LL%: I8% = LP1% - J1%: I9 = I8% + 1: X3# = 0
6870 FOR J2% = I9 TO LL%: X3# = X3# + R3#(P4%(I8%),J2%)*B#(J2%): NEXT J2%
6880 B#(I8%) = (B#(I8%) - X3#)/R3#(P4%(I8%),I8%): NEXT J1%: RETURN


1000 REM SCRN\DISTANCE.BAS DISTANCE - BASED INTERPOLATION
     Copyright (c) 1988 D.F.Watson
1010 DIM P1#(63,6), P2#(237,3), P5#(60), P8(80), P9(2,3), S1(7742), S2(3), S4(60),
     PO(4,2)
1020 DIM R1#(3,3), R3#(25,25), R4#(3,2), R5#(3,3), R7#(3), R8(2,2), S3(2,3), BX(6,3),
     R9#(25), B#(25)
1030 DIM P3%(237,3), P4%(237), P6%(60), P7%(25,2), R2%(2,2), R6%(3,2), SP%(16)
1040 DATA 0,8,1,4,5,9,6,12,13,2,7,3,10,11,14,15: REM spectrum
1050 DATA 1,0,0,1,1,0,0,1,0,0,0,0,1,0,0,1,0,1, - 1, - 1,0,1,2,5, - 1,0,2,3, - 1,5,1E37,3,1
1060 DEF FNCV(RT#) = RT#^2*LOG(RT#): REM minimum curvature
1090 P9(1,1) =  - 1E37: P9(1,2) =  - 1E37: P9(1,3) =  - 1E37: P9(2,1) = 1E37: P9(2,2) =
     1E37: P9(2,3) = 1E37
1100 SL = 0: SR = 0: SZ = 480: GR = 19: DP = 0: CL = 1: PT = 1: CR = 0: PW = 3: WD =
     640: HI = 350: LW = 5: AZ =  - 25: TL =  - 15: RD = 2: BI = 1.5: BJ = 7: IF DP = 0
     THEN SR = 0
1140 OPEN "A: \HILL.DAT" FOR INPUT AS #1
1150 REM OPEN "A: \JDAVIS.DAT" FOR INPUT AS #1
1170 FOR I1% = 1 TO 16: READ SP%(I1%): NEXT I1% 1180 INPUT#1,ND%,NC%: FOR
     I1% = 1 TO NC%: INPUT#1,P8(I1%): NEXT I1%: IF CR>0 AND NC%>31 THEN NC%
     = 31
```

Appendix

```
1185 FOR I1% = 1 TO ND%: P4%(I1%) = I1%: NEXT I1%: FOR I1% = 1 TO ND%: I2% =
     INT(RND*ND% + .5): IF I2%<1 THEN I2% = 1
1187 IF I2%>ND% THEN I2% = ND%
1188 I3% = P4%(I1%): P4%(I1%) = P4%(I2%): P4%(I2%) = I3%: NEXT I1%
1190 INPUT#1,XS,YS,DT: I2% = 1
1220 I1% = P4%(I2%)
1222 INPUT#1,P1#(I1%,1),P1#(I1%,2),P1#(I1%,3): P1#(I1%,1) = (P1#(I1%,1) - XS)/DT:
     P1#(I1%,2) = (P1#(I1%,2) - YS)/DT: IF (P1#(I1%,1) - .5)^2 + (P1#(I1%,2) - .5)^2<RD
     THEN GOTO 1270
1230 ND% = ND% - 1: IF I2%>ND% THEN GOTO 1360
1260 GOTO 1222
1270 IF P9(1,1)<P1#(I1%,1) THEN P9(1,1) = P1#(I1%,1)
1280 IF P9(2,1)>P1#(I1%,1) THEN P9(2,1) = P1#(I1%,1)
1290 IF P9(1,2)<P1#(I1%,2) THEN P9(1,2) = P1#(I1%,2)
1300 IF P9(2,2)>P1#(I1%,2) THEN P9(2,2) = P1#(I1%,2)
1310 IF P9(1,3)<P1#(I1%,3) THEN P9(1,3) = P1#(I1%,3)
1320 IF P9(2,3)>P1#(I1%,3) THEN P9(2,3) = P1#(I1%,3)
1350 I2% = I2% + 1: IF I2%< = ND% THEN GOTO 1220
1360 IF ND%<4 THEN STOP
1370 XF = P9(1,1) - P9(2,1): YF = P9(1,2) - P9(2,2): VS = P9(1,3) - P9(2,3): R8(2,1) = AZ:
     R8(2,2) = TL
1380 FOR I1% = 1 TO 6: READ BX(I1%,1),BX(I1%,2),BX(I1%,3): NEXT I1%
1390 BN% = 1: IF DP<1 THEN GOTO 1450
1400 BN% = 2: BX(2,1) = 0: BX(2,3) = 1: IF R8(2,1)>0 THEN GOTO 1420
1410 BX(2,1) = 1: BX(3,1) = 1: BX(4,1) = 1: BX(5,1) = 0: BX(6,1) = 0
1420 IF DP<2 OR CR>0 THEN GOTO 1450
1430 NC1% = INT(SZ/10): NC% = NC1% + 1: U0 = .9999/NC1%: U1 = .0000511 - U0
1440 FOR I1% = 1 TO NC%: U1 = U1 + U0: P8(I1%) = U1: NEXT I1%: GOTO 1460
1450 FOR I1% = 1 TO NC%: P8(I1%) = (P8(I1%) - P9(2,3))/VS + .00001*(RND - .5):
     NEXT I1%
1460 FOR I1% = 1 TO ND%: P1#(I1%,1) = P1#(I1%,1) + .0001*(RND - .5): P1#(I1%,2) =
     P1#(I1%,2) + .0001*(RND - .5) : P1#(I1%,3) = (P1#(I1%,3) - P9(2,3))/VS: NEXT I1%
1470 FOR I1% = 1 TO 3: READ R1#(I1%,1), R1#(I1%,2), R1#(I1%,3), R6%(I1%,1),
     R6%(I1%,2)
1480 P1#(I1% + ND%,1) = R1#(I1%,1)*XF + P9(2,1): P1#(I1% + ND%,2) = R1#(I1%,2)*YF
     + P9(2,2): P1#(I1% + ND%,3) = 0: NEXT I1%
1490 IF DP<1 THEN GOTO 1550
1500 FOR I1% = 1 TO 2: R8(I1%,1) = COS(R8(2,I1%)/57.29578): R8(I1%,2) =
     SIN(R8(2,I1%)/57.29578): NEXT I1%
1510 FOR I1% = 2 TO 6
1520 YY = (BX(I1%,2) - .5)*R8(1,1) + (BX(I1%,1) - .5)*R8(1,2) + .5: BX(I1%,1) =
     (BX(I1%,1) - .5)*R8(1,1) - (BX(I1%,2) - .5)*R8(1,2) + .5
1530 BX3 = (BX(I1%,3) - .5)*R8(2,1) + (YY - .5)*R8(2,2) + .5: BX(I1%,2) = (YY -
     .5)*R8(2,1) - (BX(I1%,3) - .5)*R8(2,2) + .5: BX(I1%,3) = BX3: NEXT I1%
1540 REM end data input, begin gradient estimation
1550 SOUND 208,5: SOUND 165,5: SOUND 185,5: SOUND 123,5
1560 IF CL<1 THEN GOTO 3030
1570 PRINT " AT " TIME$ " BEGIN GRADIENT ESTIMATION  -  PLEASE WAIT"
1580 FOR I0% = 1 TO ND%
1590 FOR I1% = 1 TO 3: P3%(1,I1%) = ND% + I1%: P3%(2,I1%) = ND% + I1%:
     P2#(1,I1%) = R1#(I1%,3): P2#(2,I1%) = R1#(I1%,3) : NEXT I1%
1600 FOR I1% = 1 TO 237: P4%(I1%) = I1%: NEXT I1%: R2%(1,1) = 1: R2%(1,2) = 1:
     R2%(2,1) = 3: R2%(2,2) = 4 : K1% = I0%: K2% = I0%: K3% = ND% + 4: K4% = - .5
1610 FOR I1% = K1% TO K2%
1620 IF I1% = K3% THEN GOTO 2000
1630 K5 = K4
1640 K6 = 1.5 + K5: IF K3%>ND% THEN K5 =  - K5
1650 U0 = 0: K7% = 0: L6% = K6 - 2
1660 FOR I2% = 1 TO R2%(1,K6)
1670 L6% = L6% + 2: IF P3%(L6%,1) = 0 THEN GOTO 1670
1680 U1# = P2#(L6%,3) - (P1#(I1%,1) - P2#(L6%,1))^2: IF U1#<0 THEN GOTO 1840
1690 U1# = U1# - (P1#(I1%,2) - P2#(L6%,2))^2: IF U1#<0 THEN GOTO 1840
1700 R2%(2,K6) = R2%(2,K6) - 2: P4%(R2%(2,K6)) = L6%: U0 = U0 - 1
1710 FOR I3% = 1 TO 3
1720 L1% = 1: IF L1% = I3% THEN L1% = L1% + 1
1730 L2% = L1% + 1: IF L2% = I3% THEN L2% = L2% + 1
```

254

```
1740 IF K7%<1 THEN GOTO 1810
1750 K9% = K7%
1760 FOR I4% = 1 TO K9%
1770 IF P3%(L6%,L1%)<>P7%(I4%,1) THEN GOTO 1800
1780 IF P3%(L6%,L2%)<>P7%(I4%,2) THEN GOTO 1800
1790 P7%(I4%,1) = P7%(K9%,1): P7%(I4%,2) = P7%(K9%,2): K7% = K7% - 1: GOTO
     1820
1800 NEXT I4%
1810 K7% = K7% + 1: P7%(K7%,1) = P3%(L6%,L1%): P7%(K7%,2) = P3%(L6%,L2%)
1820 NEXT I3%
1830 P3%(L6%,1) = 0
1840 NEXT I2%
1850 IF K7%<1 THEN GOTO 2000
1860 REM make new triangles
1870 FOR I2% = 1 TO K7%
1880 IF K5>0 THEN GOTO 1900
1890 IF P7%(I2%,1)<>I0% AND P7%(I2%,2)<>I0% THEN GOTO 1970
1900 FOR I3% = 1 TO 2
1910 R3#(I3%,1) = P1#(P7%(I2%,I3%),1) - P1#(I1%,1): R3#(I3%,2) =
     P1#(P7%(I2%,I3%),2) - P1#(I1%,2)
1920 R3#(I3%,3) = R3#(I3%,1)*(P1#(P7%(I2%,I3%),1) + P1#(I1%,1))/2 + R3#(I3%,2)
     *(P1#(P7%(I2%,I3%),2) + P1#(I1%,2))/2
1930 NEXT I3%
1940 U1# = R3#(1,1)*R3#(2,2) - R3#(2,1)*R3#(1,2): P2#(P4%(R2%(2,K6)),1) =
     (R3#(1,3)*R3#(2,2) - R3#(2,3)*R3#(1,2))/U1#: P2#(P4%(R2%(2,K6)),2) =
     (R3#(1,1)*R3#(2,3) - R3#(2,1)*R3#(1,3))/U1#
1950 P2#(P4%(R2%(2,K6)),3) = (P1#(I1%,1) - P2#(P4%(R2%(2,K6)),1))^2 + (P1#(I1%,2) -
     P2#(P4%(R2%(2,K6)),2))^2
1960 P3%(P4%(R2%(2,K6)),1) = P7%(I2%,1): P3%(P4%(R2%(2,K6)),2) = P7%(I2%,2) :
     P3%(P4%(R2%(2,K6)),3) = I1%: R2%(2,K6) = R2%(2,K6) + 2: U0 = U0 + 1
1970 NEXT I2%
1980 R2%(1,K6) = R2%(1,K6) + U0: IF K5>0 THEN GOTO 2000
1990 K5 = - K5: GOTO 1640
2000 NEXT I1%
2010 IF K2%>K1% GOTO 2040
2020 K1% = 1: K2% = ND%: K3% = I0%: GOTO 1610
2030 REM order triangles positively
2040 FOR I1% = 1 TO 2
2050 L8% = I1% - 2
2060 FOR I2% = 1 TO R2%(1,I1%)
2070 L8% = L8% + 2: IF P3%(L8%,1) = 0 THEN GOTO 2070
2080 IF P3%(L8%,1)>ND% THEN GOTO 2140
2090 FOR I3% = 1 TO 2
2100 R3#(1,I3%) = P1#(P3%(L8%,2),I3%) - P1#(P3%(L8%,1),I3%): R3#(2,I3%) =
     P1#(P3%(L8%,3),I3%) - P1#(P3%(L8%,1),I3%)
2110 NEXT I3%
2120 IF R3#(1,1)*R3#(2,2) - R3#(1,2)*R3#(2,1)> = 0 THEN GOTO 2140
2130 K8 = P3%(L8%,3): P3%(L8%,3) = P3%(L8%,2): P3%(L8%,2) = K8
2140 NEXT I2%
2150 NEXT I1%
2160 IF SL>0 THEN GOTO 2470
2170 REM calculate natural neighbor coordinates
2180 R5#(1,1) = P1#(I0%,1): R5#(1,2) = P1#(I0%,2): R5#(2,1) = R5#(1,1) + .0001 :
     R5#(2,2) = R5#(1,2): R5#(3,1) = R5#(1,1): R5#(3,2) = R5#(1,2) + .0001
2190 FOR I1% = 1 TO 3
2200 L4% = 0: L5% = 0: L6% = 0
2210 FOR I2% = 1 TO R2%(1,2)
2220 L6% = L6% + 2: IF P3%(L6%,1) = 0 THEN GOTO 2220
2230 IF P3%(L6%,1)>ND% THEN GOTO 2410
2240 U1# = P2#(L6%,3) - (R5#(I1%,1) - P2#(L6%,1))^2: IF U1#<0 THEN GOTO 2410
2250 U1# = U1# - (R5#(I1%,2) - P2#(L6%,2))^2: IF U1#<0 THEN GOTO 2410
2260 FOR I3% = 1 TO 3
2270 FOR I4% = 1 TO 2
2280 R3#(I4%,1) = P1#(P3%(L6%,R6%(I3%,I4%)),1) - R5#(I1%,1): R3#(I4%,2) =
     P1#(P3%(L6%,R6%(I3%,I4%)),2) - R5#(I1%,2)
```

Appendix

```
2290 R3#(I4%,3) = R3#(I4%,1)*(P1#(P3%(L6%,R6%(I3%,I4%)),1) + R5#(I1%,1))/2 +
     R3#(I4%,2) *(P1#(P3%(L6%,R6%(I3%,I4%)),2) + R5#(I1%,2))/2: NEXT I4%
2300 U1# = R3#(1,1)*R3#(2,2) - R3#(1,2)*R3#(2,1)
2310 R4#(R6%(I3%,2),1) = (R3#(1,3)*R3#(2,2) - R3#(2,3)*R3#(1,2))/U1#:
     R4#(R6%(I3%,2),2) = (R3#(1,1)*R3#(2,3) - R3#(2,1)*R3#(1,3))/U1#: NEXT I3%: L3%=0
2320 FOR I3% = 1 TO 3: R3#(3,I3%) = ((R4#(R6%(I3%,1),1) -
     P2#(L6%,1))*(R4#(R6%(I3%,2),2) - P2#(L6%,2)) - (R4#(R6%(I3%,2),1) -
     P2#(L6%,1))*(R4#(R6%(I3%,1),2) - P2#(L6%,2)))/2: IF R3#(3,I3%)>0 THEN L3% =
     L3% + 1
2330 NEXT I3%
2340 IF L3%>2 THEN L4% = 1
2350 FOR I3% = 1 TO 3: IF L5%<1 THEN GOTO 2390
2360 FOR I4% = 1 TO L5%: IF P3%(L6%,I3%)<>P6%(I4%) THEN GOTO 2380 2370
     P5#(I4%) = P5#(I4%) + R3#(3,I3%): GOTO 2400
2380 NEXT I4%
2390 L5% = L5% + 1: P6%(L5%) = P3%(L6%,I3%): P5#(L5%) = R3#(3,I3%)
2400 NEXT I3%
2410 NEXT I2%: IF L4%<1 THEN GOTO 2470
2420 U4 = 0
2430 FOR I2% = 1 TO L5%: U4 = U4 + P5#(I2%): NEXT I2%: R7#(I1%) = 0
2440 FOR I2% = 1 TO L5%: P5#(I2%) = P5#(I2%)/U4: R7#(I1%) = R7#(I1%) +
     P5#(I2%)*P1#(P6%(I2%),3): NEXT I2%: NEXT I1%
2450 P1#(I0%,6) = P1#(I0%,3) - R7#(1): P1#(I0%,4) = (R7#(1) - R7#(2))/.0001 :
     P1#(I0%,5) = (R7#(1) - R7#(3))/.0001 : GOTO 3080
2460 REM cross product gradients
2470 IF SL>1 THEN GOTO 2610
2480 P1#(I0%,4) = 0: P1#(I0%,5) = 0: P1#(I0%,6) = 0: P7# = 0: L7% = - 1
2490 FOR I1% = 1 TO R2%(1,1)
2500 L7% = L7% + 2: IF P3%(L7%,1) = 0 THEN GOTO 2500
2510 IF P3%(L7%,1)>ND% THEN GOTO 2570
2520 FOR I2% = 1 TO 2: FOR I3% = 1 TO 3: R3#(I2%,I3%) = P1#(P3%(L7%,1),I3%) -
     P1#(P3%(L7%,I2% + 1),I3%): NEXT I3% : NEXT I2%
2530 R3#(3,1) = R3#(1,2)*R3#(2,3) - R3#(2,2)*R3#(1,3): R3#(3,2) = R3#(1,3)*R3#(2,1) -
     R3#(2,3)*R3#(1,1): R3#(3,3) = R3#(1,1)*R3#(2,2) - R3#(2,1)*R3#(1,2): U3 = 1
2540 IF R3#(3,3)<0 THEN U3 = - 1
2550 U2 = (R3#(3,1)^2 + R3#(3,2)^2 + R3#(3,3)^2)^.5: P7# = P7# + U2
2560 FOR I2% = 1 TO 3: P1#(I0%,I2% + 3) = P1#(I0%,I2% + 3) + R3#(3,I2%)*U3: NEXT
     I2%
2570 NEXT I1%
2580 U2 = P1#(I0%,4)^2 + P1#(I0%,5)^2 + P1#(I0%,6)^2: P7# = 1 - U2^.5/P7#
2590 P1#(I0%,4) = P1#(I0%,4)/P1#(I0%,6): P1#(I0%,5) = P1#(I0%,5)/P1#(I0%,6):
     P1#(I0%,6) = P7#
2600 GOTO 3080
2610 IF SL>2 THEN GOTO 2810 : REM least squares gradients
2620 P1#(I0%,4) = 0: P1#(I0%,5) = 0: P1#(I0%,6) = .2: L7% = - 1: P6%(1) = I0%: L5% =1
2630 R4#(1,1) = P1#(I0%,3): R4#(2,1) = P1#(I0%,1)*P1#(I0%,3): R4#(3,1) =
     P1#(I0%,2)*P1#(I0%,3)
2640 R3#(2,1) = P1#(I0%,1): R3#(2,2) = P1#(I0%,1)^2: R3#(3,1) = P1#(I0%,2) : R3#(3,2)
     = P1#(I0%,1)*P1#(I0%,2): R3#(3,3) = P1#(I0%,2)^2
2650 FOR I1% = 1 TO R2%(1,1)
2660 L7% = L7% + 2
2670 IF P3%(L7%,1) = 0 THEN GOTO 2660
2680 IF P3%(L7%,1)>ND% THEN GOTO 2760
2690 FOR I2% = 1 TO 3
2700 FOR I3% = 1 TO L5%
2710 IF P3%(L7%,I2%) = P6%(I3%) THEN GOTO 2750
2720 NEXT I3%
2730 L5% = L5% + 1: P6%(L5%) = P3%(L7%,I2%): R4#(1,1) = R4#(1,1) +
     P1#(P6%(L5%),3) : R4#(2,1) = R4#(2,1) + P1#(P6%(L5%),1)*P1#(P6%(L5%),3):
     R4#(3,1) = R4#(3,1) + P1#(P6%(L5%),2)*P1#(P6%(L5%),3)
2740 R3#(2,1) = R3#(2,1) + P1#(P6%(L5%),1): R3#(2,2) = R3#(2,2) +
     P1#(P6%(L5%),1)^2: R3#(3,1) = R3#(3,1)  + P1#(P6%(L5%),2): R3#(3,2) = R3#(3,2) +
     P1#(P6%(L5%),1)*P1#(P6%(L5%),2): R3#(3,3) = R3#(3,3) + P1#(P6%(L5%),2)^2
2750 NEXT I2%
2760 NEXT I1%
```

```
2770 U1# = L5%*(R3#(2,2)*R3#(3,3) - R3#(3,2)^2) - R3#(2,1)*(R3#(2,1)*R3#(3,3) -
     R3#(3,1)*R3#(3,2)) + R3#(3,1)*(R3#(2,1)*R3#(3,2) - R3#(3,1)*R3#(2,2))
2780 P1#(I0%,4) = - (L5%*(R4#(2,1)*R3#(3,3) - R3#(3,2)*R4#(3,1)) -
     R4#(1,1)*(R3#(2,1)*R3#(3,3) - R3#(3,1)*R3#(3,2)) + R3#(3,1)*(R3#(2,1)*R4#(3,1) -
     R3#(3,1)*R4#(2,1)))/U1#
2790 P1#(I0%,5) = - (L5%*(R3#(2,2)*R4#(3,1) - R3#(3,2)*R4#(2,1)) -
     R3#(2,1)*(R3#(2,1)*R4#(3,1) - R3#(3,1)*R4#(2,1)) + R4#(1,1)*(R3#(2,1)*R3#(3,2) -
     R3#(3,1)*R3#(2,2)))/U1#
2800 GOTO 3080 : REM spline gradients
2810 IF SL>3 THEN GOTO 3035
2815 P1#(I0%,4) = 0: P1#(I0%,5) = 0: P1#(I0%,6) = .2: L7% = - 1: L5% = 1: P6%(1) =
     I0%
2820 R5#(1,1) = P1#(I0%,1): R5#(1,2) = P1#(I0%,2): R5#(2,1) = R5#(1,1) + .0001 :
     R5#(2,2) = R5#(1,2): R5#(3,1) = R5#(1,1): R5#(3,2) = R5#(1,2) + .0001
2830 FOR I1% = 1 TO R2%(1,1)
2840 L7% = L7% + 2: IF P3%(L7%,1) = 0 THEN GOTO 2840
2850 IF P3%(L7%,1)>ND% THEN GOTO 2890
2860 FOR I2% = 1 TO 3: FOR I3% = 1 TO L5%: IF P3%(L7%,I2%) = P6%(I3%) THEN
     GOTO 2880
2870 NEXT I3%: L5% = L5% + 1: P6%(L5%) = P3%(L7%,I2%)
2880 NEXT I2%
2890 NEXT I1%
2900 LL% = L5% + 3
2910 FOR I1% = 1 TO L5%
2920 R3#(I1%,1) = 1: R3#(LL%,I1% + 3) = 1: R3#(I1%,2) = P1#(P6%(I1%),1): R3#(L5% +
     1,I1% + 3) = R3#(I1%,2)
2930 R3#(I1%,3) = P1#(P6%(I1%),2): R3#(L5% + 2,I1% + 3) = R3#(I1%,3): R9#(I1%) =
     P1#(P6%(I1%),3): NEXT I1%
2940 FOR I1% = 1 TO 3: FOR I2% = 1 TO 3: R3#(I1% + L5%,I2%) = 0: NEXT I2%:
     R9#(I1% + L5%) = 0: NEXT I1%
2950 FOR I1% = 1 TO L5%: R3#(I1%,I1% + 3) = 0: IJ = I1% + 1: IF IJ>L5% THEN GOTO
     2970
2960 FOR I2% = IJ TO L5%: RT# = ((R3#(I1%,2) - R3#(I2%,2))^2 + (R3#(I1%,3) -
     R3#(I2%,3))^2)^.5: R3#(I2%,I1% + 3) = FNCV(RT#): R3#(I1%,I2% + 3) = R3#(I2%,I1%
     + 3): NEXT I2%
2970 NEXT I1%: GOSUB 6690
2980 TR# = 0: TS# = 0
2990 FOR I1% = 4 TO LL%
3000 RT# = ((P1#(P6%(I1% - 3),1) - R5#(2,1))^2 + (P1#(P6%(I1% - 3),2) -
     R5#(2,2))^2)^.5: TR# = TR# + B#(I1%)*FNCV(RT#)
3010 RS# = ((P1#(P6%(I1% - 3),1) - R5#(3,1))^2 + (P1#(P6%(I1% - 3),2) -
     R5#(3,2))^2)^.5: TS# = TS# + B#(I1%)*FNCV(RS#) : NEXT I1%
3020 R5#(2,3) = B#(1) + B#(2)*R5#(2,1) + B#(3)*R5#(2,2) + TR#: P1#(I0%,4) =
     (P1#(I0%,3) - R5#(2,3))/.0001
3030 R5#(3,3) = B#(1) + B#(2)*R5#(3,1) + B#(3)*R5#(3,2) + TS#: P1#(I0%,5) =
     (P1#(I0%,3) - R5#(3,3))/.0001: GOTO 3080
3033 REM hyperboloid gradients
3035 R3#(5,1) = 1: FOR I1% = 1 TO 4: R9#(I1%) = 0: FOR I2% = 1 TO 4: R3#(I1%,I2%) =
     0: NEXT I2%: NEXT I1%: LL% = 4
3036 P1#(I0%,4) = 0: P1#(I0%,5) = 0: P1#(I0%,6) = .2: L7% = - 1: L5% = 0: FOR I1% = 1
     TO R2%(1,1)
3037 L7% = L7% + 2: IF P3%(L7%,1) = 0 THEN GOTO 3037
3038 IF P3%(L7%,1)>ND% THEN GOTO 3070
3039 FOR I2% = 1 TO 3: IF L5%<1 THEN GOTO 3044
3041 FOR I3% = 1 TO L5%: IF P3%(L7%,I2%) = P6%(I3%) THEN GOTO 3060
3042 NEXT I3%
3044 L5% = L5% + 1: P6%(L5%) = P3%(L7%,I2%)
3050 R3#(5,2) = P1#(P6%(L5%),1): R3#(5,3) = P1#(P6%(L5%),2): R3#(5,4) =
     P1#(P6%(L5%),1)*P1#(P6%(L5%),2)
3055 FOR I3% = 1 TO 4: FOR I4% = 1 TO 4: R3#(I3%,I4%) = R3#(I3%,I4%) +
     R3#(5,I3%)*R3#(5,I4%): NEXT I4% : R9#(I3%) = R9#(I3%) +
     R3#(5,I3%)*P1#(P6%(L5%),3): NEXT I3%
3060 NEXT I2%
3070 NEXT I1%: GOSUB 6690: P1#(I0%,4) = - B#(2) - B#(4)*P1#(I0%,2): P1#(I0%,5) = -
     B#(3) - B#(4)*P1#(I0%,1): P1#(I0%,6) = .2
3080 NEXT I0%
```

Appendix

```
3820 REM make triangular grid
3830 PRINT " AT " TIME$ " BEGIN SURFACE INTERPOLATION  -  PLEASE WAIT":
     SOUND 123,5: SOUND 185,5: SOUND 208,5 : SOUND 165,5
3840 I0% = GR + 1: IS = 1: I2% = 2*I0% + 1: I3% = 1 + INT(GR/3^.5): I4% = I3%*I2% +
     I0%: SX = 1/GR: SY = 1/(2*I3%): S2(3) = .000001
3850 WX# = S2(3): S2(1) = 1.00001: AA# = SX*SY*.5*DT*DT: SM# = 0: SA# = 0: WY# =
     1: U8 = 1
3860 FOR L1% = 1 TO I4%
3870 S1(L1%) = - 999: WXO# = WX#: WYO# = WY#
4490 REM .....calculate distance weighted coordinates of WX#,WY# **********
4500 IF PW<8 THEN GOTO 4550
4510 P5#(1) = 1E37
4520 FOR L2% = 1 TO ND%: U4 = (P1#(L2%,1) - WX#)^2 + (P1#(L2%,2) - WY#)^2: IF
     U4>P5#(1) THEN GOTO 4540
4530 P6%(1) = L2%: P5#(1) = U4
4540 NEXT L2%: S1(L1%) = P1#(P6%(1),3): GOTO 4590
4550 FOR L2% = 1 TO ND%: P5#(L2%) = 1/(((P1#(L2%,1) - WX#)^2 + (P1#(L2%,2) -
     WY#)^2)^.5)^PW: P6%(L2%) = L2%: NEXT L2%: K5% = ND%
4560 S1(L1%) = 0: U4 = 0
4570 FOR L2% = 1 TO K5%: U4 = U4 + P5#(L2%): NEXT L2%
4580 FOR L2% = 1 TO K5%: P5#(L2%) = P5#(L2%)/U4: S1(L1%) = S1(L1%) +
     P5#(L2%)*P1#(P6%(L2%),3): NEXT L2%
4590 IF CL<1 THEN GOTO 4940
4600 IF PW<8 THEN GOTO 4620
4610 S1(L1%) = (P1#(P6%(1),4)*P1#(P6%(1),1) + P1#(P6%(1),5)*P1#(P6%(1),2) +
     P1#(P6%(1),3) - P1#(P6%(1),4)*WX# - P1#(P6%(1),5)*WY#): GOTO 4940
4620 FOR L2% = 1 TO K5%: S4(L2%) = 0
4630 IF P5#(L2%)<.00001 OR P5#(L2%)>1 THEN GOTO 4690
4640 IF ABS(P1#(P6%(L2%),6))<.00001 THEN GOTO 4690
4650 RS# = ABS(P1#(P6%(L2%),6)) + BI: RT# = RS#*BJ: RB# = 1/RT#: BD# =
     P5#(L2%)^RT#: BB# = BD#*2: IF BD#>.5 THEN BB# = (1 - BD#)*2
4660 BB# = BB#^RS#/2: IF BD#>.5 THEN BB# = 1 - BB#
4670 HP# = BB#^RB#
4680 S4(L2%) = ((P1#(P6%(L2%),4)*P1#(P6%(L2%),1) +
     P1#(P6%(L2%),5)*P1#(P6%(L2%),2) + P1#(P6%(L2%),3) - P1#(P6%(L2%),4)*WX# -
     P1#(P6%(L2%),5)*WY#) - S1(L1%))*HP#
4690 NEXT L2%
4700 FOR L2% = 1 TO K5%: S1(L1%) = S1(L1%) + S4(L2%): NEXT L2%
4940 IS = IS + SX*U8: IF IS>0 THEN GOTO 4960
4950 WXO# = WXO# + SX/2: IS = 1.001 - SX
4960 WX# = WXO# + SX*U8: WY# = WYO#: IF U8*WX#<S2(2 - U8) THEN GOTO 4980
4970 U8 = - U8: WX# = S2(2 + U8): WY# = WY# - SY: IS = U8
4980 NEXT L1%
5060 REM end interpolation to triangular grid and begin display construction
5080 SCREEN 9: COLOR 2,0: CLS: SOUND 208,5: SOUND 165,5: SOUND 185,5:
     SOUND 123,5: IF DP>0 THEN SZ = INT(SZ*2/3)
5090 IF SR>0 AND SZ>240 THEN SZ = 240
5100 ZS = INT(SZ*.71): SW = (WD - SZ)\2: ZW = (HI - LW)\2 + ZS\2: IF DP>0 THEN ZW
     = ZW - LW - 10
5110 IF SR>0 AND CR<1 THEN SW = (WD - 2*SZ)\3 - 10
5120 SV = SW + SZ + 60: IF SR<1 THEN GOTO 5130
5125 FOR L1% = 2 TO 6: BX(L1%,1) = (BX(L1%,1) - .5)*.99875 - (BX(L1%,3) - .5)*.05 +
     .5: BX(L1%,3) = (BX(L1%,3) - .5)*.99875 + (BX(L1%,1) - .5)*.05 + .5: NEXT L1%
5130 FOR L1% = BN% + 1 TO BN% + 4: X1 = SW + BX(L1% - 1,1)*SZ: Y1 = ZW -
     BX(L1% - 1,2)*ZS: X2 = SW + BX(L1%,1)*SZ: Y2 = ZW - BX(L1%,2)*ZS: LINE(X1,Y1)
     - (X2,Y2),14: NEXT L1%: IF SR<1 OR CR>0 THEN GOTO 5210
5140 FOR L1% = 2 TO 6: BX(L1%,1) = (BX(L1%,1) - .5)*.995 + (BX(L1%,3) - .5)*.1 + .5:
     NEXT L1%
5150 FOR L1% = BN% + 1 TO BN% + 4: X1 = SV + BX(L1% - 1,1)*SZ: Y1 = ZW -
     BX(L1% - 1,2)*ZS: X2 = SV + BX(L1%,1)*SZ: Y2 = ZW - BX(L1%,2)*ZS: LINE(X1,Y1)
     - (X2,Y2),14: NEXT L1%
5200 REM plot contours
5210 SW = SW + 1: SZ = SZ - 2: ZS = ZS - 2: ZW = ZW - 1: ZX = SX: ZY = SY: ZZ = .5:
     R3#(2,2) = 1 + 2*SY: R3#(2,1) = 0
5220 FOR L1% = 1 TO I3%
```

```
5230 R3#(1,2) = R3#(2,2) - 2*SY - 2*ZY: R3#(1,1) = R3#(2,1): N1% = (2*L1% - .5 +
     ZZ)*I0% + L1% + .5/ZZ: R3#(1,3) = S1(N1%): R3#(2,2) = R3#(2,2) - 2*SY - ZY:
     R3#(2,1) = .5 - ZZ: N2% = (2*L1% - .5 + ZZ)*I0% + L1%
5240 R3#(2,3) = S1(N2%): R3#(3,2) = R3#(2,2): R3#(3,1) = R3#(2,1) + ZX/2: N3% = N2%
     - .5/ZZ: R3#(3,3) = S1(N3%): AR# = AA#/2
5250 GOSUB 5820
5260 R3#(1,2) = R3#(1,2) + 2*ZY: N2% = (2*L1% - .5/ZZ - .5 - ZZ)*I0% + L1%: R3#(1,3) =
     S1(N2%)
5270 GOSUB 5820
5280 AR# = AA#
5290 FOR L2% = 1 TO GR
5300 R3#(2,2) = R3#(1,2): R3#(2,1) = R3#(1,1) + ZX: N4% = N2% + .5/ZZ: R3#(2,3) =
     S1(N4%)
5310 GOSUB 5820
5320 R3#(1,2) = R3#(3,2): R3#(1,1) = R3#(3,1) + ZX
5330 IF L2% = GR THEN R3#(1,1) = R3#(1,1) - ZX/2
5340 N6% = N3% - .5/ZZ: R3#(1,3) = S1(N6%): IF L2% = GR THEN AR# = AA#/2
5350 GOSUB 5820
5360 R3#(2,2) = R3#(2,2) - 2*ZY: N5% = N1% + .5/ZZ: R3#(2,3) = S1(N5%)
5370 GOSUB 5820
5380 R3#(1,2) = R3#(2,2): R3#(1,1) = R3#(2,1) - ZX: R3#(1,3) = S1(N1%): IF L2% = GR
     THEN AR# = AA#
5390 GOSUB 5820
5400 N1% = N5%: N2% = N4%: N3% = N6%: R3#(1,2) = R3#(1,2) + 2*ZY: R3#(1,1) =
     R3#(2,1) : R3#(1,3) = S1(N4%): R3#(3,1) = R3#(3,1) + ZX: R3#(3,3) = S1(N6%)
5410 NEXT L2%
5420 ZY = - ZY: ZX = - ZX: ZZ = - ZZ
5430 NEXT L1%
5440 REM end plotting contours
5450 REM plot data points
5460 IF PT<1 OR CR>0 THEN GOTO 5790
5480 FOR L1% = 1 TO ND%
5490 IF P1#(L1%,1)< - .0001 OR P1#(L1%,1)>1.0001 OR P1#(L1%,2)< - .0001 OR
     P1#(L1%,2)>1.0001 THEN GOTO 5610
5500 PP1# = P1#(L1%,1): PP2# = P1#(L1%,2): PP3# = P1#(L1%,3): IF DP<1 THEN
     GOTO 5526
5510 YY = (PP2# - .5)*R8(1,1) + (PP1# - .5)*R8(1,2) + .5: PP1# = (PP1# - .5)*R8(1,1) -
     (PP2# - .5)*R8(1,2) + .5
5520 ZZ3 = (PP3# - .5)*R8(2,1) + (YY - .5)*R8(2,2) + .5: PP2# = (YY - .5)*R8(2,1) - (PP3#
     - .5)*R8(2,2) + .5: PP3# = ZZ3: IF SR<1 THEN GOTO 5526
5525 PPP# = PP1#: PP1# = (PP1# - .5)*.99875 - (PP3# - .5)*.05 + .5
5526 S3(1,1) = PP1#*SZ + SW - 2: S3(1,2) = ZW - PP2#*ZS - 2
5527 FOR L2% = 1 TO 3: FOR L3% = 1 TO 3: PSET(S3(1,1) + L2%,S3(1,2) + L3%),4:
     NEXT L3%: NEXT L2%
5528 IF SR<1 THEN GOTO 5610
5529 PP1# = (PPP# - .5)*.99875 + (PP3# - .5)*.05 + .5
5530 S3(1,1) = PP1#*SZ + SV - 2: S3(1,2) = ZW - PP2#*ZS - 2
5531 FOR L2% = 1 TO 3: FOR L3% = 1 TO 3: PSET(S3(1,1) + L2%,S3(1,2) + L3%),4:
     NEXT L3%: NEXT L2%
5610 NEXT L1%
5790 SOUND 247,5: SOUND 153,5: SOUND 185,5: SOUND 123,10: TM$ = TIME$: IF
     SR>0 THEN GOTO 5802
5800 SM# = SM#/SA#: LOCATE 12,1: PRINT "Average": PRINT "height of": PRINT
     "surface": PRINT USING "#####.###";SM#
5802 LOCATE 19,1: PRINT " ALL": PRINT "DONE!": PRINT " At": PRINT TM$
5804 PRINT "Press": INPUT "ENTER ",PAWS: END
5810 REM curvemaker
5820 ST# = R3#(1,3)
5830 IF ST#>R3#(2,3) THEN ST# = R3#(2,3)
5840 IF ST#>R3#(3,3) THEN ST# = R3#(3,3)
5850 IF ST#< = - 999 THEN RETURN
5860 SM# = SM# + AR#*((R3#(1,3) + R3#(2,3) + R3#(3,3))/3*VS + P9(2,3)): SA# = SA# +
     AR#: IF CR<1 THEN GOTO 5910
5870 PO(1,1) = - 1E37: PO(2,1) = PO(1,1): PO(1,2) = 1E37: PO(2,2) = PO(1,2): S3(1,1) =
     R3#(1,3) : S3(1,2) = R3#(2,3): S3(1,3) = R3#(3,3): IF DP<1 THEN GOTO 5910
```

Appendix

```
5880 FOR L3% = 1 TO 3: R1#(L3%,1) = R3#(L3%,1): R1#(L3%,2) = R3#(L3%,2):
     R1#(L3%,3) = R3#(L3%,3)
5890 YY = (R3#(L3%,2) - .5)*R8(1,1) + (R3#(L3%,1) - .5)*R8(1,2): R3#(L3%,1) =
     (R3#(L3%,1) - .5)*R8(1,1) - (R3#(L3%,2) - .5)*R8(1,2) + .5
5900 R33# = (R3#(L3%,3) - .5)*R8(2,1) + YY*R8(2,2) + .5: R3#(L3%,2) = YY*R8(2,1) -
     (R3#(L3%,3) - .5)*R8(2,2) + .5: R3#(L3%,3) = R33#: NEXT L3%
5910 FOR L3% = 1 TO 3
5920 IF PO(1,1)<R3#(L3%,1) THEN PO(1,1) = R3#(L3%,1)
5930 IF PO(2,1)<R3#(L3%,2) THEN PO(2,1) = R3#(L3%,2)
5940 IF PO(1,2)>R3#(L3%,1) THEN PO(1,2) = R3#(L3%,1)
5950 IF PO(2,2)>R3#(L3%,2) THEN PO(2,2) = R3#(L3%,2)
5960 NEXT L3%
5970 X1 = SW + INT(PO(1,2)*SZ): XH = X1 - (X1\2)*2 - .5: X2 = SW + INT(PO(1,1)*SZ):
     Y1 = ZW - INT(PO(2,2)*ZS): Y2 = ZW - INT(PO(2,1)*ZS): YG = Y2 - (Y2\2)*2 - .5
5980 ZJ# = ((R3#(2,1) - R3#(1,1))*(R3#(3,2) - R3#(1,2)) - ((R3#(3,1) - R3#(1,1))*(R3#(2,2)
     - R3#(1,2)))
5990 FOR L3% = Y2 TO Y1
6000 YG = - YG: YY = (ZW - L3%)/ZS: XG = XH
6010 FOR L4% = X1 TO X2
6020 XG = - XG: XX = (L4% - SW)/SZ: ZI = 1
6030 FOR L5% = 1 TO 3
6040 L8% = 1: IF L8% = L5% THEN L8% = L8% + 1
6050 L9% = L8% + 1: IF L9% = L5% THEN L9% = L9% + 1
6060 S3(2,L5%) = (((R3#(L8%,1) - XX)*(R3#(L9%,2) - YY)) - ((R3#(L9%,1) -
     XX)*(R3#(L8%,2) - YY)))/ZJ#*ZI: IF S3(2,L5%)<0 THEN GOTO 6150
6070 ZI = - ZI: NEXT L5%
6080 ZC = S3(1,1)*S3(2,1) + S3(1,2)*S3(2,2) + S3(1,3)*S3(2,3): IF ZC<P8(1) THEN
     GOTO 6150
6090 FOR L5% = 2 TO NC%
6100 IF P8(L5%)<ZC THEN GOTO 6140
6110 LC% = INT(L5%/2 + XG/YG*.25 + .75): IF LC%<1 THEN GOTO 6150
6120 IF LC%>16 THEN LC% = 1
6130 PSET(L4%,L3%),SP%(LC%): GOTO 6150
6140 NEXT L5%
6150 NEXT L4%: NEXT L3%
6160 IF DP<1 THEN GOTO 6200
6170 FOR L3% = 1 TO 3
6180 R3#(L3%,1) = R1#(L3%,1): R3#(L3%,2) = R1#(L3%,2): R3#(L3%,3) = R1#(L3%,3)
6190 NEXT L3%
6200 RETURN
6210 L8% = 0: IF DP<2 THEN GOTO 6310
6220 L8% = 1
6230 REM rotate triangle for orthogonal profiles
6240 FOR L3% = 1 TO 3
6250 Z# = R3#(L3%,3): R3#(L3%,3) = 1 - R3#(L3%,L8%): R3#(L3%,L8%) = Z# 6260
     NEXT L3%
6270 ST# = R3#(1,3)
6280 IF ST#>R3#(2,3) THEN ST# = R3#(2,3)
6290 IF ST#>R3#(3,3) THEN ST# = R3#(3,3)
6300 IF ST# < = - 999 THEN RETURN
6310 TP# = R3#(1,3)
6320 IF TP#<R3#(2,3) THEN TP# = R3#(2,3)
6330 IF TP#<R3#(3,3) THEN TP# = R3#(3,3)
6340 IF TP# = ST# THEN RETURN
6350 FOR L0% = 1 TO NC%
6360 REM slice the triangle
6370 IF P8(L0%)>TP# OR P8(L0%)<ST# THEN GOTO 6610
6380 L5% = 1: L4% = 0
6390 L4% = L4% + 1
6400 L7% = 1: IF L7% = L4% THEN L7% = L7% + 1
6410 L3% = L7% + 1: IF L3% = L4% THEN L3% = L3% + 1
6420 IF R3#(L7%,3) = R3#(L3%,3) THEN GOTO 6460
6430 F = (P8(L0%) - R3#(L7%,3))/(R3#(L3%,3) - R3#(L7%,3)): IF F<0 OR F>1 THEN
     GOTO 6460
6440 S3(L5%,1) = R3#(L7%,1) + (R3#(L3%,1) - R3#(L7%,1))*F: S3(L5%,2) = R3#(L7%,2)
     + (R3#(L3%,2) - R3#(L7%,2))*F
```

```
6450 L5% = L5% + 1
6460 IF L5%<3 THEN GOTO 6390
6470 S3(1,3) = P8(L0%): S3(2,3) = P8(L0%): IF DP<2 THEN GOTO 6500
6480 REM reverse rotate the intersection trace
6490 FOR L3% = 1 TO 2: Z# = S3(L3%,3): S3(L3%,3) = S3(L3%,L8%): S3(L3%,L8%) = 1
     - Z#: NEXT L3%
6500 IF DP = 0 THEN GOTO 6552
6510 REM apply perspective
6520 FOR L3% = 1 TO 2
6530 YY = (S3(L3%,2) - .5)*R8(1,1) + (S3(L3%,1) - .5)*R8(1,2): S3(L3%,1) = (S3(L3%,1) -
     .5)*R8(1,1) - (S3(L3%,2) - .5)*R8(1,2) + .5
6540 S33 = (S3(L3%,3) - .5)*R8(2,1) + YY*R8(2,2) + .5: S3(L3%,2) = YY*R8(2,1) -
     (S3(L3%,3) - .5)*R8(2,2) + .5: S3(L3%,3) = S33
6550 NEXT L3%
6552 IF SR<1 THEN GOTO 6580
6554 S3(1,1) = (S3(1,1) - .5)*.99875 - (S3(1,3) - .5)*.05 + .5: S3(1,3) = (S3(1,3) -
     .5)*.99875 + (S3(1,1) - .5)*.05 + .5: S3(2,1) = (S3(2,1) - .5)*.99875 - (S3(2,3) - .5)*.05 +
     .5: S3(2,3) = (S3(2,3) - .5)*.99875 + (S3(2,1) - .5)*.05 + .5
6560 REM draw a line from (S3(1,1),S3(1,2)) to (S3(2,1),S3(2,2))
6580 LINE((SW + S3(1,1)*SZ),(ZW - S3(1,2)*ZS)) - ((SW + S3(2,1)*SZ,(ZW -
     S3(2,2)*ZS)),14: IF SR<1 OR DP<1 THEN GOTO 6610
6590 S3(1,1) = (S3(1,1) - .5)*.995 + (S3(1,3) - .5)*.1 + .5: S3(2,1) = (S3(2,1) - .5)*.995 +
     (S3(2,3) - .5)*.1 + .5
6600 LINE((SV + S3(1,1)*SZ),(ZW - S3(1,2)*ZS)) - ((SV + S3(2,1)*SZ,(ZW -
     S3(2,2)*ZS)),14
6610 NEXT L0%
6620 IF L8%<1 THEN RETURN
6630 REM reverse rotate the triangle
6640 FOR L3% = 1 TO 3
6650 Z# = R3#(L3%,3): R3#(L3%,3) = R3#(L3%,L8%): R3#(L3%,L8%) = 1 - Z#
6660 NEXT L3%
6670 L8% = L8% + 1: IF L8%<3 THEN GOTO 6240 ELSE RETURN
6680 REM solve R3#(LL%,LL%)xB#(LL%) = R9#(LL%)
6690 FOR J2% = 1 TO LL%: P4%(J2%) = J2%: B#(J2%) = 0
6700 FOR J3% = 1 TO LL%: IF ABS(R3#(J2%,J3%))> = B#(J2%) THEN B#(J2%) =
     ABS(R3#(J2%,J3%))
6710 NEXT J3%: NEXT J2%: LM1% = LL% - 1: LP1% = LL% + 1
6720 FOR J2% = 1 TO LM1%: X3# = 0
6730 FOR J3% = J2% TO LL%: X4# = ABS(R3#(P4%(J3%),J2%))/B#(P4%(J3%)): IF
     X4#< = X3# THEN GOTO 6750
6740 X3# = X4#: K3% = J3%
6750 NEXT J3%: IF K3% = J2% GOTO 6770
6760 K2% = P4%(J2%): P4%(J2%) = P4%(K3%): P4%(K3%) = K2%
6770 X3# = R3#(P4%(J2%),J2%): K4% = J2% + 1
6780 FOR J3% = K4% TO LL%: X4# =   - R3#(P4%(J3%),J2%)/X3#: R3#(P4%(J3%),J2%)
     =  - X4#
6790 FOR J4% = K4% TO LL%: R3#(P4%(J3%),J4%) = R3#(P4%(J3%),J4%) +
     X4#*R3#(P4%(J2%),J4%): NEXT J4%: NEXT J3% : NEXT J2%
6800 B#(1) = R9#(P4%(1))
6810 FOR J1% = 2 TO LL%
6820 I8% = J1% - 1: X3# = 0
6830 FOR J2% = 1 TO I8%: X3# = X3# + R3#(P4%(J1%),J2%)*B#(J2%): NEXT J2%
6840 B#(J1%) = R9#(P4%(J1%)) - X3#: NEXT J1%
6850 B#(LL%) = B#(LL%)/R3#(P4%(LL%),LL%)
6860 FOR J1% = 2 TO LL%: I8% = LP1% - J1%: I9 = I8% + 1: X3# = 0
6870 FOR J2% = I9 TO LL%: X3# = X3# + R3#(P4%(I8%),J2%)*B#(J2%): NEXT J2%
6880 B#(I8%) = (B#(I8%) - X3#)/R3#(P4%(I8%),I8%): NEXT J1%: RETURN

1000 REM SCRN\FITFUNCT.BAS FITTED FUNCTION INTERPOLATION
     Copyright (c) 1988 D.F.Watson
1010 DIM P1#(55,3), P8(80), P9(2,3), S1(5934), S2(3), RO(55), PO(4,2)
1020 DIM R1#(3,3), R3#(55,55), R8(2,2), S3(2,3), BX(6,3), R9#(55), B#(55)
1030 DIM P4%(55), R6%(3,2), SP%(16)
1040 DATA 0,8,1,4,5,9,6,12,13,2,7,3,10,11,14,15: REM spectrum
1050 DATA 1,0,0,1,1,0,0,1,0,0,0,0,1,0,0,1,0,1, - 1, - 1,0,1,2,5, - 1,0,2,3, - 1,5,1E37,3,1
```

Appendix

```
1060 DEF FNCV(RT#) = RT#^2*LOG(RT#): REM minimum curvature: REM DEF
     FNCV(RT#) = RT#^2*(LOG(RT#) - 1): REM biharmonic spline: REM DEF FNCV(RT#)
     = RT#: REM dummy function
1070 REM DEF FNCV(RT#) = NG + SI*(1 - EXP( - RT#/RG)): NG = .25: SI = 1: RG = 1/3:
     REM exponential semivariogram
1080 REM DEF FNCV(RT#) = NG + SI*(1.5*RT#/RG - (RT#/RG)^3/2): NG = .25: SI = 1:
     RG = 1: REM spherical semivariogram
1090 P9(1,1) = - 1E37: P9(1,2) = - 1E37: P9(1,3) = - 1E37: P9(2,1) = 1E37: P9(2,2) =
     1E37: P9(2,3) = 1E37
1100 SR = 0: SZ = 480: GR = 19: DP = 0: TB = 3: PT = 1: CR = 0: DG = 3: WD = 640: HI
     = 350: LW = 5: AZ = - 25: TL = - 15: RD = 2: IF DP = 0 THEN SR = 0
1140 OPEN "A:\HILL.DAT" FOR INPUT AS #1
1150 REM OPEN "A:\JDAVIS.DAT" FOR INPUT AS #1
1170 FOR I1% = 1 TO 16: READ SP%(I1%): NEXT I1%
1180 INPUT#1,ND%,NC%: FOR I1% = 1 TO NC%: INPUT#1,P8(I1%): NEXT I1%: IF
     CR>0 AND NC%>31 THEN NC% = 31
1190 INPUT#1,XS,YS,DT: I1% = 1
1220 INPUT#1,P1#(I1%,1),P1#(I1%,2),P1#(I1%,3): RO(I1%) = .1*RND: P1#(I1%,1) =
     (P1#(I1%,1) - XS)/DT: P1#(I1%,2) = (P1#(I1%,2) - YS)/DT: IF (P1#(I1%,1) - .5)^2 +
     (P1#(I1%,2) - .5)^2<RD THEN GOTO 1270
1230 ND% = ND% - 1: IF I1%>ND% THEN GOTO 1360
1260 GOTO 1220
1270 IF P9(1,1)<P1#(I1%,1) THEN P9(1,1) = P1#(I1%,1)
1280 IF P9(2,1)>P1#(I1%,1) THEN P9(2,1) = P1#(I1%,1)
1290 IF P9(1,2)<P1#(I1%,2) THEN P9(1,2) = P1#(I1%,2)
1300 IF P9(2,2)>P1#(I1%,2) THEN P9(2,2) = P1#(I1%,2)
1310 IF P9(1,3)<P1#(I1%,3) THEN P9(1,3) = P1#(I1%,3)
1320 IF P9(2,3)>P1#(I1%,3) THEN P9(2,3) = P1#(I1%,3)
1350 I1% = I1% + 1: IF I1%< = ND% THEN GOTO 1220
1360 IF ND%<4 THEN STOP
1370 XF = P9(1,1) - P9(2,1): YF = P9(1,2) - P9(2,2): VS = P9(1,3) - P9(2,3): R8(2,1) = AZ:
     R8(2,2) = TL
1380 FOR I1% = 1 TO 6: READ BX(I1%,1),BX(I1%,2),BX(I1%,3): NEXT I1% 1390 BN% =
     1: IF DP<1 THEN GOTO 1450
1400 BN% = 2: BX(2,3) = 0: BX(2,3) = 1: IF R8(2,1)>0 THEN GOTO 1420
1410 BX(2,1) = 1: BX(3,1) = 1: BX(4,1) = 1: BX(5,1) = 0: BX(6,1) = 0
1420 IF DP<2 OR CR>0 THEN GOTO 1450
1430 NC1% = INT(SZ/10): NC% = NC1% + 1: U0 = .9999/NC1%: U1 = .0000511 - U0
1440 FOR I1% = 1 TO NC%: U1 = U1 + U0: P8(I1%) = U1: NEXT I1%: GOTO 1460
1450 FOR I1% = 1 TO NC%: P8(I1%) = (P8(I1%) - P9(2,3))/VS + .00001*(RND - .5):
     NEXT I1%
1460 FOR I1% = 1 TO ND%: P1#(I1%,1) = P1#(I1%,1) + .0001*(RND - .5): P1#(I1%,2) =
     P1#(I1%,2) + .0001*(RND - .5): P1#(I1%,3) = (P1#(I1%,3) - P9(2,3))/VS: NEXT I1%
1470 FOR I1% = 1 TO 3: READ R1#(I1%,1), R1#(I1%,2), R1#(I1%,3), R6%(I1%,1),
     R6%(I1%,2)
1480 P1#(I1% + ND%,1) = R1#(I1%,1)*XF + P9(2,1): P1#(I1% + ND%,2) = R1#(I1%,2)*YF
     + P9(2,2): P1#(I1% + ND%,3) = 0: NEXT I1%
1490 IF DP<1 THEN GOTO 3480
1500 FOR I1% = 1 TO 2: R8(I1%,1) = COS(R8(2,I1%)/57.29578): R8(I1%,2) =
     SIN(R8(2,I1%)/57.29578): NEXT I1%
1510 FOR I1% = 2 TO 6
1520 YY = (BX(I1%,2) - .5)*R8(1,1) + (BX(I1%,1) - .5)*R8(1,2) + .5: BX(I1%,1) =
     (BX(I1%,1) - .5)*R8(1,1) - (BX(I1%,2) - .5)*R8(1,2) + .5
1530 BX3 = (BX(I1%,3) - .5)*R8(2,1) + (YY - .5)*R8(2,2) + .5: BX(I1%,2) = (YY -
     .5)*R8(2,1) - (BX(I1%,3) - .5)*R8(2,2) + .5: BX(I1%,3) = BX3: NEXT I1%
1540 REM end data input
3480 PRINT " AT " TIME$ " BEGIN SURFACE DEFINITION - PLEASE WAIT"
3490 REM minimum curvature splines
3500 IF TB>3 THEN GOTO 3600
3510 LL% = ND% + 3
3520 FOR I1% = 1 TO ND%
3530 R3#(I1%,1) = 1: R3#(LL%,I1% + 3) = 1: R3#(I1%,2) = P1#(I1%,1): R3#(ND% +
     1,I1% + 3) = R3#(I1%,2)
3540 R3#(I1%,3) = P1#(I1%,2): R3#(ND% + 2,I1% + 3) = R3#(I1%,3): R9#(I1%) =
     P1#(I1%,3): NEXT I1%
```

```
3550 FOR I1% = 1 TO 3: FOR I2% = 1 TO 3: R3#(I1% + ND%,I2%) = 0: NEXT I2%:
     R9#(I1% + ND%) = 0: NEXT I1%
3560 FOR I1% = 1 TO ND%: R3#(I1%,I1% + 3) = RO(I1%): IJ = I1% + 1: IF IJ>ND%
     THEN GOTO 3580
3570 FOR I2% = IJ TO ND%: RR# = ((R3#(I1%,2) - R3#(I2%,2))^2 + (R3#(I1%,3) -
     R3#(I2%,3))^2)^.5: R3#(I2%,I1% + 3) = FNCV(RR#): R3#(I1%,I2% + 3) = R3#(I2%,I1%
     + 3): NEXT I2%
3580 NEXT I1%: GOSUB 6690: GOTO 3830
3590 REM polynomial trend surface
3600 IF TB>4 THEN GOTO 3750
3610 B#(1) = 1: LL% = (DG + 1)*(DG + 2)/2: FOR I1% = 1 TO LL%: FOR I2% = 1 TO
     LL%: R3#(I1%,I2%) = 0: NEXT I2%: R9#(I1%) = 0 : NEXT I1%
3620 FOR I1% = 1 TO ND%: I5% = 1
3630 FOR I2% = 1 TO DG
3640 FOR I3% = 1 TO I2%: I5% = I5% + 1: I6% = I5% - I2%: B#(I5%) =
     B#(I6%)*P1#(I1%,1): NEXT I3%
3650 I5% = I5% + 1: B#(I5%) = B#(I6%)*P1#(I1%,2): NEXT I2%
3660 FOR I2% = 1 TO LL%: R9#(I2%) = R9#(I2%) + B#(I2%)*P1#(I1%,3)
3670 FOR I3% = 1 TO LL%: R3#(I2%,I3%) = R3#(I2%,I3%) + B#(I2%)*B#(I3%): NEXT
     I3%
3680 NEXT I2%
3690 NEXT I1%: GOSUB 6690
3700 FOR L2% = 1 TO ND%: R9#(1) = 1: L5% = 1: RO(L2%) = 0
3710 FOR L3% = 1 TO DG
3720 FOR L4% = 1 TO L3%: L5% = L5% + 1: L6% = L5% - L3%: R9#(L5%) =
     R9#(L6%)*P1#(L2%,1): NEXT L4% : L5% = L5% + 1: R9#(L5%) =
     R9#(L6%)*P1#(L2%,2): NEXT L3%
3730 FOR L3% = 1 TO LL%: RO(L2%) = RO(L2%) + B#(L3%)*R9#(L3%): NEXT L3%:
     RO(L2%) = (RO(L2%) - P1#(L2%,3))*VS: NEXT L2%: GOTO 3830
3740 REM multiquadric collocation
3750 LL% = ND%
3760 FOR I1% = 1 TO LL%
3770 FOR I2% = 1 TO LL%
3780 IF I1% = I2% THEN GOTO 3800
3790 R3#(I1%,I2%) = ((P1#(I1%,1) - P1#(I2%,1))^2 + (P1#(I1%,2) - P1#(I2%,2))^2 +
     RO(I2%)^2)^.5: IF TB<8 THEN GOTO 3810
3795 R3#(I1%,I2%) = 2*((P1#(I1%,1) - P1#(I2%,1))^2 + (P1#(I1%,2) - P1#(I2%,2))^2)^.5 +
     RO(I2%) - R3#(I1%,I2%) : GOTO 3810
3800 R3#(I1%,I2%) = RO(I1%): IF TB = 8 THEN R3#(I1%,I2%) = 0
3810 NEXT I2%: R9#(I1%) = P1#(I1%,3): NEXT I1%: GOSUB 6690
3820 REM make triangular grid
3830 PRINT " AT " TIME$ " BEGIN SURFACE INTERPOLATION  -  PLEASE WAIT":
     SOUND 123,5: SOUND 185,5: SOUND 208,5 : SOUND 165,5
3840 I0% = GR + 1: IS = 1: I2% = 2*I0% + 1: I3% = 1 + INT(GR/3^.5): I4% = I3%*I2% +
     I0%: SX = 1/GR: SY = 1/(2*I3%): S2(3) = .000001
3850 WX# = S2(3): S2(1) = 1.00001: AA# = SX*SY*.5*DT*DT: SM# = 0: SA# = 0: WY# =
     1: U8 = 1
3860 FOR L1% = 1 TO I4%
3870 S1(L1%) =  - 999: WXO# = WX#: WYO# = WY#
4710 REM minimum curvature splines
4720 S1(L1%) = 0: IF TB>3 THEN GOTO 4760
4730 FOR L2% = 4 TO LL%: RR# = ((P1#(L2% - 3,1) - WX#)^2 + (P1#(L2% - 3,2) -
     WY#)^2)^.5: S1(L1%) = S1(L1%) + B#(L2%)*FNCV(RR#): NEXT L2%
4740 S1(L1%) = S1(L1%) + B#(1) + B#(2)*WX# + B#(3)*WY#: GOTO 4940
4750 REM polynomial trend surface
4760 IF TB>4 THEN GOTO 4830
4770 R9#(1) = 1: L5% = 1
4780 FOR L2% = 1 TO DG
4790 FOR L3% = 1 TO L2%: L5% = L5% + 1: L6% = L5% - L2%: R9#(L5%) =
     R9#(L6%)*WX#: NEXT L3%
4800 L5% = L5% + 1: R9#(L5%) = R9#(L6%)*WY#: NEXT L2%
4810 FOR L2% = 1 TO LL%: S1(L1%) = S1(L1%) + B#(L2%)*R9#(L2%): NEXT L2%:
     GOTO 4940
4830 REM multiquadric collocation
4840 FOR L2% = 1 TO LL%: IF TB = 8 THEN GOTO 4844
```

Appendix

```
4842 S1(L1%) = S1(L1%) + B#(L2%)*((P1#(L2%,1) - WX#)^2 + (P1#(L2%,2) - WY#)^2 +
     RO(L2%)^2)^.5: GOTO 4848
4844 S1(L1%) = S1(L1%) + B#(L2%)*(2*((P1#(L2%,1) - WX#)^2 + (P1#(L2%,2) -
     WY#)^2)^.5 + RO(L2%) - ((P1#(L2%,1) - WX#)^2 + (P1#(L2%,2) - WY#)^2 +
     RO(L2%)^2)^.5)
4848 NEXT L2%
4940 IS = IS + SX*U8: IF IS>0 THEN GOTO 4960
4950 WXO# = WXO# + SX/2: IS = 1.001 - SX
4960 WX# = WXO# + SX*U8: WY# = WYO#: IF U8*WX#<S2(2 - U8) THEN GOTO 4980
4970 U8 = - U8: WX# = S2(2 + U8): WY# = WY# - SY: IS = U8
4980 NEXT L1%
5060 REM end interpolation to triangular grid and begin display construction
5080 SCREEN 9: COLOR 2,0: CLS: SOUND 208,5: SOUND 165,5: SOUND 185,5:
     SOUND 123,5: IF DP>0 THEN SZ = INT(SZ*2/3)
5090 IF SR>0 AND SZ>240 THEN SZ = 240
5100 ZS = INT(SZ*.71): SW = (WD - SZ)\2: ZW = (HI - LW)\2 + ZS\2: IF DP>0 THEN ZW
     = ZW - LW - 10
5110 IF SR>0 AND CR<1 AND DP>0 THEN SW = (WD - 2*SZ)\3 - 10
5120 SV = SW + SZ + 60: IF SR<1 THEN GOTO 5130
5125 FOR L1% = 2 TO 6: BX(L1%,1) = (BX(L1%,1) - .5)*.99875 - (BX(L1%,3) - .5)*.05 +
     .5: BX(L1%,3) = (BX(L1%,3) - .5)*.99875 + (BX(L1%,1) - .5)*.05 + .5: NEXT L1%
5130 FOR L1% = BN% + 1 TO BN% + 4: X1 = SW + BX(L1% - 1,1)*SZ: Y1 = ZW -
     BX(L1% - 1,2)*ZS: X2 = SW + BX(L1%,1)*SZ: Y2 = ZW - BX(L1%,2)*ZS: LINE(X1,Y1)
     - (X2,Y2),14: NEXT L1%: IF SR<1 OR CR>0 THEN GOTO 5210
5140 FOR L1% = 2 TO 6: BX(L1%,1) = (BX(L1%,1) - .5)*.995 + (BX(L1%,3) - .5)*.1 + .5:
     NEXT L1%
5150 FOR L1% = BN% + 1 TO BN% + 4: X1 = SV + BX(L1% - 1,1)*SZ: Y1 = ZW -
     BX(L1% - 1,2)*ZS: X2 = SV + BX(L1%,1)*SZ: Y2 = ZW - BX(L1%,2)*ZS: LINE(X1,Y1) -
     (X2,Y2),14: NEXT L1%
5200 REM plot contours
5210 SW = SW + 1: SZ = SZ - 2: ZS = ZS - 2: ZW = ZW - 1: ZX = SX: ZY = SY: ZZ = .5:
     R3#(2,2) = 1 + 2*SY: R3#(2,1) = 0
5220 FOR L1% = 1 TO I3%
5230 R3#(1,2) = R3#(2,2) - 2*SY - 2*ZY: R3#(1,1) = R3#(2,1): N1% = (2*L1% - .5 +
     ZZ)*I0% + L1% + .5/ZZ: R3#(1,3) = S1(N1%): R3#(2,2) = R3#(2,2) - 2*SY - ZY:
     R3#(2,1) = .5 - ZZ: N2% = (2*L1% - .5 + ZZ)*I0% + L1%
5240 R3#(2,3) = S1(N2%): R3#(3,2) = R3#(2,2): R3#(3,1) = R3#(2,1) + ZX/2: N3% = N2%
     - .5/ZZ: R3#(3,3) = S1(N3%): AR# = AA#/2
5250 GOSUB 5820
5260 R3#(1,2) = R3#(1,2) + 2*ZY: N2% = (2*L1% - .5/ZZ - .5 - ZZ)*I0% + L1%: R3#(1,3) =
     S1(N2%)
5270 GOSUB 5820
5280 AR# = AA#
5290 FOR L2% = 1 TO GR
5300 R3#(2,2) = R3#(1,2): R3#(2,1) = R3#(1,1) + ZX: N4% = N2% + .5/ZZ: R3#(2,3) =
     S1(N4%)
5310 GOSUB 5820
5320 R3#(1,2) = R3#(3,2): R3#(1,1) = R3#(3,1) + ZX
5330 IF L2% = GR THEN R3#(1,1) = R3#(1,1) - ZX/2
5340 N6% = N3% - .5/ZZ: R3#(1,3) = S1(N6%): IF L2% = GR THEN AR# = AA#/2
5350 GOSUB 5820
5360 R3#(2,2) = R3#(2,2) - 2*ZY: N5% = N1% + .5/ZZ: R3#(2,3) = S1(N5%)
5370 GOSUB 5820
5380 R3#(1,2) = R3#(2,2): R3#(1,1) = R3#(2,1) - ZX: R3#(1,3) = S1(N1%): IF L2% = GR
     THEN AR# = AA#
5390 GOSUB 5820
5400 N1% = N5%: N2% = N4%: N3% = N6%: R3#(1,2) = R3#(1,2) + 2*ZY: R3#(1,1) =
     R3#(2,1) : R3#(1,3) = S1(N4%): R3#(3,1) = R3#(3,1) + ZX: R3#(3,3) = S1(N6%)
5410 NEXT L2%
5420 ZY = - ZY: ZX = - ZX: ZZ = - ZZ
5430 NEXT L1%
5440 REM end plotting contours
5450 REM plot data points
5460 IF PT<1 OR CR>0 THEN GOTO 5790
5480 FOR L1% = 1 TO ND%
```

```
5490 IF P1#(L1%,1)< - .0001 OR P1#(L1%,1)>1.0001 OR P1#(L1%,2)< - .0001 OR
     P1#(L1%,2)>1.0001 THEN GOTO 5610
5500 PP1# = P1#(L1%,1): PP2# = P1#(L1%,2): PP3# = P1#(L1%,3): IF DP<1 THEN
     GOTO 5526
5510 YY = (PP2# - .5)*R8(1,1) + (PP1# - .5)*R8(1,2) + .5: PP1# = (PP1# - .5)*R8(1,1) -
     (PP2# - .5)*R8(1,2) + .5
5520 ZZ3 = (PP3# - .5)*R8(2,1) + (YY - .5)*R8(2,2) + .5: PP2# = (YY - .5)*R8(2,1) - (PP3#
     - .5)*R8(2,2) + .5: PP3# = ZZ3: IF SR<1 THEN GOTO 5526
5525 PPP# = PP1#: PP1# = (PP1# - .5)*.99875 - (PP3# - .5)*.05 + .5
5526 S3(1,1) = PP1#*SZ + SW - 2: S3(1,2) = ZW - PP2#*ZS - 2
5527 FOR L2% = 1 TO 3: FOR L3% = 1 TO 3: PSET(S3(1,1) + L2%,S3(1,2) + L3%),4:
     NEXT L3%: NEXT L2%
5528 IF SR<1 THEN GOTO 5610
5529 PP1# = (PPP# - .5)*.99875 + (PP3# - .5)*.05 + .5
5530 S3(1,1) = PP1#*SZ + SW - 2: S3(1,2) = ZW - PP2#*ZS - 2
5531 FOR L2% = 1 TO 3: FOR L3% = 1 TO 3: PSET(S3(1,1) + L2%,S3(1,2) + L3%),4:
     NEXT L3%: NEXT L2%
5610 NEXT L1%
5790 SOUND 247,5: SOUND 153,5: SOUND 185,5: SOUND 123,10: TM$ = TIME$: IF
     SR>0 THEN GOTO 5802
5800 SM# = SM#/SA#: LOCATE 12,1: PRINT "Average": PRINT "height of": PRINT
     "surface": PRINT USING "#####.###";SM#
5802 LOCATE 19,1: PRINT " ALL": PRINT "DONE!": PRINT " At": PRINT TM$
5804 PRINT "Press": INPUT "ENTER ",PM$: END
5810 REM curvemaker
5820 ST# = R3#(1,3)
5830 IF ST#>R3#(2,3) THEN ST# = R3#(2,3)
5840 IF ST#>R3#(3,3) THEN ST# = R3#(3,3)
5850 IF ST#< = - 999 THEN RETURN
5860 SM# = SM# + AR#*((R3#(1,3) + R3#(2,3) + R3#(3,3))/3*VS + P9(2,3)): SA# = SA# +
     AR#: IF CR<1 THEN GOTO 6210
5870 PO(1,1) = - 1E37: PO(2,1) = PO(1,1): PO(1,2) = 1E37: PO(2,2) = PO(1,2): S3(1,3) =
     R3#(1,3) : S3(1,2) = R3#(2,3): S3(1,3) = R3#(3,3): IF DP<1 THEN GOTO 5910
5880 FOR L3% = 1 TO 3: R1#(L3%,1) = R3#(L3%,1): R1#(L3%,2) = R3#(L3%,2):
     R1#(L3%,3) = R3#(L3%,3)
5890 YY = (R3#(L3%,2) - .5)*R8(1,1) + (R3#(L3%,1) - .5)*R8(1,2): R3#(L3%,1) =
     (R3#(L3%,1) - .5)*R8(1,1) - (R3#(L3%,2) - .5)*R8(1,2) + .5
5900 R33# = (R3#(L3%,3) - .5)*R8(2,1) + YY*R8(2,2) + .5: R3#(L3%,2) = YY*R8(2,1) -
     (R3#(L3%,3) - .5)*R8(2,2) + .5: R3#(L3%,3) = R33#: NEXT L3%
5910 FOR L3% = 1 TO 3
5920 IF PO(1,1)<R3#(L3%,1) THEN PO(1,1) = R3#(L3%,1)
5930 IF PO(2,1)<R3#(L3%,2) THEN PO(2,1) = R3#(L3%,2)
5940 IF PO(1,2)>R3#(L3%,1) THEN PO(1,2) = R3#(L3%,1)
5950 IF PO(2,2)>R3#(L3%,2) THEN PO(2,2) = R3#(L3%,2)
5960 NEXT L3%
5970 X1 = SW + INT(PO(1,2)*SZ): XH = X1 - (X1\2)*2 - .5: X2 = SW + INT(PO(1,1)*SZ):
     Y1 = ZW - INT(PO(2,2)*ZS): Y2 = ZW - INT(PO(2,1)*ZS): YG = Y2 - (Y2\2)*2 - .5
5980 ZJ# = ((R3#(2,1) - R3#(1,1))*(R3#(3,2) - R3#(1,2))) - ((R3#(3,1) - R3#(1,1))*(R3#(2,2)
     - R3#(1,2)))
5990 FOR L3% = Y2 TO Y1
6000 YG = - YG: YY = (ZW - L3%)/ZS: XG = XH
6010 FOR L4% = X1 TO X2
6020 XG = - XG: XX = (L4% - SW)/SZ: ZI = 1
6030 FOR L5% = 1 TO 3
6040 L8% = 1: IF L8% = L5% THEN L8% = L8% + 1
6050 L9% = L8% + 1: IF L9% = L5% THEN L9% = L9% + 1
6060 S3(2,L5%) = (((R3#(L8%,1) - XX)*(R3#(L9%,2) - YY)) - ((R3#(L9%,1) -
     XX)*(R3#(L8%,2) - YY)))/ZJ#*ZI: IF S3(2,L5%)<0 THEN GOTO 6150
6070 ZI = - ZI: NEXT L5%
6080 ZC = S3(1,1)*S3(2,1) + S3(1,2)*S3(2,2) + S3(1,3)*S3(2,3): IF ZC<P8(1) THEN
     GOTO 6150
6090 FOR L5% = 2 TO NC%
6100 IF P8(L5%)<ZC THEN GOTO 6140
6110 LC% = INT(L5%/2 + XG/YG*.25 + .75): IF LC%<1 THEN GOTO 6150 6120 IF
     LC%>16 THEN LC% = 1
6130 PSET(L4%,L3%),SP%(LC%): GOTO 6150
```

Appendix

```
6140 NEXT L5%
6150 NEXT L4%: NEXT L3%
6160 IF DP<1 THEN GOTO 6200
6170 FOR L3% = 1 TO 3
6180 R3#(L3%,1) = R1#(L3%,1): R3#(L3%,2) = R1#(L3%,2): R3#(L3%,3) = R1#(L3%,3)
6190 NEXT L3%
6200 RETURN
6210 L8% = 0: IF DP<2 THEN GOTO 6310
6220 L8% = 1
6230 REM rotate triangle for orthogonal profiles
6240 FOR L3% = 1 TO 3
6250 Z# = R3#(L3%,3): R3#(L3%,3) = 1 - R3#(L3%,L8%): R3#(L3%,L8%) = Z#
6260 NEXT L3%
6270 ST# = R3#(1,3)
6280 IF ST#>R3#(2,3) THEN ST# = R3#(2,3)
6290 IF ST#>R3#(3,3) THEN ST# = R3#(3,3)
6300 IF ST# < = - 999 THEN RETURN
6310 TP# = R3#(1,3)
6320 IF TP#<R3#(2,3) THEN TP# = R3#(2,3)
6330 IF TP#<R3#(3,3) THEN TP# = R3#(3,3)
6340 IF TP# = ST# THEN RETURN
6350 FOR L0% = 1 TO NC%
6360 REM slice the triangle
6370 IF P8(L0%)>TP# OR P8(L0%)<ST# THEN GOTO 6610
6380 L5% = 1: L4% = 0
6390 L4% = L4% + 1
6400 L7% = 1: IF L7% = L4% THEN L7% = L7% + 1
6410 L3% = L7% + 1: IF L3% = L4% THEN L3% = L3% + 1
6420 IF R3#(L7%,3) = R3#(L3%,3) THEN GOTO 6460
6430 F = (P8(L0%) - R3#(L7%,3))/(R3#(L3%,3) - R3#(L7%,3)): IF F<0 OR F>1 THEN
     GOTO 6460
6440 S3(L5%,1) = R3#(L7%,1) + (R3#(L3%,1) - R3#(L7%,1))*F: S3(L5%,2) = R3#(L7%,2)
     + (R3#(L3%,2) - R3#(L7%,2))*F
6450 L5% = L5% + 1
6460 IF L5%<3 THEN GOTO 6390
6470 S3(1,3) = P8(L0%): S3(2,3) = P8(L0%): IF DP<2 THEN GOTO 6500
6480 REM reverse rotate the intersection trace
6490 FOR L3% = 1 TO 2: Z# = S3(L3%,3): S3(L3%,3) = S3(L3%,L8%): S3(L3%,L8%) = 1
     - Z#: NEXT L3%
6500 IF DP = 0 THEN GOTO 6552
6510 REM apply perspective
6520 FOR L3% = 1 TO 2
6530 YY = (S3(L3%,2) - .5)*R8(1,1) + (S3(L3%,1) - .5)*R8(1,2): S3(L3%,1) = (S3(L3%,1) -
     .5)*R8(1,1) - (S3(L3%,2) - .5)*R8(1,2) + .5
6540 S33 = (S3(L3%,3) - .5)*R8(2,1) + YY*R8(2,2) + .5: S3(L3%,2) = YY*R8(2,1) -
     (S3(L3%,3) - .5)*R8(2,2) + .5: S3(L3%,3) = S33
6550 NEXT L3%
6552 IF SR<1 THEN GOTO 6580
6554 S3(1,1) = (S3(1,1) - .5)*.99875 - (S3(1,3) - .5)*.05 + .5: S3(1,3) = (S3(1,3) -
     .5)*.99875 + (S3(1,1) - .5)*.05 + .5: S3(2,1) = (S3(2,1) - .5)*.99875 - (S3(2,3) - .5)*.05 +
     .5: S3(2,3) = (S3(2,3) - .5)*.99875 + (S3(2,1) - .5)*.05 + .5
6560 REM draw a line from (S3(1,1),S3(1,2)) to (S3(2,1),S3(2,2))
6580 LINE((SW + S3(1,1)*SZ),(ZW - S3(1,2)*ZS)) - ((SW + S3(2,1)*SZ),(ZW -
     S3(2,2)*ZS)),14: IF SR<1 OR DP<1 THEN GOTO 6610
6590 S3(1,1) = (S3(1,1) - .5)*.995 + (S3(1,3) - .5)*.1 + .5: S3(2,1) = (S3(2,1) - .5)*.995 +
     (S3(2,3) - .5)*.1 + .5
6600 LINE((SV + S3(1,1)*SZ),(ZW - S3(1,2)*ZS)) - ((SV + S3(2,1)*SZ),(ZW -
     S3(2,2)*ZS)),14
6610 NEXT L0%
6620 IF L8%<1 THEN RETURN
6630 REM reverse rotate the triangle
6640 FOR L3% = 1 TO 3
6650 Z# = R3#(L3%,3): R3#(L3%,3) = R3#(L3%,L8%): R3#(L3%,L8%) = 1 - Z#
6660 NEXT L3%
6670 L8% = L8% + 1: IF L8%<3 THEN GOTO 6240 ELSE RETURN
6680 REM solve R3#(LL%,LL%)xB#(LL%) = R9#(LL%)
```

```
6690 FOR J2% = 1 TO LL%: P4%(J2%) = J2%: B#(J2%) = 0
6700 FOR J3% = 1 TO LL%: IF ABS(R3#(J2%,J3%))> = B#(J2%) THEN B#(J2%) =
     ABS(R3#(J2%,J3%))
6710 NEXT J3%: NEXT J2%: LM1% = LL% - 1: LP1% = LL% + 1
6720 FOR J2% = 1 TO LM1%: X3# = 0
6730 FOR J3% = J2% TO LL%: X4# = ABS(R3#(P4%(J3%),J2%))/B#(P4%(J3%)): IF
     X4#< = X3# THEN GOTO 6750
6740 X3# = X4#: K3% = J3%
6750 NEXT J3%: IF K3% = J2% GOTO 6770
6760 K2% = P4%(J2%): P4%(J2%) = P4%(K3%): P4%(K3%) = K2% 6770 X3# =
     R3#(P4%(J2%),J2%): K4% = J2% + 1
6780 FOR J3% = K4% TO LL%: X4# = - R3#(P4%(J3%),J2%)/X3#: R3#(P4%(J3%),J2%)
     = - X4#
6790 FOR J4% = K4% TO LL%: R3#(P4%(J3%),J4%) = R3#(P4%(J3%),J4%) +
     X4#*R3#(P4%(J2%),J4%): NEXT J4%: NEXT J3% : NEXT J2%
6800 B#(1) = R9#(P4%(1))
6810 FOR J1% = 2 TO LL%
6820 I8% = J1% - 1: X3# = 0
6830 FOR J2% = 1 TO I8%: X3# = X3# + R3#(P4%(J1%),J2%)*B#(J2%): NEXT J2%
6840 B#(J1%) = R9#(P4%(J1%)) - X3#: NEXT J1%
6850 B#(LL%) = B#(LL%)/R3#(P4%(LL%),LL%)
6860 FOR J1% = 2 TO LL%: I8% = LP1% - J1%: I9 = I8% + 1: X3# = 0
6870 FOR J2% = I9 TO LL%: X3# = X3# + R3#(P4%(I8%),J2%)*B#(J2%): NEXT J2%
6880 B#(I8%) = (B#(I8%) - X3#)/R3#(P4%(I8%),I8%): NEXT J1%: RETURN


1000 REM SCRN\RECTANGL.BAS RECTANGLE - BASED INTERPOLATION
     Copyright (c) 1988 D.F.Watson
1010 DIM P1#(63,6), P5#(50), P8(80), P9(2,3), S1(2033), S2(3), S4(50), PO(4,2)
1020 DIM R1#(3,3), R3#(50,50), R4#(3,2), R5#(3,3), R8(2,2), S3(2,3), BX(6,3), R9#(50),
     B#(55)
1030 DIM P4%(50), P6(50), R6(3,2), SP%(16)
1040 DATA 0,8,1,4,5,9,6,12,13,2,7,3,10,11,14,15: REM spectrum
1050 DATA 1,0,0,1,1,0,0,1,0,0,0,0,0,1,0,0,1,0,1, - 1, - 1,0,1,2,5, - 1,0,2,3, - 1,5,1E37,3,1
1060 DEF FNCV(RT#) = RT#^2*LOG(RT#): REM minimum curvature: REM DEF
     FNCV(RT#) = RT#^2*(LOG(RT#) - 1): REM biharmonic spline: REM DEF FNCV(RT#)
     = RT#: REM dummy function
1070 REM DEF FNCV(RT#) = NG + SI*(1 -  EXP( - RT#/RG)): NG = .25: SI = 1: RG = 1/3:
     REM exponential semivariogram
1080 REM DEF FNCV(RT#) = NG + SI*(1.5*RT#/RG - (RT#/RG)^3/2): NG = .25: SI = 1:
     RG = 1: REM spherical semivariogram
1090 P9(1,1) = - - 1E37: P9(1,2) = - 1E37: P9(1,3) = - 1E37: P9(2,1) = 1E37: P9(2,2) =
     1E37: P9(2,3) = 1E37
1100 SL = 2: SR = 0: SZ = 480: GR = 19: DP = 0: CL = 1: PT = 1: CR = 0: TB = 7: WD =
     640: HI = 350: LW = 5: AZ = - 25: TL = - 15: RD = 2: BI = 1.5: BJ = 7: IF DP = 0
     THEN SR = 0
1130 OPEN "A:\SHARPG.DAT" FOR INPUT AS #1
1170 FOR I1% = 1 TO 16: READ SP%(I1%): NEXT I1%: IF TB>6 THEN CL = 0
1180 INPUT#1,ND%,NC%: FOR I1% = 1 TO NC%: INPUT#1,P8(I1%): NEXT I1%: IF
     CR>0 AND NC%>31 THEN NC% = 31
1190 INPUT#1,XS,YS,DT: I1% = 1: JJ = 1: INPUT#1,FX,FY,DS,NS: P1#(I1%,1) = FX:
     P1#(I1%,2) = FY
1210 INPUT#1,P1#(I1%,3): P1#(I1%,1) = (P1#(I1%,1) - XS)/DT: P1#(I1%,2) = (P1#(I1%,2)
     - YS)/DT: IF ABS(P1#(I1%,1) - .5)<RD AND ABS(P1#(I1%,2) - .5)<RD THEN GOTO
     1270
1230 ND% = ND% - 1: IF I1%>ND% THEN GOTO 1360
1240 P1#(I1% + 1,1) = FX + JJ*DS: P1#(I1% + 1,2) = FY: JJ = JJ + 1: IF JJ< = NS THEN
     GOTO 1210
1250 FY = FY + DS: P1#(I1% + 1,2) = FY: JJ = 1: P1#(I1% + 1,2) = FY
1260 GOTO 1210
1270 IF P9(1,1)<P1#(I1%,1) THEN P9(1,1) = P1#(I1%,1)
1280 IF P9(2,1)>P1#(I1%,1) THEN P9(2,1) = P1#(I1%,1)
1290 IF P9(1,2)<P1#(I1%,2) THEN P9(1,2) = P1#(I1%,2)
1300 IF P9(2,2)>P1#(I1%,2) THEN P9(2,2) = P1#(I1%,2)
1310 IF P9(1,3)<P1#(I1%,3) THEN P9(1,3) = P1#(I1%,3)
1320 IF P9(2,3)>P1#(I1%,3) THEN P9(2,3) = P1#(I1%,3)
```

Appendix

```
1330 P1#(I1% + 1,2) = FY + JJ*DS: P1#(I1% + 1,1) = FX: JJ = JJ + 1: IF JJ< = NS THEN
     GOTO 1350
1340 FX = FX + DS: P1#(I1% + 1,1) = FX: JJ = 1: P1#(I1% + 1,2) = FY 1350 I1% = I1% +
     1: IF I1%< = ND% THEN GOTO 1210
1360 IF ND%<4 THEN STOP
1370 XF = P9(1,1) - P9(2,1): YF = P9(1,2) - P9(2,2): VS = P9(1,3) - P9(2,3): R8(2,1) = AZ:
     R8(2,2) = TL: IF INT(ND%^.5 + .1)<NS THEN NS = INT(ND%^.5)
1380 FOR I1% = 1 TO 6: READ BX(I1%,1),BX(I1%,2),BX(I1%,3): NEXT I1% 1390 BN% =
     1: IF DP<1 THEN GOTO 1450
1400 BN% = 2: BX(2,1) = 0: BX(2,3) = 1: IF R8(2,1)>0 THEN GOTO 1420
1410 BX(2,1) = 1: BX(3,1) = 1: BX(4,1) = 1: BX(5,1) = 0: BX(6,1) = 0
1420 IF DP<2 OR CR>0 THEN GOTO 1450
1430 NC1% = INT(SZ/10): NC% = NC1% + 1: U0 = .9999/NC1%: U1 = .0000511 - U0
1440 FOR I1% = 1 TO NC%: U1 = U1 + U0: P8(I1%) = U1: NEXT I1%: GOTO 1460
1450 FOR I1% = 1 TO NC%: P8(I1%) = (P8(I1%) - P9(2,3))/VS + .00001*(RND - .5):
     NEXT I1%
1460 FOR I1% = 1 TO ND%: P1#(I1%,1) = P1#(I1%,1) + .0001*(RND - .5): P1#(I1%,2) =
     P1#(I1%,2) + .0001*(RND - .5) : P1#(I1%,3) = (P1#(I1%,3) - P9(2,3))/VS: NEXT I1%
1470 FOR I1% = 1 TO 3: READ R1#(I1%,1),R1#(I1%,2),R1#(I1%,3),R6#(I1%,1),R6#(I1%,2)
1480 P1#(I1% + ND%,1) = R1#(I1%,1)*XF + P9(2,1): P1#(I1% + ND%,2) = R1#(I1%,2)*YF
     + P9(2,2): P1#(I1% + ND%,3) = 0: NEXT I1%
1490 IF DP<1 THEN GOTO 1550
1500 FOR I1% = 1 TO 2: R8(I1%,1) = COS(R8(2,I1%)/57.29578): R8(I1%,2) =
     SIN(R8(2,I1%)/57.29578): NEXT I1%
1510 FOR I1% = 2 TO 6
1520 YY = (BX(I1%,2) - .5)*R8(1,1) + (BX(I1%,1) - .5)*R8(1,2) + .5: BX(I1%,1) =
     (BX(I1%,1) - .5)*R8(1,1) - (BX(I1%,2) - .5)*R8(1,2) + .5
1530 BX3 = (BX(I1%,3) - .5)*R8(2,1) + (YY - .5)*R8(2,2) + .5: BX(I1%,2) = (YY -
     .5)*R8(2,1) - (BX(I1%,3) - .5)*R8(2,2) + .5: BX(I1%,3) = BX3: NEXT I1%
1540 REM end data input, begin gradient estimation
1550 SOUND 208,5: SOUND 165,5: SOUND 185,5: SOUND 123,5
1560 IF CL<1 THEN GOTO 3830
1570 PRINT " AT " TIME$ " BEGIN GRADIENT ESTIMATION - PLEASE WAIT"
1580 FOR I0% = 1 TO ND%
2610 IF SL>2 THEN GOTO 2810 : REM least squares gradients
2620 P1#(I0%,4) = 0: P1#(I0%,5) = 0: P1#(I0%,6) = .2: L5% = 1
2641 R4#(1,1) = 0: R4#(2,1) = 0: R4#(3,1) = 0: R3#(2,1) = 0: R3#(2,2) = 0: R3#(3,1) = 0:
     R3#(3,2) = 0: R3#(3,3) = 0
2642 FOR I2% = - NS TO NS STEP NS: FOR I3% = - 1 TO 1 STEP 1: P6(L5%) = I0% +
     I2% + I3%: IF P6(L5%)<1 OR P6(L5%)>ND% THEN GOTO 2745
2730 R4#(1,1) = R4#(1,1) + P1#(P6(L5%),3): R4#(2,1) = R4#(2,1) + P1#(P6(L5%),1)
     *P1#(P6(L5%),3): R4#(3,1) = R4#(3,1) + P1#(P6(L5%),2)*P1#(P6(L5%),3)
2740 R3#(2,1) = R3#(2,1) + P1#(P6(L5%),1): R3#(2,2) = R3#(2,2) + P1#(P6(L5%),1)^2:
     R3#(3,1) = R3#(3,1) + P1#(P6(L5%),2): R3#(3,2) = R3#(3,2) +
     P1#(P6(L5%),1)*P1#(P6(L5%),2): R3#(3,3) = R3#(3,3) + P1#(P6(L5%),2)^2: L5% =
     L5% + 1
2745 NEXT I3%: NEXT I2%: L5% = L5% - 1
2770 U1# = L5%*(R3#(2,2)*R3#(3,3) - R3#(3,2)^2) - R3#(2,1)*(R3#(2,1)*R3#(3,3) -
     R3#(3,1)*R3#(3,2)) + R3#(3,1)*(R3#(2,1)*R3#(3,2) - R3#(3,1)*R3#(2,2))
2780 P1#(I0%,4) = - (L5%*(R4#(2,1)*R3#(3,3) - R3#(3,2)*R4#(3,1)) -
     R4#(1,1)*(R3#(2,1)*R3#(3,3) - R3#(3,1)*R3#(3,2)) + R3#(3,1)*(R3#(2,1)*R4#(3,1) -
     R3#(3,1)*R4#(2,1))/U1#
2790 P1#(I0%,5) = - (L5%*(R3#(2,2)*R4#(3,1) - R3#(3,2)*R4#(2,1)) -
     R3#(2,1)*(R3#(2,1)*R4#(3,1) - R3#(3,1)*R4#(2,1)) + R4#(1,1)*(R3#(2,1)*R3#(3,2) -
     R3#(3,1)*R3#(2,2))/U1#: GOTO 3080
2810 IF SL>3 THEN GOTO 3035 : REM spline gradients
2815 P1#(I0%,4) = 0: P1#(I0%,5) = 0: P1#(I0%,6) = .2: L5% = 1
2820 R5#(1,1) = P1#(I0%,1): R5#(1,2) = P1#(I0%,2): R5#(2,1) = R5#(1,1) + .0001 :
     R5#(2,2) = R5#(1,2): R5#(3,1) = R5#(1,1): R5#(3,2) = R5#(1,2) + .0001
2822 FOR I2% = - NS TO NS STEP NS: FOR I3% = - 1 TO 1 STEP 1: P6(L5%) = I0% +
     I2% + I3%: IF P6(L5%)<1 OR P6(L5%)>ND% THEN GOTO 2826
2824 L5% = L5% + 1
2826 NEXT I3%: NEXT I2%: L5% = L5% - 1
2900 LL% = L5% + 3
2910 FOR I1% = 1 TO L5%
```

```
2920 R3#(I1%,1) = 1: R3#(LL%,I1% + 3) = 1: R3#(I1%,2) = P1#(P6(I1%),1): R3#(L5% +
     1,I1% + 3) = R3#(I1%,2)
2930 R3#(I1%,3) = P1#(P6(I1%),2): R3#(L5% + 2,I1% + 3) = R3#(I1%,3): R9#(I1%) =
     P1#(P6(I1%),3): NEXT I1%
2940 FOR I1% = 1 TO 3: FOR I2% = 1 TO 3: R3#(I1% + L5%,I2%) = 0: NEXT I2%:
     R9#(I1% + L5%) = 0: NEXT I1%
2950 FOR I1% = 1 TO L5%: R3#(I1%,I1% + 3) = 0: IJ = I1% + 1: IF IJ>L5% THEN GOTO
     2970
2960 FOR I2% = IJ TO L5%: RT# = ((R3#(I1%,2) - R3#(I2%,2))^2 + (R3#(I1%,3) -
     R3#(I2%,3))^2)^.5: R3#(I2%,I1% + 3) = FNCV(RT#): R3#(I1%,I2% + 3) = R3#(I2%,I1%
     + 3): NEXT I2%
2970 NEXT I1%: GOSUB 6690
2980 TR# = 0: TS# = 0
2990 FOR I1% = 4 TO LL%
3000 RT# = ((P1#(P6(I1% - 3),1) - R5#(2,1))^2 + (P1#(P6(I1% - 3),2) - R5#(2,2))^2)^.5:
     TR# = TR# + B#(I1%)*FNCV(RT#)
3010 RS# = ((P1#(P6(I1% - 3),1) - R5#(3,1))^2 + (P1#(P6(I1% - 3),2) - R5#(3,2))^2)^.5:
     TS# = TS# + B#(I1%)*FNCV(RS#) : NEXT I1%
3020 R5#(2,3) = B#(1) + B#(2)*R5#(2,1) + B#(3)*R5#(2,2) + TR#: P1#(I0%,4) =
     (P1#(I0%,3) - R5#(2,3))/.0001
3030 R5#(3,3) = B#(1) + B#(2)*R5#(3,1) + B#(3)*R5#(3,2) + TS#: P1#(I0%,5) =
     (P1#(I0%,3) - R5#(3,3))/.0001: GOTO 3080
3033 REM hyperboloid gradients
3035 R3#(5,1) = 1: FOR I1% = 1 TO 4: R9#(I1%) = 0: FOR I2% = 1 TO 4: R3#(I1%,I2%) =
     0: NEXT I2%: NEXT I1%: LL% = 1
3040 FOR I1% = - NS TO NS STEP NS: FOR I2% = - 1 TO 1 STEP 1: P6%(LL%) = I0%
     + I1% + I2%: IF P6%(LL%)<1 OR P6%(LL%)>ND% THEN GOTO 3060
3050 R3#(5,2) = P1#(P6%(LL%),1): R3#(5,3) = P1#(P6%(LL%),2): R3#(5,4) =
     P1#(P6%(LL%),1)*P1#(P6%(LL%),2)
3053 FOR I3% = 1 TO 4: FOR I4% = 1 TO 4
3055 R3#(I3%,I4%) = R3#(I3%,I4%) + R3#(5,I3%)*R3#(5,I4%)
3056 NEXT I4%
3057 R9#(I3%) = R9#(I3%) + R3#(5,I3%)*P1#(P6%(LL%),3)
3058 NEXT I3%
3059 LL% = LL% + 1
3060 NEXT I2%
3070 NEXT I1%: LL% = 4
3071 GOSUB 6690
3072 P1#(I0%,4) = - B#(2) - B#(4)*P1#(I0%,2)
3073 P1#(I0%,5) = - B#(3) - B#(4)*P1#(I0%,1)
3074 P1#(I0%,6) = .2
3080 NEXT I0%
3820 REM make triangular grid
3830 PRINT " AT " TIME$ " BEGIN SURFACE INTERPOLATION  -  PLEASE WAIT":
     SOUND 123,5: SOUND 185,5: SOUND 208,5 : SOUND 165,5
3831 IF TB<7 THEN GOTO 3840 : REM minimum curvature spline
3833 FOR I1% = 1 TO 3: FOR I2% = 1 TO 3: R3#(I1% + ND%,I2%) = 0: NEXT I2%:
     R9#(I1% + ND%) = 0: NEXT I1%: LL% = ND% + 3
3835 FOR I1% = 1 TO ND%: R3#(I1%,1) = 1: R3#(LL%,I1% + 3) = 1: R3#(I1%,2) =
     P1#(I1%,1): R3#(ND% + 1,I1% + 3) = R3#(I1%,2) : R3#(I1%,3) = P1#(I1%,2):
     R3#(ND% + 2,I1% + 3) = R3#(I1%,3): R9#(I1%) = P1#(I1%,3): R3#(I1%,I1% + 3) = 0:
     IF I1% = 1 THEN GOTO 3838
3837 FOR I2% = I1% TO ND%: RT# = ((P1#(I2%,1) - R3#(I1% - 1,2))^2 + (P1#(I2%,2) -
     R3#(I1% - 1,3))^2)^.5: R3#(I2%,I1% + 2) = FNCV(RT#): R3#(I1% - 1,I2% + 3) =
     R3#(I2%,I1% + 2): NEXT I2%
3838 NEXT I1%: GOSUB 6690
3840 I0% = GR + 1: IS = 1: I2% = 2*I0% + 1: I3% = 1 + INT(GR/3^.5): I4% = I3%*I2% +
     I0%: SX = 1/GR: SY = 1/(2*I3%): S2(3) = .000001
3850 WX# = S2(3): S2(1) = 1.00001: AA# = SX*SY*.5*DT*DT: SM# = 0: SA# = 0: WY# =
     1: U8 = 1: K5% = 4: OX = 0: OY = 0
3860 FOR L1% = 1 TO I4%: REM loop to 4980
3865 IF TB>6 THEN GOTO 4870
3870 S1(L1%) = - 999: RX# = 1 + (WX# - P9(2,1))*DT/DS: QX = INT(RX#): RY# = 1 +
     (WY# - P9(2,2))*DT/DS: QY = INT(RY#): IF QX<1 OR QX> = NS OR QY<1 OR QY> =
     NS THEN GOTO 4940
4558 S1(L1%) = 0: IF QX = OX AND QY = OY THEN GOTO 4562
```

Appendix

```
4560 P6(1) = (QX - 1)*NS + QY: P6(2) = P6(1) + NS: P6(4) = P6(1) + 1: P6(3) = P6(4) +
     NS
4562 RX# = RX# - QX: RY# = RY# - QY
4563 P5#(1) = (1 - RX#)*(1 - RY#): P5#(2) = RX#*(1 - RY#): P5#(3) = RX#*RY#: P5#(4) =
     RY#*(1 - RX#)
4580 FOR L2% = 1 TO K5%: S1(L1%) = S1(L1%) + P5#(L2%)*P1#(P6(L2%),3): NEXT
     L2%: OX = QX: OY = QY
4590 SII = S1(L1%): IF CL<1 THEN GOTO 4940: REM blend bilinear and gradients
4620 FOR L2% = 1 TO K5%: S4(L2%) = 0
4630 IF P5#(L2%)<.00001 OR P5#(L2%)>1 THEN GOTO 4690
4640 IF ABS(P1#(P6(L2%),6))<.00001 THEN GOTO 4690
4650 RS# = ABS(P1#(P6(L2%),6)) + BI: RT# = RS#*BJ: RB# = 1/RT#: BD# =
     P5#(L2%)^RT#: BB# = BD#*2: IF BD#>.5 THEN BB# = (1 - BD#)*2
4660 BB# = BB#^RS#/2: IF BD#>.5 THEN BB# = 1 - BB#
4670 HP# = BB#^RB#
4680 S4(L2%) = ((P1#(P6(L2%),4)*P1#(P6(L2%),1) + P1#(P6(L2%),5)*P1#(P6(L2%),2) +
     P1#(P6(L2%),3) - P1#(P6(L2%),4)*WX# - P1#(P6(L2%),5)*WY#) - S1(L1%))*HP#
4690 NEXT L2%
4700 FOR L2% = 1 TO K5%: S1(L1%) = S1(L1%) + S4(L2%): NEXT L2%: GOTO 4940
4870 TR# = 0:   REM minimum curvature splines
4880 FOR L2% = 4 TO LL%: L3% = L2% - 3: RT# = ((P1#(L3%,1) - WX#)^2 +
     (P1#(L3%,2) - WY#)^2)^.5: TR# = TR# + B#(L2%)*FNCV(RT#): NEXT L2%: S1(L1%) =
     B#(1) + B#(2)*WX# + B#(3)*WY# + TR#
4940 IS = IS + SX*U8: IF IS>0 THEN GOTO 4960
4950 WX# = WX# + SX/2: IS = 1.001 - SX
4960 WX# = WX# + SX*U8: IF U8*WX#<S2(2 - U8) THEN GOTO 4980
4970 U8 = - U8: WX# = S2(2 + U8): WY# = WY# - SY: IS = U8
4980 NEXT L1%
5060 REM end interpolation to triangular grid and begin display construction
5080 SCREEN 9: COLOR 2,0: CLS: SOUND 208,5: SOUND 165,5: SOUND 185,5:
     SOUND 123,5: IF DP>0 THEN SZ = INT(SZ*2/3)
5090 IF SR>0 AND SZ>240 THEN SZ = 240
5100 ZS = INT(SZ*.71): SW = (WD - SZ)\2: ZW = (HI - LW)\2 + ZS\2: IF DP>0 THEN ZW
     = ZW - LW - 10
5110 IF SR>0 AND CR<1 AND DP>0 THEN SW = (WD - 2*SZ)\3 - 12
5120 SV = SW + SZ + 60: IF SR<1 THEN GOTO 5130
5125 FOR L1% = 2 TO 6: BX(L1%,1) = (BX(L1%,1) - .5)*.99875 - (BX(L1%,3) - .5)*.05 +
     .5: BX(L1%,3) = (BX(L1%,3) - .5)*.99875 + (BX(L1%,1) - .5)*.05 + .5: NEXT L1%
5130 FOR L1% = BN% + 1 TO BN% + 4: X1 = SW + BX(L1% - 1,1)*SZ: Y1 = ZW -
     BX(L1% - 1,2)*ZS: X2 = SW + BX(L1%,1)*SZ: Y2 = ZW - BX(L1%,2)*ZS: LINE(X1,Y1)
     - (X2,Y2),14: NEXT L1%: IF SR<1 OR SR<1 OR CR>0 THEN GOTO 5210
5140 FOR L1% = 2 TO 6: BX(L1%,1) = (BX(L1%,1) - .5)*.995 + (BX(L1%,3) - .5)*.1 + .5:
     NEXT L1%
5150 FOR L1% = BN% + 1 TO BN% + 4: X1 = SV + BX(L1% - 1,1)*SZ: Y1 = ZW -
     BX(L1% - 1,2)*ZS: X2 = SV + BX(L1%,1)*SZ: Y2 = ZW - BX(L1%,2)*ZS: LINE(X1,Y1) -
     (X2,Y2),14: NEXT L1%
5200 REM plot contours
5210 SW = SW + 1: SZ = SZ - 2: ZS = ZS - 2: ZW = ZW - 1: ZX = SX: ZY = SY: ZZ = .5:
     R3#(2,2) = 1 + 2*SY: R3#(2,1) = 0
5220 FOR L1% = 1 TO I3%
5230 R3#(1,2) = R3#(2,2) - 2*SY - 2*ZY: R3#(1,1) = R3#(2,1): N1% = (2*L1% - .5 +
     ZZ)*I0% + L1% + .5/ZZ: R3#(1,3) = S1(N1%): R3#(2,2) = R3#(2,2) - 2*SY - ZY:
     R3#(2,1) = .5 - ZZ: N2% = (2*L1% - .5 + ZZ)*I0% + L1%
5240 R3#(2,3) = S1(N2%): R3#(3,2) = R3#(2,2): R3#(3,1) = R3#(2,1) + ZX/2: N3% = N2%
     - .5/ZZ: R3#(3,3) = S1(N3%): AR# = AA#/2
5250 GOSUB 5820
5260 R3#(1,2) = R3#(1,2) + 2*ZY: N2% = (2*L1% - .5/ZZ - .5 - ZZ)*I0% + L1%: R3#(1,3) =
     S1(N2%) 5270 GOSUB 5820
5280 AR# = AA#
5290 FOR L2% = 1 TO GR
5300 R3#(2,2) = R3#(1,2): R3#(2,1) = R3#(1,1) + ZX: N4% = N2% + .5/ZZ: R3#(2,3) =
     S1(N4%)
5310 GOSUB 5820
5320 R3#(1,2) = R3#(3,2): R3#(1,1) = R3#(3,1) + ZX
5330 IF L2% = GR THEN R3#(1,1) = R3#(1,1) - ZX/2
5340 N6% = N3% - .5/ZZ: R3#(1,3) = S1(N6%): IF L2% = GR THEN AR# = AA#/2
```

BASIC Programs

```
5350 GOSUB 5820
5360 R3#(2,2) = R3#(2,2) - 2*ZY: N5% = N1% + .5/ZZ: R3#(2,3) = S1(N5%)
5370 GOSUB 5820
5380 R3#(1,2) = R3#(2,2): R3#(1,1) = R3#(2,1) - ZX: R3#(1,3) = S1(N1%): IF L2% = GR
   THEN AR# = AA#
5390 GOSUB 5820
5400 N1% = N5%: N2% = N4%: N3% = N6%: R3#(1,2) = R3#(1,2) + 2*ZY: R3#(1,1) =
   R3#(2,1) : R3#(1,3) = S1(N4%): R3#(3,1) = R3#(3,1) + ZX: R3#(3,3) = S1(N6%)
5410 NEXT L2%
5420 ZY = - ZY: ZX = - ZX: ZZ = - ZZ
5430 NEXT L1%
5440 REM end plotting contours
5450 REM plot data points
5460 IF PT<1 OR CR>0 THEN GOTO 5790
5480 FOR L1% = 1 TO ND%
5490 IF P1#(L1%,1)< - .0001 OR P1#(L1%,1)>1.0001 OR P1#(L1%,2)< - .0001 OR
   P1#(L1%,2)>1.0001 THEN GOTO 5610
5500 PP1# = P1#(L1%,1): PP2# = P1#(L1%,2): PP3# = P1#(L1%,3): IF DP<1 THEN
   GOTO 5526
5510 YY = (PP2# - .5)*R8(1,1) + (PP1# - .5)*R8(1,2) + .5: PP1# = (PP1# - .5)*R8(1,1) -
   (PP2# - .5)*R8(1,2) + .5
5520 ZZ3 = (PP3# - .5)*R8(2,1) + (YY - .5)*R8(2,2) + .5: PP2# = (YY - .5)*R8(2,1) - (PP3#
   - .5)*R8(2,2) + .5: PP3# = ZZ3: IF SR<1 THEN GOTO 5526
5525 PPP# = PP1#: PP1# = (PP1# - .5)*.99875 - (PP3# - .5)*.05 + .5
5526 S3(1,1) = PP1#*SZ + SW - 2: S3(1,2) = ZW - PP2#*ZS - 2
5527 FOR L2% = 1 TO 3: FOR L3% = 1 TO 3: PSET(S3(1,1) + L2%,S3(1,2) + L3%),4:
   NEXT L3%: NEXT L2%
5528 IF SR<1 THEN GOTO 5610
5529 PP1# = (PPP# - .5)*.99875 + (PP3# - .5)*.05 + .5
5530 S3(1,1) = PP1#*SZ + SV - 2: S3(1,2) = ZW - PP2#*ZS - 2
5531 FOR L2% = 1 TO 3: FOR L3% = 1 TO 3: PSET(S3(1,1) + L2%,S3(1,2) + L3%),4:
   NEXT L3%: NEXT L2%
5610 NEXT L1%
5790 SOUND 247,5: SOUND 153,5: SOUND 185,5: SOUND 123,10: TM$ = TIME$: IF
   SR>0 THEN GOTO 5802
5800 SM# = SM#/SA#: LOCATE 12,1: PRINT "Average": PRINT "height of": PRINT
   "surface": PRINT USING "#####.###";SM#
5802 LOCATE 19,1: PRINT " ALL": PRINT "DONE!": PRINT " At": PRINT TM$ 5804
   PRINT "Press": INPUT "ENTER ",PM$: END
5810 REM curvemaker
5820 ST = R3#(1,3)
5830 IF ST>R3#(2,3) THEN ST = R3#(2,3)
5840 IF ST>R3#(3,3) THEN ST = R3#(3,3)
5850 IF ST< = - 999 THEN RETURN
5860 SM# = SM# + AR#*((R3#(1,3) + R3#(2,3) + R3#(3,3))/3*VS + P9(2,3)): SA# = SA# +
   AR#: IF CR<1 THEN GOTO 6210
5870 PO(1,1) = - 1E37: PO(2,1) = PO(1,1): PO(1,2) = 1E37: PO(2,2) = PO(1,2): S3(1,1) =
   R3#(1,3) : S3(1,2) = R3#(2,3): S3(1,3) = R3#(3,3): IF DP<1 THEN GOTO 5910
5880 FOR L3% = 1 TO 3: R1#(L3%,1) = R3#(L3%,1): R1#(L3%,2) = R3#(L3%,2):
   R1#(L3%,3) = R3#(L3%,3)
5890 YY = (R3#(L3%,2) - .5)*R8(1,1) + (R3#(L3%,1) - .5)*R8(1,2): R3#(L3%,1) =
   (R3#(L3%,1) - .5)*R8(1,1) - (R3#(L3%,2) - .5)*R8(1,2) + .5
5900 R33# = (R3#(L3%,3) - .5)*R8(2,1) + YY*R8(2,2) + .5: R3#(L3%,2) = YY*R8(2,1) -
   (R3#(L3%,3) - .5)*R8(2,2) + .5: R3#(L3%,3) = R33#: NEXT L3%
5910 FOR L3% = 1 TO 3
5920 IF PO(1,1)<R3#(L3%,1) THEN PO(1,1) = R3#(L3%,1)
5930 IF PO(2,1)<R3#(L3%,2) THEN PO(2,1) = R3#(L3%,2)
5940 IF PO(1,2)>R3#(L3%,1) THEN PO(1,2) = R3#(L3%,1)
5950 IF PO(2,2)>R3#(L3%,2) THEN PO(2,2) = R3#(L3%,2)
5960 NEXT L3%
5970 X1 = SW + INT(PO(1,2)*SZ): XH = X1 - (X1\2)*2 - .5: X2 = SW + INT(PO(1,1)*SZ):
   Y1 = ZW - INT(PO(2,2)*ZS): Y2 = ZW - INT(PO(2,1)*ZS): YG = Y2 - (Y2\2)*2 - .5
5980 ZJ# = ((R3#(2,1) - R3#(1,1))*(R3#(3,2) - R3#(1,2))) - ((R3#(3,1) - R3#(1,1))*(R3#(2,2)
   - R3#(1,2)))
5990 FOR L3% = Y2 TO Y1
6000 YG = - YG: YY = (ZW - L3%)/ZS: XG = XH
```

Appendix

```
6010 FOR L4% = X1 TO X2
6020 XG = - XG: XX = (L4% - SW)/SZ: ZI = 1
6030 FOR L5% = 1 TO 3
6040 L8% = 1: IF L8% = L5% THEN L8% = L8% + 1
6050 L9% = L8% + 1: IF L9% = L5% THEN L9% = L9% + 1
6060 S3(2,L5%) = (((R3#(L8%,1) - XX)*(R3#(L9%,2) - YY)) - ((R3#(L9%,1) -
     XX)*(R3#(L8%,2) - YY)))/ZJ#*ZI: IF S3(2,L5%)<0 THEN GOTO 6150
6070 ZI = - ZI: NEXT L5%
6080 ZC = S3(1,1)*S3(2,1) + S3(1,2)*S3(2,2) + S3(1,3)*S3(2,3): IF ZC<P8(1) THEN
     GOTO 6150
6090 FOR L5% = 2 TO NC%
6100 IF P8(L5%)<ZC THEN GOTO 6140
6110 LC = INT(L5%/2 + XG/YG*.25 + .75): IF LC<1 THEN GOTO 6150
6120 IF LC>16 THEN LC = 1
6130 PSET(L4%,L3%),SP%(LC): GOTO 6150
6140 NEXT L5%
6150 NEXT L4%: NEXT L3%
6160 IF DP<1 THEN GOTO 6200
6170 FOR L3% = 1 TO 3
6180 R3#(L3%,1) = R1#(L3%,1): R3#(L3%,2) = R1#(L3%,2): R3#(L3%,3) = R1#(L3%,3)
6190 NEXT L3%
6200 RETURN
6210 L8% = 0: IF DP<2 THEN GOTO 6310
6220 L8% = 1
6230 REM rotate triangle for orthogonal profiles
6240 FOR L3% = 1 TO 3
6250 Z# = R3#(L3%,3): R3#(L3%,3) = 1 - R3#(L3%,L8%): R3#(L3%,L8%) = Z# 6260
     NEXT L3%
6270 ST = R3#(1,3)
6280 IF ST>R3#(2,3) THEN ST = R3#(2,3)
6290 IF ST>R3#(3,3) THEN ST = R3#(3,3)
6300 IF ST < = - 999 THEN RETURN
6310 TP = R3#(1,3)
6320 IF TP<R3#(2,3) THEN TP = R3#(2,3)
6330 IF TP<R3#(3,3) THEN TP = R3#(3,3)
6340 IF TP = ST THEN RETURN
6350 FOR L0% = 1 TO NC%
6360 REM slice the triangle
6370 IF P8(L0%)>TP OR P8(L0%)<ST THEN GOTO 6610
6380 L5% = 1: L4% = 0
6390 L4% = L4% + 1
6400 L7% = 1: IF L7% = L4% THEN L7% = L7% + 1
6410 L3% = L7% + 1: IF L3% = L4% THEN L3% = L3% + 1
6420 IF R3#(L7%,3) = R3#(L3%,3) THEN GOTO 6460
6430 F = (P8(L0%) - R3#(L7%,3))/(R3#(L3%,3) - R3#(L7%,3)): IF F<0 OR F>1 THEN
     GOTO 6460
6440 S3(L5%,1) = R3#(L7%,1) + (R3#(L3%,1) - R3#(L7%,1))*F: S3(L5%,2) = R3#(L7%,2)
     + (R3#(L3%,2) - R3#(L7%,2))*F
6450 L5% = L5% + 1 6460 IF L5%<3 THEN GOTO 6390
6470 S3(1,3) = P8(L0%): S3(2,3) = P8(L0%): IF DP<2 THEN GOTO 6500
6480 REM reverse rotate the intersection trace
6490 FOR L3% = 1 TO 2: Z# = S3(L3%,3): S3(L3%,3) = S3(L3%,L8%): S3(L3%,L8%) = 1
     - Z#: NEXT L3%
6500 IF DP = 0 THEN GOTO 6552
6510 REM apply perspective
6520 FOR L3% = 1 TO 2
6530 YY = (S3(L3%,2) - .5)*R8(1,1) + (S3(L3%,1) - .5)*R8(1,2): S3(L3%,1) = (S3(L3%,1) -
     .5)*R8(1,1) - (S3(L3%,2) - .5)*R8(1,2) + .5
6540 S33 = (S3(L3%,3) - .5)*R8(2,1) + YY*R8(2,2) + .5: S3(L3%,2) = YY*R8(2,1) -
     (S3(L3%,3) - .5)*R8(2,2) + .5: S3(L3%,3) = S33
6550 NEXT L3%
6552 IF SR<1 THEN GOTO 6580
6554 S3(1,1) = (S3(1,1) - .5)*.99875 - (S3(1,3) - .5)*.05 + .5: S3(1,3) = (S3(1,3) -
     .5)*.99875 + (S3(1,1) - .5)*.05 + .5: S3(2,1) = (S3(2,1) - .5)*.99875 - (S3(2,3) - .5)*.05 +
     .5: S3(2,3) = (S3(2,3) - .5)*.99875 + (S3(2,1) - .5)*.05 + .5
6560 REM draw a line from (S3(1,1),S3(1,2)) to (S3(2,1),S3(2,2))
```

```
6580 LINE((SW + S3(1,1)*SZ),(ZW - S3(1,2)*ZS)) - ((SW + S3(2,1)*SZ),(ZW -
     S3(2,2)*ZS)),14: IF SR<1 OR DP<1 THEN GOTO 6610
6590 S3(1,1) = (S3(1,1) - .5)*.995 + (S3(1,3) - .5)*.1 + .5: S3(2,1) = (S3(2,1) - .5)*.995 +
     (S3(2,3) - .5)*.1 + .5
6600 LINE((SV + S3(1,1)*SZ),(ZW - S3(1,2)*ZS)) - ((SV + S3(2,1)*SZ),(ZW -
     S3(2,2)*ZS)),14
6610 NEXT L0%
6620 IF L8%<1 THEN RETURN
6630 REM reverse rotate the triangle
6640 FOR L3% = 1 TO 3
6650 Z# = R3#(L3%,3): R3#(L3%,3) = R3#(L3%,L8%): R3#(L3%,L8%) = 1 - Z#
6660 NEXT L3%
6670 L8% = L8% + 1: IF L8%<3 THEN GOTO 6240 ELSE RETURN
6680 REM solve R3#(LL%,LL%)xB#(LL%) = R9#(LL%)
6690 FOR J2% = 1 TO LL%: P4%(J2%) = J2%: B#(J2%) = 0
6700 FOR J3% = 1 TO LL%: IF ABS(R3#(J2%,J3%))> = B#(J2%) THEN B#(J2%) =
     ABS(R3#(J2%,J3%))
6710 NEXT J3%: NEXT J2%: LM1 = LL% - 1: LP1 = LL% + 1
6720 FOR J2% = 1 TO LM1: X3# = 0
6730 FOR J3% = J2% TO LL%: X4# = ABS(R3#(P4%(J3%),J2%))/B#(P4%(J3%)): IF
     X4#< = X3# THEN GOTO 6750
6740 X3# = X4#: K3% = J3%
6750 NEXT J3%: IF K3% = J2% GOTO 6770
6760 K2% = P4%(J2%): P4%(J2%) = P4%(K3%): P4%(K3%) = K2%
6770 X3# = R3#(P4%(J2%),J2%): K4% = J2% + 1
6780 FOR J3% = K4% TO LL%: X4# = - R3#(P4%(J3%),J2%)/X3#: R3#(P4%(J3%),J2%)
     = - X4#
6790 FOR J4% = K4% TO LL%: R3#(P4%(J3%),J4%) = R3#(P4%(J3%),J4%) +
     X4#*R3#(P4%(J2%),J4%): NEXT J4%: NEXT J3% : NEXT J2%
6800 B#(1) = R9#(P4%(1))
6810 FOR J1% = 2 TO LL%
6820 I8 = J1% - 1: X3# = 0
6830 FOR J2% = 1 TO I8: X3# = X3# + R3#(P4%(J1%),J2%)*B#(J2%): NEXT J2% 6840
     B#(J1%) = R9#(P4%(J1%)) - X3#: NEXT J1%
6850 B#(LL%) = B#(LL%)/R3#(P4%(LL%),LL%)
6860 FOR J1% = 2 TO LL%: I8 = LP1 - J1%: I9 = I8 + 1: X3# = 0
6870 FOR J2% = I9 TO LL%: X3# = X3# + R3#(P4%(I8),J2%)*B#(J2%): NEXT J2% 6880
     B#(I8) = (B#(I8) - X3#)/R3#(P4%(I8),I8): NEXT J1%: RETURN


1000 REM PLOT\CON2R.BAS NEIGHBORHOOD - BASED INTERPOLATION
     Copyright (c) 1988 D.F.Watson
1010 DIM P1#(299,6), P2#(593,3), P5#(50), P8(80), P9(2,3), S1(1067), S2(3), S4(50),
     PO(4,2)
1020 DIM R1#(3,3), R3#(3,3), R4#(3,2), R5#(3,3), R7#(3), R8(2,2), S3(2,3), BX(6,3)
1030 DIM P3%(593,3), P4%(593), P6%(50), P7%(25,2), R2%(2,2), R6%(3,2)
1050 DATA 1,0,0,1,1,0,0,1,0,0,0,0,1,0,0,1,0,1, - 1, - 1,0,1,2,5, - 1,0,2,3, - 1,5,1E37,3,1
1090 P9(1,1) = - 1E37: P9(1,2) = - 1E37: P9(1,3) = - 1E37: P9(2,1) = 1E37: P9(2,2) =
     1E37: P9(2,3) = 1E37
1100 SR = 0: SZ = 5100: GR = 19: DP = 0: CL = 1: PT = 1: DZ = 15: WD = 10000: HI =
     6800: AZ = - 25: TL = - 15: RD = 2: BI = 1.5: BJ = 7: IF DP = 0 THEN SR = 0
1140 OPEN "A: \HILL.DAT" FOR INPUT AS #1
1150 REM OPEN "A: \JDAVIS.DAT" FOR INPUT AS #1 1180 INPUT#1,ND%,NC%: RM =
     13/ND%: FOR I1% = 1 TO NC%: INPUT#1,P8(I1%): NEXT I1%: IF CR>0 AND
     NC%>31 THEN NC% = 31
1185 FOR I1% = 1 TO ND%: P4%(I1%) = I1%: NEXT I1%: FOR I1% = 1 TO ND%: I2% =
     INT(RND*ND% + .5): IF I2%<1 THEN I2% = 1
1187 IF I2%>ND% THEN I2% = ND%
1188 I3% = P4%(I1%): P4%(I1%) = P4%(I2%): P4%(I2%) = I3%: NEXT I1%
1190 INPUT#1,XS,YS,DT: I2% = 1
1220 I1% = P4%(I2%)
1222 INPUT#1,P1#(I1%,1),P1#(I1%,2),P1#(I1%,3): P1#(I1%,1) = (P1#(I1%,1) - XS)/DT:
     P1#(I1%,2) = (P1#(I1%,2) - YS)/DT: IF (P1#(I1%,1) - .5)^2 + (P1#(I1%,2) - .5)^2<RD
     THEN GOTO 1270
1230 ND% = ND% - 1: IF I1%>ND% THEN GOTO 1360
1260 GOTO 1222
```

Appendix

```
1270 IF P9(1,1)<P1#(I1%,1) THEN P9(1,1) = P1#(I1%,1)
1280 IF P9(2,1)>P1#(I1%,1) THEN P9(2,1) = P1#(I1%,1)
1290 IF P9(1,2)<P1#(I1%,2) THEN P9(1,2) = P1#(I1%,2)
1300 IF P9(2,2)>P1#(I1%,2) THEN P9(2,2) = P1#(I1%,2)
1310 IF P9(1,3)<P1#(I1%,3) THEN P9(1,3) = P1#(I1%,3)
1320 IF P9(2,3)>P1#(I1%,3) THEN P9(2,3) = P1#(I1%,3)
1350 I2% = I2% + 1: IF I2%< = ND% THEN GOTO 1220
1360 IF ND%<4 THEN STOP
1370 XF = P9(1,1) - P9(2,1): YF = P9(1,2) - P9(2,2): VS = P9(1,3) - P9(2,3): R8(2,1) = AZ:
     R8(2,2) = TL
1380 FOR I1% = 1 TO 6: READ BX(I1%,1),BX(I1%,2),BX(I1%,3): NEXT I1%
1390 BN% = 1: IF DP<1 THEN GOTO 1450
1400 BN% = 2: BX(2,1) = 0: BX(2,3) = 1: IF R8(2,1)>0 THEN GOTO 1420
1410 BX(2,1) = 1: BX(3,1) = 1: BX(4,1) = 1: BX(5,1) = 0: BX(6,1) = 0
1420 IF DP<2 OR CR>0 THEN GOTO 1450
1430 NC1% = INT(SZ/120): NC% = NC1% + 1: U0 = .9999/NC1%: U1 = .0000511 - U0
1440 FOR I1% = 1 TO NC%: U1 = U1 + U0: P8(I1%) = U1: NEXT I1%: GOTO 1460
1450 FOR I1% = 1 TO NC%: P8(I1%) = (P8(I1%) - P9(2,3))/VS + .00001*(RND - .5):
     NEXT I1%
1460 FOR I1% = 1 TO ND%: P1#(I1%,1) = P1#(I1%,1) + .0001*(RND - .5): P1#(I1%,2) =
     P1#(I1%,2) + .0001*(RND - .5) : P1#(I1%,3) = (P1#(I1%,3) - P9(2,3))/VS: NEXT I1%
1470 FOR I1% = 1 TO 3: READ R1#(I1%,1), R1#(I1%,2), R1#(I1%,3), R6%(I1%,1),
     R6%(I1%,2)
1480 P1#(I1% + ND%,1) = R1#(I1%,1)*XF + P9(2,1): P1#(I1% + ND%,2) = R1#(I1%,2)*YF
     + P9(2,2): P1#(I1% + ND%,3) = 0: NEXT I1%
1490 IF DP<1 THEN GOTO 1560
1500 FOR I1% = 1 TO 2: R8(I1%,1) = COS(R8(2,I1%)/57.29578): R8(I1%,2) =
     SIN(R8(2,I1%)/57.29578): NEXT I1%
1510 FOR I1% = 2 TO 6
1520 YY = (BX(I1%,2) - .5)*R8(1,1) + (BX(I1%,1) - .5)*R8(1,2) + .5: BX(I1%,1) =
     (BX(I1%,1) - .5)*R8(1,1) - (BX(I1%,2) - .5)*R8(1,2) + .5
1530 BX3 = (BX(I1%,3) - .5)*R8(2,1) + (YY - .5)*R8(2,2) + .5: BX(I1%,2) = (YY -
     .5)*R8(2,1) - (BX(I1%,3) - .5)*R8(2,2) + .5: BX(I1%,3) = BX3: NEXT I1%
1540 REM end data input, begin gradient estimation
1560 IF CL<1 THEN GOTO 3110
1570 PRINT " AT " TIME$ " BEGIN GRADIENT ESTIMATION  -  PLEASE WAIT"
1580 FOR I0% = 1 TO ND%
1590 FOR I1% = 1 TO 3: P3%(1,I1%) = ND% + I1%: P3%(2,I1%) = ND% + I1%:
     P2#(1,I1%) = R1#(I1%,3): P2#(2,I1%) = R1#(I1%,3) : NEXT I1%
1600 FOR I1% = 1 TO 593: P4%(I1%) = I1%: NEXT I1%: R2%(1,1) = 1: R2%(1,2) = 1:
     R2%(2,1) = 3: R2%(2,2) = 4 : K1% = I0%: K2% = I0%: K3% = ND% + 4: K4 =  - .5
1610 FOR I1% = K1% TO K2%
1620 IF I1% = K3% THEN GOTO 2000
1630 K5 = K4
1640 K6 = 1.5 + K5: IF K3%>ND% THEN K5 =  - K5
1650 U0 = 0: K7% = 0: L6% = K6 - 2
1660 FOR I2% = 1 TO R2%(1,K6)
1670 L6% = L6% + 2: IF P3%(L6%,1) = 0 THEN GOTO 1670
1680 U1# = P2#(L6%,3) - (P1#(I1%,1) - P2#(L6%,1))^2: IF U1#<0 THEN GOTO 1840
1690 U1# = U1# - (P1#(I1%,2) - P2#(L6%,2))^2: IF U1#<0 THEN GOTO 1840
1700 R2%(2,K6) = R2%(2,K6) - 2: P4%(R2%(2,K6)) = L6%: U0 = U0 - 1
1710 FOR I3% = 1 TO 3
1720 L1% = 1: IF L1% = I3% THEN L1% = L1% + 1
1730 L2% = L1% + 1: IF L2% = I3% THEN L2% = L2% + 1
1740 IF K7%<1 THEN GOTO 1810
1750 K9% = K7%
1760 FOR I4% = 1 TO K9%
1770 IF P3%(L6%,L1%)<>P7%(I4%,1) THEN GOTO 1800
1780 IF P3%(L6%,L2%)<>P7%(I4%,2) THEN GOTO 1800
1790 P7%(I4%,1) = P7%(K9%,1): P7%(I4%,2) = P7%(K9%,2): K7% = K7% - 1: GOTO
     1820
1800 NEXT I4%
1810 K7% = K7% + 1: P7%(K7%,1) = P3%(L6%,L1%): P7%(K7%,2) = P3%(L6%,L2%)
1820 NEXT I3%
1830 P3%(L6%,1) = 0
1840 NEXT I2%
```

```
1850 IF K7%<1 THEN GOTO 2000
1860 REM make new triangles
1870 FOR I2% = 1 TO K7% 1880 IF K5>0 THEN GOTO 1900
1890 IF P7%(I2%,1)<>I0% AND P7%(I2%,2)<>I0% THEN GOTO 1970
1900 FOR I3% = 1 TO 2
1910 R3#(I3%,1) = P1#(P7%(I2%,I3%),1) - P1#(I1%,1): R3#(I3%,2) =
     P1#(P7%(I2%,I3%),2) - P1#(I1%,2)
1920 R3#(I3%,3) = R3#(I3%,1)*(P1#(P7%(I2%,I3%),1) + P1#(I1%,1))/2 + R3#(I3%,2)
     *(P1#(P7%(I2%,I3%),2) + P1#(I1%,2))/2
1930 NEXT I3%
1940 U1# = R3#(1,1)*R3#(2,2) - R3#(2,1)*R3#(1,2): P2#(P4%(R2%(2,K6)),1) =
     (R3#(1,3)*R3#(2,2) - R3#(2,3)*R3#(1,2))/U1#: P2#(P4%(R2%(2,K6)),2) =
     (R3#(1,1)*R3#(2,3) - R3#(2,1)*R3#(1,3))/U1#
1950 P2#(P4%(R2%(2,K6)),3) = (P1#(I1%,1) - P2#(P4%(R2%(2,K6)),1))^2 + (P1#(I1%,2) -
     P2#(P4%(R2%(2,K6)),2))^2
1960 P3%(P4%(R2%(2,K6)),1) = P7%(I2%,1): P3%(P4%(R2%(2,K6)),2) = P7%(I2%,2) :
     P3%(P4%(R2%(2,K6)),3) = I1%: R2%(2,K6) = R2%(2,K6) + 2: U0 = U0 + 1
1970 NEXT I2%
1980 R2%(1,K6) = R2%(1,K6) + U0: IF K5>0 THEN GOTO 2000
1990 K5 = - K5: GOTO 1640
2000 NEXT I1%
2010 IF K2%>K1% GOTO 2040
2020 K1% = 1: K2% = ND%: K3% = I0%: GOTO 1610
2030 REM order triangles positively
2040 FOR I1% = 1 TO 2
2050 L8% = I1% - 2
2060 FOR I2% = 1 TO R2%(1,I1%)
2070 L8% = L8% + 2: IF P3%(L8%,1) = 0 THEN GOTO 2070
2080 IF P3%(L8%,1)>ND% THEN GOTO 2140
2090 FOR I3% = 1 TO 2
2100 R3#(1,I3%) = P1#(P3%(L8%,2),I3%) - P1#(P3%(L8%,1),I3%): R3#(2,I3%) =
     P1#(P3%(L8%,3),I3%) - P1#(P3%(L8%,1),I3%)
2110 NEXT I3%
2120 IF R3#(1,1)*R3#(2,2) - R3#(1,2)*R3#(2,1)> = 0 THEN GOTO 2140
2130 K8 = P3%(L8%,3): P3%(L8%,3) = P3%(L8%,2): P3%(L8%,2) = K8
2140 NEXT I2%
2150 NEXT I1%
2170 REM calculate natural neighbor coordinates
2180 R5#(1,1) = P1#(I0%,1): R5#(1,2) = P1#(I0%,2): R5#(2,1) = R5#(1,1) + .0001 :
     R5#(2,2) = R5#(1,2): R5#(3,1) = R5#(1,1): R5#(3,2) = R5#(1,2) + .0001
2190 FOR I1% = 1 TO 3
2200 L4% = 0: L5% = 0: L6% = 0
2210 FOR I2% = 1 TO R2%(1,2)
2220 L6% = L6% + 2: IF P3%(L6%,1) = 0 THEN GOTO 2220
2230 IF P3%(L6%,1)>ND% THEN GOTO 2410
2240 U1# = P2#(L6%,3) - (R5#(I1%,1) - P2#(L6%,1))^2: IF U1#<0 THEN GOTO 2410
2250 U1# = U1# - (R5#(I1%,2) - P2#(L6%,2))^2: IF U1#<0 THEN GOTO 2410
2260 FOR I3% = 1 TO 3
2270 FOR I4% = 1 TO 2
2280 R3#(I4%,1) = P1#(P3%(L6%,R6%(I3%,I4%)),1) - R5#(I1%,1): R3#(I4%,2) =
     P1#(P3%(L6%,R6%(I3%,I4%)),2) - R5#(I1%,2)
2290 R3#(I4%,3) = R3#(I4%,1)*(P1#(P3%(L6%,R6%(I3%,I4%)),1) + R5#(I1%,1))/2 +
     R3#(I4%,2) *(P1#(P3%(L6%,R6%(I3%,I4%)),2) + R5#(I1%,2))/2: NEXT I4%
2300 U1# = R3#(1,1)*R3#(2,2) - R3#(1,2)*R3#(2,1)
2310 R4#(R6%(I3%,2),1) = (R3#(1,3)*R3#(2,2) - R3#(2,3)*R3#(1,2))/U1#:
     R4#(R6%(I3%,2),2) = (R3#(1,1)*R3#(2,3) - R3#(2,1)*R3#(1,3))/U1#: NEXT I3%: L3%=0
2320 FOR I3% = 1 TO 3: R3#(3,I3%) = ((R4#(R6%(I3%,1),1) -
     P2#(L6%,1))*(R4#(R6%(I3%,2),2) - P2#(L6%,2)) - (R4#(R6%(I3%,2),1) -
     P2#(L6%,1))*(R4#(R6%(I3%,1),2) - P2#(L6%,2)))/2: IF R3#(3,I3%)>0 THEN L3% =
     L3% + 1
2330 NEXT I3%
2340 IF L3%>2 THEN L4% = 1
2350 FOR I3% = 1 TO 3: IF L5%<1 THEN GOTO 2390
2360 FOR I4% = 1 TO L5%: IF P3%(L6%,I3%)<>P6%(I4%) THEN GOTO 2380 2370
     P5#(I4%) = P5#(I4%) + R3#(3,I3%): GOTO 2400
2380 NEXT I4%
```

Appendix

```
2390 L5% = L5% + 1: P6%(L5%) = P3%(L6%,I3%): P5#(L5%) = R3#(3,I3%)
2400 NEXT I3%
2410 NEXT I2%: IF L4%<1 THEN GOTO 2480
2420 U4 = 0
2430 FOR I2% = 1 TO L5%: U4 = U4 + P5#(I2%): NEXT I2%: R7#(I1%) = 0
2440 FOR I2% = 1 TO L5%: P5#(I2%) = P5#(I2%)/U4: R7#(I1%) = R7#(I1%) +
     P5#(I2%)*P1#(P6%(I2%),3): NEXT I2%: NEXT I1%
2450 P1#(I0%,6) = P1#(I0%,3) - R7#(1): P1#(I0%,4) = (R7#(1) - R7#(2))/.0001:
     P1#(I0%,5) = (R7#(1) - R7#(3))/.0001 : GOTO 3080
2460 REM cross product gradients
2480 P1#(I0%,4) = 0: P1#(I0%,5) = 0: P1#(I0%,6) = 0: P7# = 0: L7% = - 1
2490 FOR I1% = 1 TO R2%(1,1)
2500 L7% = L7% + 2: IF P3%(L7%,1) = 0 THEN GOTO 2500
2510 IF P3%(L7%,1)>ND% THEN GOTO 2570
2520 FOR I2% = 1 TO 2: FOR I3% = 1 TO 3: R3#(I2%,I3%) = P1#(P3%(L7%,1),I3%) -
     P1#(P3%(L7%,I2% + 1),I3%): NEXT I3% : NEXT I2%
2530 R3#(3,1) = R3#(1,2)*R3#(2,3) - R3#(2,2)*R3#(1,3): R3#(3,2) = R3#(1,3)*R3#(2,1) -
     R3#(2,3)*R3#(1,1): R3#(3,3) = R3#(1,1)*R3#(2,2) - R3#(2,1)*R3#(1,2): U3 = 1
2540 IF R3#(3,3)<0 THEN U3 = - 1
2550 U2 = (R3#(3,1)^2 + R3#(3,2)^2 + R3#(3,3)^2)^.5: P7# = P7# + U2
2560 FOR I2% = 1 TO 3: P1#(I0%,I2% + 3) = P1#(I0%,I2% + 3) + R3#(3,I2%)*U3: NEXT
     I2%
2570 NEXT I1%
2580 U2 = P1#(I0%,4)^2 + P1#(I0%,5)^2 + P1#(I0%,6)^2: P7# = 1 - U2^.5/P7#
2590 P1#(I0%,4) = P1#(I0%,4)/P1#(I0%,6): P1#(I0%,5) = P1#(I0%,5)/P1#(I0%,6):
     P1#(I0%,6) = P7#
3080 NEXT I0%
3100 REM natural neighbor sort
3110 FOR I0% = 1 TO 3: P3%(1,I0%) = ND% + I0%: P2#(1,I0%) = R1#(I0%,3): NEXT I0%
3120 FOR I0% = 1 TO 593: P4%(I0%) = I0%: NEXT I0%: NT% = 1: L0% = 2
3130 FOR I0% = 1 TO ND%: L3% = 0
3140 FOR I1% = 1 TO NT%
3150 U1# = P2#(I1%,3) - (P1#(I0%,1) - P2#(I1%,1))^2: IF U1#<0 THEN GOTO 3300
3160 U1# = U1# - (P1#(I0%,2) - P2#(I1%,2))^2: IF U1#<0 THEN GOTO 3300
3170 L0% = L0% - 1: P4%(L0%) = I1%
3180 FOR I2% = 1 TO 3
3190 L1% = 1: IF L1% = I2% THEN L1% = L1% + 1
3200 L2% = L1% + 1: IF L2% = I2% THEN L2% = L2% + 1
3210 IF L3%<1 THEN GOTO 3280
3220 L5% = L3%
3230 FOR I3% = 1 TO L5%
3240 IF P3%(I1%,L1%)<>P7%(I3%,1) THEN GOTO 3270
3250 IF P3%(I1%,L2%)<>P7%(I3%,2) THEN GOTO 3270
3260 P7%(I3%,1) = P7%(L5%,1): P7%(I3%,2) = P7%(L5%,2): L3% = L3% - 1: GOTO
     3290
3270 NEXT I3%
3280 L3% = L3% + 1: P7%(L3%,1) = P3%(I1%,L1%): P7%(L3%,2) = P3%(I1%,L2%)
3290 NEXT I2%
3300 NEXT I1%
3310 FOR I1% = 1 TO L3%
3320 FOR I2% = 1 TO 2
3330 R3#(I2%,1) = P1#(P7%(I1%,I2%),1) - P1#(I0%,1): R3#(I2%,2) =
     P1#(P7%(I1%,I2%),2) - P1#(I0%,2): R3#(I2%,3) = R3#(I2%,1)*(P1#(P7%(I1%,I2%),1) +
     P1#(I0%,1))/2
3340 R3#(I2%,3) = R3#(I2%,3) + R3#(I2%,2)*(P1#(P7%(I1%,I2%),2) + P1#(I0%,2))/2
3350 NEXT I2%
3360 U1# = R3#(1,1)*R3#(2,2) - R3#(2,1)*R3#(1,2): P2#(P4%(L0%),1) =
     (R3#(1,3)*R3#(2,2) - R3#(2,3)*R3#(1,2))/U1#: P2#(P4%(L0%),2) = (R3#(1,1)*R3#(2,3)
     - R3#(2,1)*R3#(1,3))/U1#
3370 P2#(P4%(L0%),3) = (P1#(I0%,1) - P2#(P4%(L0%),1))^2 + (P1#(I0%,2) -
     P2#(P4%(L0%),2))^2
3380 P3%(P4%(L0%),1) = P7%(I1%,1): P3%(P4%(L0%),2) = P7%(I1%,2):
     P3%(P4%(L0%),3) = I0%: L0% = L0% + 1
3390 NEXT I1%
3400 NT% = NT% + 2: NEXT I0%
3410 FOR I0% = 1 TO NT%: IF P3%(I0%,1)>ND% THEN GOTO 3460
```

BASIC PROGRAMS

```
3420 FOR I1% = 1 TO 2
3430 R3#(1,I1%) = P1#(P3%(I0%,2),I1%) - P1#(P3%(I0%,1),I1%): R3#(2,I1%) =
     P1#(P3%(I0%,3),I1%) - P1#(P3%(I0%,1),I1%): NEXT I1%
3440 IF R3#(1,1)*R3#(2,2) - R3#(1,2)*R3#(2,1)> = 0 THEN GOTO 3460
3450 L4% = P3%(I0%,3): P3%(I0%,3) = P3%(I0%,2): P3%(I0%,2) = L4%
3460 NEXT I0%
3820 REM make triangular grid
3830 PRINT " AT " TIME$ " BEGIN SURFACE INTERPOLATION  -  PLEASE WAIT":
     SOUND 123,5: SOUND 185,5: SOUND 208,5 : SOUND 165,5
3840 I0% = GR + 1: IS = 1: I2% = 2*I0% + 1: I3% = 1 + INT(GR/3^.5): I4% = I3%*I2% +
     I0%: SX = 1/GR: SY = 1/(2*I3%): S2(3) = .000001
3850 WX# = S2(3): S2(1) = 1.00001: AA# = SX*SY*.5*DT*DT: WY# = 1: U8 = 1
3860 FOR L1% = 1 TO I4%
3870 S1(L1%) =  - 999: WXO# = WX#: WYO# = WY#
3880 REM .....calculate natural neighbor coordinates of WX#,WY# *********
3890 K4% = 0: K5% = 0
3900 FOR L2% = 1 TO NT%
3910 IF P3%(L2%,1)>ND% THEN GOTO 4280
3915 IF P2#(L2%,3)>RM THEN GOTO 4280
3920 U1# = P2#(L2%,3) - (WX# - P2#(L2%,1))^2: IF U1#<0 THEN GOTO 4280
3930 U1# = U1# - (WY# - P2#(L2%,2))^2: IF U1#<0 THEN GOTO 4280
3940 FOR L3% = 1 TO 3
3950 L4% = 1: IF L4% = L3% THEN L4% = L4% + 1
3960 L5% = L4% + 1: IF L5% = L3% THEN L5% = L5% + 1
3970 IF ABS((P1#(P3%(L2%,L5%),1) - P1#(P3%(L2%,L4%),1))*(WY# -
     P1#(P3%(L2%,L4%),2)) - (WX# - P1#(P3%(L2%,L4%),1))*(P1#(P3%(L2%,L5%),2) -
     P1#(P3%(L2%,L4%),2)))>.000001 THEN GOTO 4030
4000 DEN# = ABS(P1#(P3%(L2%,L4%),2) - P1#(P3%(L2%,L5%),2)) +
     ABS(P1#(P3%(L2%,L5%),1) - P1#(P3%(L2%,L4%),1))
4010 WX# = WX# + (P1#(P3%(L2%,L4%),2) - P1#(P3%(L2%,L5%),2))*.0001/DEN#: WY#
     = WY# + (P1#(P3%(L2%,L5%),1) - P1#(P3%(L2%,L4%),1))*.0001/DEN#
4030 NEXT L3%
4040 FOR L3% = 1 TO 3
4050 FOR L4% = 1 TO 2
4060 R3#(L4%,1) = P1#(P3%(L2%,R6%(L3%,L4%)),1) - WX#: R3#(L4%,2) =
     P1#(P3%(L2%,R6%(L3%,L4%)),2) - WY#
4070 R3#(L4%,3) = R3#(L4%,1)*(P1#(P3%(L2%,R6%(L3%,L4%)),1) + WX#)/2 +
     R3#(L4%,2) *(P1#(P3%(L2%,R6%(L3%,L4%)),2) + WY#)/2
4080 NEXT L4%
4090 U1# = R3#(1,1)*R3#(2,2) - R3#(1,2)*R3#(2,1)
4100 R4#(R6%(L3%,2),1) = (R3#(1,3)*R3#(2,2) - R3#(2,3)*R3#(1,2))/U1#
4110 R4#(R6%(L3%,2),2) = (R3#(1,1)*R3#(2,3) - R3#(2,1)*R3#(1,3))/U1# 4120 NEXT
     L3% 4130 K3% = 0
4140 FOR L3% = 1 TO 3
4150 R3#(3,L3%) = ((R4#(R6%(L3%,1),1) - P2#(L2%,1))*(R4#(R6%(L3%,2),2) -
     P2#(L2%,2)) - (R4#(R6%(L3%,2),1) - P2#(L2%,1))*(R4#(R6%(L3%,1),2) -
     P2#(L2%,2)))/2
4160 IF R3#(3,L3%)>0 THEN K3% = K3% + 1
4170 NEXT L3%
4180 IF K3%>2 THEN K4% = 1
4190 FOR L3% = 1 TO 3
4200 IF K5%<1 THEN GOTO 4260
4210 FOR L4% = 1 TO K5%
4220 IF P3%(L2%,L3%)<>P6%(L4%) THEN GOTO 4250
4230 P5#(L4%) = P5#(L4%) + R3#(3,L3%)
4240 GOTO 4270
4250 NEXT L4%
4260 K5% = K5% + 1: P6%(K5%) = P3%(L2%,L3%): P5#(K5%) = R3#(3,L3%)
4270 NEXT L3%
4280 NEXT L2%
4300 IF K4%<1 THEN GOTO 4940
4310 REM ....end determining K5% local coordinates *********
4560 S1(L1%) = 0: U4 = 0
4570 FOR L2% = 1 TO K5%: U4 = U4 + P5#(L2%): NEXT L2%
4580 FOR L2% = 1 TO K5%: P5#(L2%) = P5#(L2%)/U4: S1(L1%) = S1(L1%) +
     P5#(L2%)*P1#(P6%(L2%),3): NEXT L2%
```

277

```
4590 IF CL<1 THEN GOTO 4940
4620 FOR L2% = 1 TO K5%: S4(L2%) = 0
4630 IF P5#(L2%)<.00001 OR P5#(L2%)>1 THEN GOTO 4690
4640 IF ABS(P1#(P6%(L2%),6))<.00001 THEN GOTO 4690
4650 RS# = ABS(P1#(P6%(L2%),6)) + BI: RT# = RS#*BJ: RB# = 1/RT#: BD# =
     P5#(L2%)^RT#: BB# = BD#*2: IF BD#>.5 THEN BB# = (1 - BD#)*2
4660 BB# = BB#^RS#/2: IF BD#>.5 THEN BB# = 1 - BB#
4670 HP# = BB#^RB#
4680 S4(L2%) = ((P1#(P6%(L2%),4)*P1#(P6%(L2%),1) +
     P1#(P6%(L2%),5)*P1#(P6%(L2%),2) + P1#(P6%(L2%),3) - P1#(P6%(L2%),4)*WX# -
     P1#(P6%(L2%),5)*WY#) - S1(L1%))*HP#
4690 NEXT L2%
4700 FOR L2% = 1 TO K5%: S1(L1%) = S1(L1%) + S4(L2%): NEXT L2%
4940 IS = IS + SX*U8: IF IS>0 THEN GOTO 4960
4950 WXO# = WXO# + SX/2: IS = 1.001 - SX
4960 WX# = WXO# + SX*U8: WY# = WYO#: IF U8*WX#<S2(2 - U8) THEN GOTO 4980
4970 U8 = - U8: WX# = S2(2 + U8): WY# = WY# - SY: IS = U8
4980 NEXT L1%
5060 REM end interpolation to triangular grid and begin display construction
5070 SOUND 208,5: SOUND 165,5: SOUND 185,5: SOUND 123,5: PRINT " AT " TIME$;
5071 INPUT " PREPARE PLOTTER AND PRESS ENTER",PM$
5090 LW = (HI - SZ)/2: IF DP>0 THEN SZ = INT(SZ*2/3)
5091 IF DP>0 AND SZ>HI/2 THEN SZ = HI/2
5100 SW = (WD - SZ)/2: ZW = (HI - SZ)/2
5110 IF SR>0 THEN SW = (WD - 2*SZ)/3
5120 SW = SW + 1: SV = 2*SW + SZ: SR = SR*2: SZ = SZ - 2: SZ = SZ - 2: ZW = ZW -
     1: SU = SW: IF SR>0 THEN SU = (2*SW + SV)/3
5160 REM plot open
5170 LPRINT "IN;IP0,0," + STR$(WD) + "," + STR$(HI) + ";SP1;": IF SR<2 THEN GOTO
     5180
5175 FOR L1% = 2 TO 6: BX1 = (BX(L1%,1) - .5)*.99875 - (BX(L1%,3) - .5)*.05 + .5:
     BX(L1%,3) = (BX(L1%,3) - .5)*.99875 + (BX(L1%,1) - .5)*.05 + .5: BX(L1%,1) = BX1:
     NEXT L1%: LPRINT "PU;PA0,6800;"
5180 LPRINT "PU;PA" + STR$(BX(BN%,1)*SZ + SW) + "," + STR$(BX(BN%,2)*SZ +
     ZW) + ";PD;" : ZX = SX: ZY = SY: ZZ = .5: R3#(2,2) = 1 + 2*SY: R3#(2,1) = 0
5190 FOR L1% = BN% + 1 TO BN% + 4: LPRINT "PA" + STR$(BX(L1%,1)*SZ + SW) +
     "," + STR$(BX(L1%,2)*SZ + ZW) + ";": NEXT L1% : LPRINT "PU;"
5192 FOR L1% = 2 TO 6: BX(L1%,1) = (BX(L1%,1) - .5)*.995 + (BX(L1%,3) - .5)*.1 + .5:
     NEXT L1%: SM = 0: SA = 0
5200 REM plot contours
5220 FOR L1% = 1 TO I3%
5230 R3#(1,2) = R3#(2,2) - 2*SY - 2*ZY: R3#(1,1) = R3#(2,1): N1% = (2*L1% - .5 +
     ZZ)*I0% + L1% + .5/ZZ: R3#(1,3) = S1(N1%): R3#(2,2) = R3#(2,2) - 2*SY - ZY:
     R3#(2,1) = .5 - ZZ: N2% = (2*L1% - .5 + ZZ)*I0% + L1%
5240 R3#(2,3) = S1(N2%): R3#(3,2) = R3#(2,2): R3#(3,1) = R3#(2,1) + ZX/2: N3% = N2%
     - .5/ZZ: R3#(3,3) = S1(N3%): AR# = AA#/2
5250 GOSUB 5820
5260 R3#(1,2) = R3#(1,2) + 2*ZY: N2% = (2*L1% - .5/ZZ - .5 - ZZ)*I0% + L1%: R3#(1,3) =
     S1(N2%)
5270 GOSUB 5820
5280 AR# = AA#
5290 FOR L2% = 1 TO GR
5300 R3#(2,2) = R3#(1,2): R3#(2,1) = R3#(1,1) + ZX: N4% = N2% + .5/ZZ: R3#(2,3) =
     S1(N4%)
5310 GOSUB 5820
5320 R3#(1,2) = R3#(3,2): R3#(1,1) = R3#(3,1) + ZX
5330 IF L2% = GR THEN R3#(1,1) = R3#(1,1) - ZX/2
5340 N6% = N3% - .5/ZZ: R3#(1,3) = S1(N6%): IF L2% = GR THEN AR# = AA#/2
5350 GOSUB 5820
5360 R3#(2,2) = R3#(2,2) - 2*ZY: N5% = N1% + .5/ZZ: R3#(2,3) = S1(N5%)
5370 GOSUB 5820
5380 R3#(1,2) = R3#(2,2): R3#(1,1) = R3#(2,1) - ZX: R3#(1,3) = S1(N1%): IF L2% = GR
     THEN AR# = AA#
5390 GOSUB 5820
5400 N1% = N5%: N2% = N4%: N3% = N6%: R3#(1,2) = R3#(1,2) + 2*ZY: R3#(1,1) =
     R3#(2,1) : R3#(1,3) = S1(N4%): R3#(3,1) = R3#(3,1) + ZX: R3#(3,3) = S1(N6%)
```

```
5410 NEXT L2% 5420 ZY = - ZY: ZX = - ZX: ZZ = - ZZ
5430 NEXT L1%: REM end plotting contours
5450 REM plot data points
5460 IF PT<1 THEN GOTO 5770
5480 DZ% = DZ: FOR L1% = 1 TO ND%
5490 IF P1#(L1%,1)< - .0001 OR P1#(L1%,1)>1.0001 OR P1#(L1%,2)< - .0001 OR
     P1#(L1%,2)>1.0001 THEN GOTO 5610
5500 PP1# = P1#(L1%,1): PP2# = P1#(L1%,2): PP3# = P1#(L1%,3): IF DP<1 THEN
     GOTO 5530
5510 YY = (PP2# - .5)*R8(1,1) + (PP1# - .5)*R8(1,2) + .5: PP1# = (PP1# - .5)*R8(1,1) -
     (PP2# - .5)*R8(1,2) + .5
5520 ZZ3 = (PP3# - .5)*R8(2,1) + (YY - .5)*R8(2,2) + .5: PP2# = (YY - .5)*R8(2,1) - (PP3#
     - .5)*R8(2,2) + .5: PP3# = ZZ3: IF SR<2 THEN GOTO 5528
5525 PP1# = (PP1# - .5)*.99875 - (PP3# - .5)*.05 + .5
5528 IF SR<>1 THEN GOTO 5530
5529 PP1# = (PP1# - .5)*.99875 + (PP3# - .5)*.05 + .5
5530 S3(1,1) = PP1#*SZ + SW: S3(1,2) = PP2#*SZ + ZW
5540 LPRINT "PA" + STR$(S3(1,1) - DZ%) + "," + STR$(S3(1,2) + DZ%) + ";": LPRINT
     "PD;PA" + STR$(S3(1,1) + DZ%) + "," + STR$(S3(1,2) + DZ%) + ";": LPRINT "PA" +
     STR$(S3(1,1) + DZ%) + "," + STR$(S3(1,2) - DZ%) + ";:"
5550 LPRINT "PA" + STR$(S3(1,1) - DZ%) + "," + STR$(S3(1,2) - DZ%) + ";": LPRINT
     "PA" + STR$(S3(1,1) - DZ%) + "," + STR$(S3(1,2) + DZ%) + ";": LPRINT "PA" +
     STR$(S3(1,1) + DZ%) + "," + STR$(S3(1,2) - DZ%) + ";PU;"
5610 NEXT L1%
5770 SR = SR - 1: SW = SV: IF SR = 1 THEN GOTO 5180
5775 LPRINT "PU;PA" + STR$(SU) + "," + STR$(LW - 500) + ";SI0.2,0.3;SL0.2;"
5777 LPRINT "LBFigure 1.0.1  -  Example Isolines" + CHR$(3): LPRINT "PU;PA0,6800;":
     REM plot close
5790 SOUND 123,5: SOUND 185,5: SOUND 208,5: SOUND 165,5: TM$ = TIME$
5800 SM = SM/SA: PRINT " AVERAGE HEIGHT OF SURFACE IS" SM: PRINT " AT "
     TM$ " RUN COMPLETE": END
5810 REM curvemaker
5820 ST = R3#(1,3)
5830 IF ST>R3#(2,3) THEN ST = R3#(2,3)
5840 IF ST>R3#(3,3) THEN ST = R3#(3,3)
5850 IF ST< = - 999 THEN RETURN
5860 SM = SM + AR#*((R3#(1,3) + R3#(2,3) + R3#(3,3))/3*VS + P9(2,3)): SA = SA + AR#
6210 L8% = 0: IF DP<2 THEN GOTO 6310
6220 L8% = 1 6230 REM rotate triangle for orthogonal profiles
6240 FOR L3% = 1 TO 3
6250 Z# = R3#(L3%,3): R3#(L3%,3) = 1 - R3#(L3%,L8%): R3#(L3%,L8%) = Z# 6260
     NEXT L3%
6270 ST = R3#(1,3)
6280 IF ST>R3#(2,3) THEN ST = R3#(2,3)
6290 IF ST>R3#(3,3) THEN ST = R3#(3,3)
6300 IF ST < = - 999 THEN RETURN
6310 TP = R3#(1,3)
6320 IF TP<R3#(2,3) THEN TP = R3#(2,3)
6330 IF TP<R3#(3,3) THEN TP = R3#(3,3)
6340 IF TP = ST THEN RETURN
6350 FOR L0% = 1 TO NC%
6360 REM slice the triangle
6370 IF P8(L0%)>TP OR P8(L0%)<ST THEN GOTO 6610
6380 L5% = 1: L4% = 0
6390 L4% = L4% + 1
6400 L7% = 1: IF L7% = L4% THEN L7% = L7% + 1
6410 L3% = L7% + 1: IF L3% = L4% THEN L3% = L3% + 1
6420 IF R3#(L7%,3) = R3#(L3%,3) THEN GOTO 6460
6430 F = (P8(L0%) - R3#(L7%,3))/(R3#(L3%,3) - R3#(L7%,3)): IF F<0 OR F>1 THEN
     GOTO 6460
6440 S3(L5%,1) = R3#(L7%,1) + (R3#(L3%,1) - R3#(L7%,1))*F: S3(L5%,2) = R3#(L7%,2)
     + (R3#(L3%,2) - R3#(L7%,2))*F
6450 L5% = L5% + 1
6460 IF L5%<3 THEN GOTO 6390
6470 S3(1,3) = P8(L0%): S3(2,3) = P8(L0%): IF DP<2 THEN GOTO 6500
6480 REM reverse rotate the intersection trace
```

Appendix

```
6490 FOR L3% = 1 TO 2: Z# = S3(L3%,3): S3(L3%,3) = S3(L3%,L8%): S3(L3%,L8%) = 1
     - Z#: NEXT L3%
6500 IF DP = 0 THEN GOTO 6570
6510 REM apply perspective
6520 FOR L3% = 1 TO 2
6530 YY = (S3(L3%,2) - .5)*R8(1,1) + (S3(L3%,1) - .5)*R8(1,2): S3(L3%,1) = (S3(L3%,1) -
     .5)*R8(1,1) - (S3(L3%,2) - .5)*R8(1,2) + .5
6540 S33 = (S3(L3%,3) - .5)*R8(2,1) + YY*R8(2,2) + .5: S3(L3%,2) = YY*R8(2,1) -
     (S3(L3%,3) - .5)*R8(2,2) + .5: S3(L3%,3) = S33
6550 NEXT L3%
6552 IF SR<2 THEN GOTO 6560
6554 S3(1,1) = (S3(1,1) - .5)*.99875 - (S3(1,3) - .5)*.05 + .5: S3(2,1) = (S3(2,1) -
     .5)*.99875 - (S3(2,3) - .5)*.05 + .5
6560 IF SR<>1 THEN GOTO 6570
6565 S3(1,1) = (S3(1,1) - .5)*.99875 + (S3(1,3) - .5)*.05 + .5: S3(2,1) = (S3(2,1) -
     .5)*.99875 + (S3(2,3) - .5)*.05 + .5
6567 REM draw a line from (S3(1,1),S3(1,2)) to (S3(2,1),S3(2,2))
6570 LPRINT "PA" + STR$(S3(1,1)*SZ + SW) + "," + STR$(S3(1,2)*SZ + ZW) + ";":
     LPRINT "PD;PA" + STR$(S3(2,1)*SZ + SW) + "," + STR$(S3(2,2)*SZ + ZW) + ";PU;"
6610 NEXT L0%
6620 IF L8%<1 THEN RETURN
6630 REM reverse rotate the triangle
6640 FOR L3% = 1 TO 3
6650 Z# = R3#(L3%,3): R3#(L3%,3) = R3#(L3%,L8%): R3#(L3%,L8%) = 1 - Z#
6660 NEXT L3%
6670 L8% = L8% + 1: IF L8%<3 THEN GOTO 6240 ELSE RETURN


1000 REM  **PLOT\NEIGHBOR.BAS**  NEIGHBORHOOD - BASED INTERPOLATION
     Copyright (c) 1988 D.F.Watson
1010 DIM P1#(55,6), P2#(237,3), P5#(25), P8(80), P9(2,3), S1(7742), S2(3), S4(25),
     P0(4,2)
1020 DIM R1#(3,3), R3#(25,25), R4#(3,2), R5#(3,3), R7#(3), R8(2,2), S3(2,3), BX(6,3),
     R9#(25), B#(25)
1030 DIM P3%(237,3), P4%(237), P6%(25), P7%(25,2), R2%(2,2), R6%(3,2)
1050 DATA 1,0,0,1,1,0,0,1,0,0,0,0,1,0,0,1,0,1, - 1, - 1,0,1,2,5, - 1,0,2,3, - 1,5,1E37,3,1
1060 DEF FNCV(RT#) = RT#^2*LOG(RT#): REM minimum curvature
1090 P9(1,1) = - 1E37: P9(1,2) = - 1E37: P9(1,3) = - 1E37: P9(2,1) = 1E37: P9(2,2) =
     1E37: P9(2,3) = 1E37
1100 SL = 0: SR = 1: SZ = 5100: GR = 19: DP = 0: CL = 1: PT = 1: DZ = 15: WD = 10000:
     HI = 6800: AZ = - 25: TL = - 15: RD = 2: BI = 1.5: BJ = 7: IF DP = 0 THEN SR = 0
1140 OPEN "A: \HILL.DAT" FOR INPUT AS #1
1150 REM OPEN "A: \JDAVIS.DAT" FOR INPUT AS #1
1180 INPUT#1,ND%,NC%: FOR I1% = 1 TO NC%: INPUT#1,P8(I1%): NEXT I1%: IF
     CR>0 AND NC%>31 THEN NC% = 31
1185 FOR I1% = 1 TO ND%: P4%(I1%) = I1%: NEXT I1%: FOR I1% = 1 TO ND%: I2% =
     INT(RND*ND% + .5): IF I2%<1 THEN I2% = 1
1187 IF I2%>ND% THEN I2% = ND%
1188 I3% = P4%(I1%): P4%(I1%) = P4%(I2%): P4%(I2%) = I3%: NEXT I1%
1190 INPUT#1,XS,YS,DT: I2% = 1
1220 I1% = P4%(I2%)
1222 INPUT#1,P1#(I1%,1),P1#(I1%,2),P1#(I1%,3): P1#(I1%,1) = (P1#(I1%,1) - XS)/DT:
     P1#(I1%,2) = (P1#(I1%,2) - YS)/DT: IF (P1#(I1%,1) - .5)^2 + (P1#(I1%,2) - .5)^2<RD
     THEN GOTO 1270
1230 ND% = ND% - 1: IF I2%>ND% THEN GOTO 1360
1260 GOTO 1222
1270 IF P9(1,1)<P1#(I1%,1) THEN P9(1,1) = P1#(I1%,1)
1280 IF P9(2,1)>P1#(I1%,1) THEN P9(2,1) = P1#(I1%,1)
1290 IF P9(1,2)<P1#(I1%,2) THEN P9(1,2) = P1#(I1%,2)
1300 IF P9(2,2)>P1#(I1%,2) THEN P9(2,2) = P1#(I1%,2)
1310 IF P9(1,3)<P1#(I1%,3) THEN P9(1,3) = P1#(I1%,3)
1320 IF P9(2,3)>P1#(I1%,3) THEN P9(2,3) = P1#(I1%,3)
1350 I2% = I2% + 1: IF I2%< = ND% THEN GOTO 1220
1360 IF ND%<4 THEN STOP
1370 XF = P9(1,1) - P9(2,1): YF = P9(1,2) - P9(2,2): VS = P9(1,3) - P9(2,3): R8(2,1) = AZ:
     R8(2,2) = TL
```

```
1380 FOR I1% = 1 TO 6: READ BX(I1%,1),BX(I1%,2),BX(I1%,3): NEXT I1%
1390 BN% = 1: IF DP<1 THEN GOTO 1450
1400 BN% = 2: BX(2,1) = 0: BX(2,3) = 1: IF R8(2,1)>0 THEN GOTO 1420
1410 BX(2,1) = 1: BX(3,1) = 1: BX(4,1) = 1: BX(5,1) = 0: BX(6,1) = 0
1420 IF DP<2 OR CR>0 THEN GOTO 1450
1430 NC1% = INT(SZ/100): NC% = NC1% + 1: U0 = .9999/NC1%: U1 = .0000511 - U0
1440 FOR I1% = 1 TO NC%: U1 = U1 + U0: P8(I1%) = U1: NEXT I1%: GOTO 1460
1450 FOR I1% = 1 TO NC%: P8(I1%) = (P8(I1%) - P9(2,3))/VS + .00001*(RND - .5):
     NEXT I1%
1460 FOR I1% = 1 TO ND%: P1#(I1%,1) = P1#(I1%,1) + .0001*(RND - .5): P1#(I1%,2) =
     P1#(I1%,2) + .0001*(RND - .5) : P1#(I1%,3) = (P1#(I1%,3) - P9(2,3))/VS: NEXT I1%
1470 FOR I1% = 1 TO 3: READ R1#(I1%,1), R1#(I1%,2), R1#(I1%,3), R6%(I1%,1),
     R6%(I1%,2)
1480 P1#(I1% + ND%,1) = R1#(I1%,1)*XF + P9(2,1): P1#(I1% + ND%,2) = R1#(I1%,2)*YF
     + P9(2,2): P1#(I1% + ND%,3) = 0: NEXT I1%
1490 IF DP<1 THEN GOTO 1550
1500 FOR I1% = 1 TO 2: R8(I1%,1) = COS(R8(2,I1%)/57.29578): R8(I1%,2) =
     SIN(R8(2,I1%)/57.29578): NEXT I1%
1510 FOR I1% = 2 TO 6
1520 YY = (BX(I1%,2) - .5)*R8(1,1) + (BX(I1%,1) - .5)*R8(1,2) + .5: BX(I1%,1) =
     (BX(I1%,1) - .5)*R8(1,1) - (BX(I1%,2) - .5)*R8(1,2) + .5
1530 BX3 = (BX(I1%,3) - .5)*R8(2,1) + (YY - .5)*R8(2,2) + .5: BX(I1%,2) = (YY -
     .5)*R8(2,1) - (BX(I1%,3) - .5)*R8(2,2) + .5: BX(I1%,3) = BX3: NEXT I1%
1540 REM end data input, begin gradient estimation
1550 SOUND 208,5: SOUND 165,5: SOUND 185,5: SOUND 123,5
1560 IF CL<1 THEN GOTO 3110
1570 PRINT " AT " TIME$ " BEGIN GRADIENT ESTIMATION  -  PLEASE WAIT"
1580 FOR I0% = 1 TO ND%
1590 FOR I1% = 1 TO 3: P3%(1,I1%) = ND% + I1%: P3%(2,I1%) = ND% + I1%:
     P2#(1,I1%) = R1#(I1%,3): P2#(2,I1%) = R1#(I1%,3) : NEXT I1%
1600 FOR I1% = 1 TO 237: P4%(I1%) = I1%: NEXT I1%: R2%(1,1) = 1: R2%(1,2) = 1:
     R2%(2,1) = 3 : R2%(2,2) = 4: K1% = I0%: K2% = I0%: K3% = ND% + 4: K4 = - .5
1610 FOR I1% = K1% TO K2%
1620 IF I1% = K3% THEN GOTO 2000
1630 K5 = K4
1640 K6 = 1.5 + K5: IF K3%>ND% THEN K5 = - K5
1650 U0 = 0: K7 = 0: L6% = K6 - 2
1660 FOR I2% = 1 TO R2%(1,K6)
1670 L6% = L6% + 2: IF P3%(L6%,1) = 0 THEN GOTO 1670
1680 U1# = P2#(L6%,3) - (P1#(I1%,1) - P2#(L6%,1))^2: IF U1#<0 THEN GOTO 1840
1690 U1# = U1# - (P1#(I1%,2) - P2#(L6%,2))^2: IF U1#<0 THEN GOTO 1840
1700 R2%(2,K6) = R2%(2,K6) - 2: P4%(R2%(2,K6)) = L6%: U0 = U0 - 1
1710 FOR I3% = 1 TO 3
1720 L1% = 1: IF L1% = I3% THEN L1% = L1% + 1
1730 L2% = L1% + 1: IF L2% = I3% THEN L2% = L2% + 1
1740 IF K7<1 THEN GOTO 1810
1750 K9 = K7
1760 FOR I4% = 1 TO K9
1770 IF P3%(L6%,L1%)<>P7%(I4%,1) THEN GOTO 1800
1780 IF P3%(L6%,L2%)<>P7%(I4%,2) THEN GOTO 1800
1790 P7%(I4%,1) = P7%(K9,1): P7%(I4%,2) = P7%(K9,2): K7 = K7 - 1: GOTO 1820
1800 NEXT I4%
1810 K7 = K7 + 1: P7%(K7,1) = P3%(L6%,L1%): P7%(K7,2) = P3%(L6%,L2%)
1820 NEXT I3%
1830 P3%(L6%,1) = 0
1840 NEXT I2%
1850 IF K7<1 THEN GOTO 2000
1860 REM make new triangles
1870 FOR I2% = 1 TO K7
1880 IF K5>0 THEN GOTO 1900
1890 IF P7%(I2%,1)<>I0% AND P7%(I2%,2)<>I0% THEN GOTO 1970
1900 FOR I3% = 1 TO 2
1910 R3#(I3%,1) = P1#(P7%(I2%,I3%),1) - P1#(I1%,1): R3#(I3%,2) =
     P1#(P7%(I2%,I3%),2) - P1#(I1%,2)
1920 R3#(I3%,3) = R3#(I3%,1)*(P1#(P7%(I2%,I3%),1) + P1#(I1%,1))/2 + R3#(I3%,2)
     *(P1#(P7%(I2%,I3%),2) + P1#(I1%,2))/2
```

Appendix

```
1930 NEXT I3%
1940 U1# = R3#(1,1)*R3#(2,2) - R3#(2,1)*R3#(1,2): P2#(P4%(R2%(2,K6)),1) =
     (R3#(1,3)*R3#(2,2) - R3#(2,3)*R3#(1,2))/U1#: P2#(P4%(R2%(2,K6)),2) =
     (R3#(1,1)*R3#(2,3) - R3#(2,1)*R3#(1,3))/U1#
1950 P2#(P4%(R2%(2,K6)),3) = (P1#(I1%,1) - P2#(P4%(R2%(2,K6)),1))^2 + (P1#(I1%,2) -
     P2#(P4%(R2%(2,K6)),2))^2
1960 P3%(P4%(R2%(2,K6)),1) = P7%(I2%,1): P3%(P4%(R2%(2,K6)),2) = P7%(I2%,2) :
     P3%(P4%(R2%(2,K6)),3) = I1%: R2%(2,K6) = R2%(2,K6) + 2: U0 = U0 + 1
1970 NEXT I2%
1980 R2%(1,K6) = R2%(1,K6) + U0: IF K5>0 THEN GOTO 2000
1990 K5 = - K5: GOTO 1640
2000 NEXT I1%
2010 IF K2%>K1% GOTO 2040
2020 K1% = 1: K2% = ND%: K3% = I0%: GOTO 1610
2030 REM order triangles positively
2040 FOR I1% = 1 TO 2
2050 L8% = I1% - 2
2060 FOR I2% = 1 TO R2%(1,I1%)
2070 L8% = L8% + 2: IF P3%(L8%,1) = 0 THEN GOTO 2070
2080 IF P3%(L8%,1)>ND% THEN GOTO 2140
2090 FOR I3% = 1 TO 2
2100 R3#(1,I3%) = P1#(P3%(L8%,2),I3%) - P1#(P3%(L8%,1),I3%): R3#(2,I3%) =
     P1#(P3%(L8%,3),I3%) - P1#(P3%(L8%,1),I3%)
2110 NEXT I3%
2120 IF R3#(1,1)*R3#(2,2) - R3#(1,2)*R3#(2,1)> = 0 THEN GOTO 2140
2130 K8% = P3%(L8%,3): P3%(L8%,3) = P3%(L8%,2): P3%(L8%,2) = K8%
2140 NEXT I2%
2150 NEXT I1%
2160 IF SL>0 THEN GOTO 2470
2170 REM calculate natural neighbor coordinates
2180 R5#(1,1) = P1#(I0%,1): R5#(1,2) = P1#(I0%,2): R5#(2,1) = R5#(1,1) + .0001 :
     R5#(2,2) = R5#(1,2): R5#(3,1) = R5#(1,1): R5#(3,2) = R5#(1,2) + .0001
2190 FOR I1% = 1 TO 3
2200 L4% = 0: L5% = 0: L6% = 0
2210 FOR I2% = 1 TO R2%(1,2)
2220 L6% = L6% + 2: IF P3%(L6%,1) = 0 THEN GOTO 2220
2230 IF P3%(L6%,1)>ND% THEN GOTO 2410
2240 U1# = P2#(L6%,3) - (R5#(I1%,1) - P2#(L6%,1))^2: IF U1#<0 THEN GOTO 2410
2250 U1# = U1# - (R5#(I1%,2) - P2#(L6%,2))^2: IF U1#<0 THEN GOTO 2410
2260 FOR I3% = 1 TO 3
2270 FOR I4% = 1 TO 2
2280 R3#(I4%,1) = P1#(P3%(L6%,R6%(I3%,I4%)),1) - R5#(I1%,1): R3#(I4%,2) =
     P1#(P3%(L6%,R6%(I3%,I4%)),2) - R5#(I1%,2)
2290 R3#(I4%,3) = R3#(I4%,1)*(P1#(P3%(L6%,R6%(I3%,I4%)),1) + R5#(I1%,1))/2 +
     R3#(I4%,2) *(P1#(P3%(L6%,R6%(I3%,I4%)),2) + R5#(I1%,2))/2: NEXT I4%
2300 U1# = R3#(1,1)*R3#(2,2) - R3#(1,2)*R3#(2,1)
2310 R4#(R6%(I3%,2),1) = (R3#(1,3)*R3#(2,2) - R3#(2,3)*R3#(1,2))/U1#:
     R4#(R6%(I3%,2),2) = (R3#(1,1)*R3#(2,3) - R3#(2,1)*R3#(1,3))/U1#: NEXT I3%: L3%=0
2320 FOR I3% = 1 TO 3: R3#(3,I3%) = ((R4#(R6%(I3%,1),1) -
     P2#(L6%,1))*(R4#(R6%(I3%,2),2) - P2#(L6%,2)) - (R4#(R6%(I3%,2),1) -
     P2#(L6%,1))*(R4#(R6%(I3%,1),2) - P2#(L6%,2)))/2: IF R3#(3,I3%)>0 THEN L3% =
     L3% + 1
2330 NEXT I3%
2340 IF L3%>2 THEN L4% = 1
2350 FOR I3% = 1 TO 3: IF L5%<1 THEN GOTO 2390
2360 FOR I4% = 1 TO L5%: IF P3%(L6%,I3%)<>P6%(I4%) THEN GOTO 2380
2370 P5#(I4%) = P5#(I4%) + R3#(3,I3%): GOTO 2400
2380 NEXT I4%
2390 L5% = L5% + 1: P6%(L5%) = P3%(L6%,I3%): P5#(L5%) = R3#(3,I3%)
2400 NEXT I3%
2410 NEXT I2%: IF L4%<1 THEN GOTO 2470
2420 U4 = 0
2430 FOR I2% = 1 TO L5%: U4 = U4 + P5#(I2%): NEXT I2%: R7#(I1%) = 0
2440 FOR I2% = 1 TO L5%: P5#(I2%) = P5#(I2%)/U4: R7#(I1%) = R7#(I1%) +
     P5#(I2%)*P1#(P6%(I2%),3): NEXT I2%: NEXT I1%
```

```
2450 P1#(I0%,6) = P1#(I0%,3) - R7#(1): P1#(I0%,4) = (R7#(1) - R7#(2))/.0001:
     P1#(I0%,5) = (R7#(1) - R7#(3))/.0001 : GOTO 3080
2460 REM cross product gradients
2470 IF SL>1 THEN GOTO 2610
2480 P1#(I0%,4) = 0: P1#(I0%,5) = 0: P1#(I0%,6) = 0: P7# = 0: L7% = - 1
2490 FOR I1% = 1 TO R2%(1,1)
2500 L7% = L7% + 2: IF P3%(L7%,1) = 0 THEN GOTO 2500
2510 IF P3%(L7%,1)>ND% THEN GOTO 2570
2520 FOR I2% = 1 TO 2: FOR I3% = 1 TO 3: R3#(I2%,I3%) = P1#(P3%(L7%,1),I3%) -
     P1#(P3%(L7%,I2% + 1),I3%): NEXT I3% : NEXT I2%
2530 R3#(3,1) = R3#(1,2)*R3#(2,3) - R3#(2,2)*R3#(1,3): R3#(3,2) = R3#(1,3)*R3#(2,1) -
     R3#(2,3)*R3#(1,1): R3#(3,3) = R3#(1,1)*R3#(2,2) - R3#(2,1)*R3#(1,2): U3 = 1
2540 IF R3#(3,3)<0 THEN U3 = - 1
2550 U2 = (R3#(3,1)^2 + R3#(3,2)^2 + R3#(3,3)^2)^.5: P7# = P7# + U2
2560 FOR I2% = 1 TO 3: P1#(I0%,I2% + 3) = P1#(I0%,I2% + 3) + R3#(3,I2%)*U3: NEXT
     I2%
2570 NEXT I1%
2580 U2 = P1#(I0%,4)^2 + P1#(I0%,5)^2 + P1#(I0%,6)^2: P7# = 1 - U2^.5/P7#
2590 P1#(I0%,4) = P1#(I0%,4)/P1#(I0%,6): P1#(I0%,5) = P1#(I0%,5)/P1#(I0%,6):
     P1#(I0%,6) = P7#
2600 GOTO 3080
2610 IF SL>2 THEN GOTO 2810 : REM least squares gradients
2620 P1#(I0%,4) = 0: P1#(I0%,5) = 0: P1#(I0%,6) = .2: L7% = - 1: P6%(1) = I0%: L5% =
     1 2630 R4#(1,1) = P1#(I0%,3): R4#(2,1) = P1#(I0%,1)*P1#(I0%,3): R4#(3,1) =
     P1#(I0%,2)*P1#(I0%,3)
2640 R3#(2,1) = P1#(I0%,1): R3#(2,2) = P1#(I0%,1)^2: R3#(3,1) = P1#(I0%,2) : R3#(3,2)
     = P1#(I0%,1)*P1#(I0%,2): R3#(3,3) = P1#(I0%,2)^2
2650 FOR I1% = 1 TO R2%(1,1)
2660 L7% = L7% + 2
2670 IF P3%(L7%,1) = 0 THEN GOTO 2660
2680 IF P3%(L7%,1)>ND% THEN GOTO 2760
2690 FOR I2% = 1 TO 3
2700 FOR I3% = 1 TO L5%
2710 IF P3%(L7%,I2%) = P6%(I3%) THEN GOTO 2750
2720 NEXT I3%
2730 L5% = L5% + 1: P6%(L5%) = P3%(L7%,I2%): R4#(1,1) = R4#(1,1) +
     P1#(P6%(L5%),3) : R4#(2,1) = R4#(2,1) + P1#(P6%(L5%),1)*P1#(P6%(L5%),3):
     R4#(3,1) = R4#(3,1) + P1#(P6%(L5%),2)*P1#(P6%(L5%),3)
2740 R3#(2,1) = R3#(2,1) + P1#(P6%(L5%),1): R3#(2,2) = R3#(2,2) +
     P1#(P6%(L5%),1)^2: R3#(3,1) = R3#(3,1) + P1#(P6%(L5%),2): R3#(3,2) = R3#(3,2) +
     P1#(P6%(L5%),1)*P1#(P6%(L5%),2): R3#(3,3) = R3#(3,3) + P1#(P6%(L5%),2)^2
2750 NEXT I2%
2760 NEXT I1%
2770 U1# = L5%*(R3#(2,2)*R3#(3,3) - R3#(3,2)^2) - R3#(2,1)*(R3#(2,1)*R3#(3,3) -
     R3#(3,1)*R3#(3,2)) + R3#(3,1)*(R3#(2,1)*R3#(3,2) - R3#(3,1)*R3#(2,2))
2780 P1#(I0%,4) = - (L5%*(R4#(2,1)*R3#(3,3) - R3#(3,2)*R4#(3,1)) -
     R4#(1,1)*(R3#(2,1)*R3#(3,3) - R3#(3,1)*R3#(3,2)) + R3#(3,1)*(R3#(2,1)*R4#(3,1) -
     R3#(3,1)*R4#(2,1)))/U1#
2790 P1#(I0%,5) = - (L5%*(R3#(2,2)*R4#(3,1) - R3#(3,2)*R4#(2,1)) -
     R3#(2,1)*(R3#(2,1)*R4#(3,1) - R3#(3,1)*R4#(2,1)) + R4#(1,1)*(R3#(2,1)*R3#(3,2) -
     R3#(3,1)*R3#(2,2)))/U1#
2800 GOTO 3080 : REM spline gradients
2810 IF SL>3 THEN GOTO 3035
2815 P1#(I0%,4) = 0: P1#(I0%,5) = 0: P1#(I0%,6) = .2: L7% = - 1: L5% = 1: P6%(1) =
     I0%
2820 R5#(1,1) = P1#(I0%,1): R5#(1,2) = P1#(I0%,2): R5#(2,1) = R5#(1,1) + .0001 :
     R5#(2,2) = R5#(1,2): R5#(3,1) = R5#(1,1): R5#(3,2) = R5#(1,2) + .0001
2830 FOR I1% = 1 TO R2%(1,1)
2840 L7% = L7% + 2: IF P3%(L7%,1) = 0 THEN GOTO 2840
2850 IF P3%(L7%,1)>ND% THEN GOTO 2890
2860 FOR I2% = 1 TO 3: FOR I3% = 1 TO L5%: IF P3%(L7%,I2%) = P6%(I3%) THEN
     GOTO 2880
2870 NEXT I3%: L5% = L5% + 1: P6%(L5%) = P3%(L7%,I2%)
2880 NEXT I2%
2890 NEXT I1%
2900 LL% = L5% + 3
```

Appendix

```
2910 FOR I1% = 1 TO L5%
2920 R3#(I1%,1) = 1: R3#(LL,I1% + 3) = 1: R3#(I1%,2) = P1#(P6%(I1%),1): R3#(L5% +
     1,I1% + 3) = R3#(I1%,2)
2930 R3#(I1%,3) = P1#(P6%(I1%),2): R3#(L5% + 2,I1% + 3) = R3#(I1%,3): R9#(I1%) =
     P1#(P6%(I1%),3): NEXT I1%
2940 FOR I1% = 1 TO 3: FOR I2% = 1 TO 3: R3#(I1% + L5%,I2%) = 0: NEXT I2%:
     R9#(I1% + L5%) = 0: NEXT I1%
2950 FOR I1% = 1 TO L5%: R3#(I1%,I1% + 3) = 0: IJ = I1% + 1: IF IJ>L5% THEN GOTO
     2970
2960 FOR I2% = IJ TO L5%: RT# = ((R3#(I1%,2) - R3#(I2%,2))^2 + (R3#(I1%,3) -
     R3#(I2%,3))^2)^.5: R3#(I2%,I1% + 3) = FNCV(RT#): R3#(I1%,I2% + 3) = R3#(I2%,I1%
     + 3): NEXT I2%
2970 NEXT I1%: GOSUB 6690
2980 TR# = 0: TS# = 0
2990 FOR I1% = 4 TO LL%
3000 RT# = ((P1#(P6%(I1% - 3),1) - R5#(2,1))^2 + (P1#(P6%(I1% - 3),2) -
     R5#(2,2))^2)^.5: TR# = TR# + B#(I1%)*FNCV(RT#)
3010 RS# = ((P1#(P6%(I1% - 3),1) - R5#(3,1))^2 + (P1#(P6%(I1% - 3),2) -
     R5#(3,2))^2)^.5: TS# = TS# + B#(I1%)*FNCV(RS#) : NEXT I1%
3020 R5#(2,3) = B#(1) + B#(2)*R5#(2,1) + B#(3)*R5#(2,2) + TR#: P1#(I0%,4) =
     (P1#(I0%,3) - R5#(2,3))/.0001
3030 R5#(3,3) = B#(1) + B#(2)*R5#(3,1) + B#(3)*R5#(3,2) + TS#: P1#(I0%,5) =
     (P1#(I0%,3) - R5#(3,3))/.0001: GOTO 3080
3033 REM hyperboloid gradients
3035 R3#(5,1) = 1: FOR I1% = 1 TO 4: R9#(I1%) = 0: FOR I2% = 1 TO 4: R3#(I1%,I2%) =
     0: NEXT I2%: NEXT I1%: LL% = 4
3036 P1#(I0%,4) = 0: P1#(I0%,5) = 0: P1#(I0%,6) = .2: L7% = - 1: L5% = 0: FOR I1% = 1
     TO R2%(1,1)
3037 L7% = L7% + 2: IF P3%(L7%,1) = 0 THEN GOTO 3037
3038 IF P3%(L7%,1)>ND% THEN GOTO 3070
3039 FOR I2% = 1 TO 3: IF L5%<1 THEN GOTO 3044
3041 FOR I3% = 1 TO L5%: IF P3%(L7%,I2%) = P6%(I3%) THEN GOTO 3060
3042 NEXT I3%
3044 L5% = L5% + 1: P6%(L5%) = P3%(L7%,I2%)
3050 R3#(5,2) = P1#(P6%(L5%),1): R3#(5,3) = P1#(P6%(L5%),2): R3#(5,4) =
     P1#(P6%(L5%),1)*P1#(P6%(L5%),2)
3055 FOR I3% = 1 TO 4: FOR I4% = 1 TO 4: R3#(I3%,I4%) = R3#(I3%,I4%) +
     R3#(5,I3%)*R3#(5,I4%): NEXT I4% : R9#(I3%) = R9#(I3%) +
     R3#(5,I3%)*P1#(P6%(L5%),3): NEXT I3%
3060 NEXT I2%
3070 NEXT I1%: GOSUB 6690: P1#(I0%,4) = - B#(2) - B#(4)*P1#(I0%,2): P1#(I0%,5) = -
     B#(3) - B#(4)*P1#(I0%,1): P1#(I0%,6) = .2
3080 NEXT I0%
3100 REM natural neighbor sort
3110 FOR I0% = 1 TO 3: P3%(1,I0%) = ND% + I0%: P2#(1,I0%) = R1#(I0%,3): NEXT I0%
3120 FOR I0% = 1 TO 237: P4%(I0%) = I0%: NEXT I0%: NT% = 1: L0% = 2
3130 FOR I0% = 1 TO ND%: L3% = 0
3140 FOR I1% = 1 TO NT%
3150 U1# = P2#(I1%,3) - (P1#(I0%,1) - P2#(I1%,1))^2: IF U1#<0 THEN GOTO 3300
3160 U1# = U1# - (P1#(I0%,2) - P2#(I1%,2))^2: IF U1#<0 THEN GOTO 3300
3170 L0% = L0% - 1: P4%(L0%) = I1%
3180 FOR I2% = 1 TO 3
3190 L1% = 1: IF L1% = I2% THEN L1% = L1% + 1
3200 L2% = L1% + 1: IF L2% = I2% THEN L2% = L2% + 1
3210 IF L3%<1 THEN GOTO 3280
3220 L5% = L3%
3230 FOR I3% = 1 TO L5%
3240 IF P3%(I1%,L1%)<>P7%(I3%,1) THEN GOTO 3270
3250 IF P3%(I1%,L2%)<>P7%(I3%,2) THEN GOTO 3270
3260 P7%(I3%,1) = P7%(L5%,1): P7%(I3%,2) = P7%(L5%,2): L3% = L3% - 1: GOTO
     3290
3270 NEXT I3%
3280 L3% = L3% + 1: P7%(L3%,1) = P3%(I1%,L1%): P7%(L3%,2) = P3%(I1%,L2%)
3290 NEXT I2%
3300 NEXT I1%
3310 FOR I1% = 1 TO L3%
```

```
3320 FOR I2% = 1 TO 2
3330 R3#(I2%,1) = P1#(P7%(I1%,I2%),1) - P1#(I0%,1): R3#(I2%,2) =
     P1#(P7%(I1%,I2%),2) - P1#(I0%,2): R3#(I2%,3) = R3#(I2%,1)*(P1#(P7%(I1%,I2%),1) +
     P1#(I0%,1))/2
3340 R3#(I2%,3) = R3#(I2%,3) + R3#(I2%,2)*(P1#(P7%(I1%,I2%),2) + P1#(I0%,2))/2
3350 NEXT I2%
3360 U1# = R3#(1,1)*R3#(2,2) - R3#(2,1)*R3#(1,2): P2#(P4%(L0%),1) =
     (R3#(1,3)*R3#(2,2) - R3#(2,3)*R3#(1,2))/U1#: P2#(P4%(L0%),2) = (R3#(1,1)*R3#(2,3)
     - R3#(2,1)*R3#(1,3))/U1#
3370 P2#(P4%(L0%),3) = (P1#(I0%,1) - P2#(P4%(L0%),1))^2 + (P1#(I0%,2) -
     P2#(P4%(L0%),2))^2
3380 P3%(P4%(L0%),1) = P7%(I1%,1): P3%(P4%(L0%),2) = P7%(I1%,2):
     P3%(P4%(L0%),3) = I0%: L0% = L0% + 1
3390 NEXT I1%
3400 NT% = NT% + 2: NEXT I0%
3410 FOR I0% = 1 TO NT%: IF P3%(I0%,1)>ND% THEN GOTO 3460
3420 FOR I1% = 1 TO 2
3430 R3#(1,I1%) = P1#(P3%(I0%,2),I1%) - P1#(P3%(I0%,1),I1%): R3#(2,I1%) =
     P1#(P3%(I0%,3),I1%) - P1#(P3%(I0%,1),I1%): NEXT I1%
3440 IF R3#(1,1)*R3#(2,2) - R3#(1,2)*R3#(2,1)> = 0 THEN GOTO 3460
3450 L4% = P3%(I0%,3): P3%(I0%,3) = P3%(I0%,2): P3%(I0%,2) = L4%
3460 NEXT I0%
3820 REM make triangular grid
3830 PRINT " AT " TIME$ " BEGIN SURFACE INTERPOLATION - PLEASE WAIT":
     SOUND 123,5: SOUND 185,5: SOUND 208,5 : SOUND 165,5
3840 I0% = GR + 1: IS = 1: I2% = 2*I0% + 1: I3% = 1 + INT(GR/3^.5): I4% = I3%*I2% +
     I0%: SX = 1/GR: SY = 1/(2*I3%): S2(3) = .000001
3850 WX# = S2(3): S2(1) = 1.00001: AA# = SX*SY*.5*DT*DT: WY# = 1: U8 = 1
3860 FOR L1% = 1 TO I4%
3870 S1(L1%) = - 999: WXO# = WX#: WYO# = WY#
3880 REM .....calculate natural neighbor coordinates of WX#,WY# *********
3890 K4% = 0: K5% = 0
3900 FOR L2% = 1 TO NT%
3910 IF P3%(L2%,1)>ND% THEN GOTO 4280
3920 U1# = P2#(L2%,3) - (WX# - P2#(L2%,1))^2: IF U1#<0 THEN GOTO 4280
3930 U1# = U1# - (WY# - P2#(L2%,2))^2: IF U1#<0 THEN GOTO 4280
3940 FOR L3% = 1 TO 3
3950 L4% = 1: IF L4% = L3% THEN L4% = L4% + 1
3960 L5% = L4% + 1: IF L5% = L3% THEN L5% = L5% + 1
3970 IF ABS((P1#(P3%(L2%,L5%),1) - P1#(P3%(L2%,L4%),1))*(WY# -
     P1#(P3%(L2%,L4%),2)) - (WX# - P1#(P3%(L2%,L4%),1))*(P1#(P3%(L2%,L5%),2) -
     P1#(P3%(L2%,L4%),2)))>.000001 THEN GOTO 4030
4000 DEN# = ABS(P1#(P3%(L2%,L4%),2) - P1#(P3%(L2%,L5%),2)) +
     ABS(P1#(P3%(L2%,L5%),1) - P1#(P3%(L2%,L4%),1))
4010 WX# = WX# + (P1#(P3%(L2%,L4%),2) - P1#(P3%(L2%,L5%),2))*.0001/DEN#: WY#
     = WY# + (P1#(P3%(L2%,L5%),1) - P1#(P3%(L2%,L4%),1))*.0001/DEN#
4030 NEXT L3%
4040 FOR L3% = 1 TO 3
4050 FOR L4% = 1 TO 2
4060 R3#(L4%,1) = P1#(P3%(L2%,R6%(L3%,L4%)),1) - WX#: R3#(L4%,2) =
     P1#(P3%(L2%,R6%(L3%,L4%)),2) - WY#
4070 R3#(L4%,3) = R3#(L4%,1)*(P1#(P3%(L2%,R6%(L3%,L4%)),1) + WX#)/2 +
     R3#(L4%,2) *(P1#(P3%(L2%,R6%(L3%,L4%)),2) + WY#)/2
4080 NEXT L4%
4090 U1# = R3#(1,1)*R3#(2,2) - R3#(1,2)*R3#(2,1)
4100 R4#(R6%(L3%,2),1) = (R3#(1,3)*R3#(2,2) - R3#(2,3)*R3#(1,2))/U1#
4110 R4#(R6%(L3%,2),2) = (R3#(1,1)*R3#(2,3) - R3#(2,1)*R3#(1,3))/U1#
4120 NEXT L3%
4130 K3% = 0
4140 FOR L3% = 1 TO 3
4150 R3#(3,L3%) = ((R4#(R6%(L3%,1),1) - P2#(L2%,1))*(R4#(R6%(L3%,2),2) -
     P2#(L2%,2)) - (R4#(R6%(L3%,2),1) - P2#(L2%,1))*(R4#(R6%(L3%,1),2) -
     P2#(L2%,2)))/2
4160 IF R3#(3,L3%)>0 THEN K3% = K3% + 1
4170 NEXT L3%
4180 IF K3%>2 THEN K4% = 1
```

285

Appendix

```
4190 FOR L3% = 1 TO 3
4200 IF K5%<1 THEN GOTO 4260
4210 FOR L4% = 1 TO K5%
4220 IF P3%(L2%,L3%)<>P6%(L4%) THEN GOTO 4250
4230 P5#(L4%) = P5#(L4%) + R3#(3,L3%)
4240 GOTO 4270
4250 NEXT L4%
4260 K5% = K5% + 1: P6%(K5%) = P3%(L2%,L3%): P5#(K5%) = R3#(3,L3%)
4270 NEXT L3%
4280 NEXT L2%
4300 IF K4%<1 THEN GOTO 4940 ELSE GOTO 4560
4310 REM ....end determining K5% local coordinates *********
4560 S1(L1%) = 0: U4 = 0
4570 FOR L2% = 1 TO K5%: U4 = U4 + P5#(L2%): NEXT L2%
4580 FOR L2% = 1 TO K5%: P5#(L2%) = P5#(L2%)/U4: S1(L1%) = S1(L1%) +
     P5#(L2%)*P1#(P6%(L2%),3): NEXT L2%
4590 IF CL<1 THEN GOTO 4940
4620 FOR L2% = 1 TO K5%: S4(L2%) = 0
4630 IF P5#(L2%)<.00001 OR P5#(L2%)>1 THEN GOTO 4690
4640 IF ABS(P1#(P6%(L2%),6))<.00001 THEN GOTO 4690
4650 RS# = ABS(P1#(P6%(L2%),6)) + BI: RT# = RS#*BJ: RB# = 1/RT#: BD# =
     P5#(L2%)^RT#: BB# = BD#*2: IF BD#>.5 THEN BB# = (1 - BD#)*2
4660 BB# = BB#^RS#/2: IF BD#>.5 THEN BB# = 1 - BB#
4670 HP# = BB#*RB#
4680 S4(L2%) = ((P1#(P6%(L2%),4)*P1#(P6%(L2%),1) +
     P1#(P6%(L2%),5)*P1#(P6%(L2%),2) + P1#(P6%(L2%),3) - P1#(P6%(L2%),4)*WX# -
     P1#(P6%(L2%),5)*WY#) - S1(L1%))*HP#
4690 NEXT L2%
4700 FOR L2% = 1 TO K5%: S1(L1%) = S1(L1%) + S4(L2%): NEXT L2%
4940 IS = IS + SX*U8: IF IS>0 THEN GOTO 4960
4950 WXO# = WXO# + SX/2: IS = 1.001 - SX
4960 WX# = WXO# + SX*U8: WY# = WYO#: IF U8*WX#<S2(2 - U8) THEN GOTO 4980
4970 U8 = - U8: WX# = S2(2 + U8): WY# = WY# - SY: IS = U8
4980 NEXT L1%
5060 REM end interpolation to triangular grid and begin display construction
5070 SOUND 208,5: SOUND 165,5: SOUND 185,5: SOUND 123,5: PRINT " AT " TIME$;
5071 INPUT " PREPARE PLOTTER AND PRESS ENTER",PM$
5090 LW = (HI - SZ)/2: IF DP>0 THEN SZ = INT(SZ*2/3)
5091 IF DP>0 AND SZ>HI/2 THEN SZ = HI/2
5100 SW = (WD - SZ)/2: ZW = (HI - SZ)/2
5110 IF SR>0 THEN SW = (WD - 2*SZ)/3
5120 SW = SW + 1: SV = 2*SW + SZ: SR = SR*2: SZ = SZ - 2: SZ = SZ - 2: ZW = ZW -
     1: SU = SW: IF SR>0 THEN SU = (2*SW + SV)/3
5160 REM plot open
5170 LPRINT "IN;IP0,0," + STR$(WD) + "," + STR$(HI) + ";SP1;": IF SR<2 THEN GOTO
     5180
5175 FOR L1% = 2 TO 6: BX1 = (BX(L1%,1) - .5)*.99875 - (BX(L1%,3) - .5)*.05 + .5:
     BX(L1%,3) = (BX(L1%,3) - .5)*.99875 + (BX(L1%,1) - .5)*.05 + .5: BX(L1%,1) = BX1:
     NEXT L1%: LPRINT "PU;PA0,6800;"
5180 LPRINT "PU;PA" + STR$(BX(BN%,1)*SZ + SW) + "," + STR$(BX(BN%,2)*SZ +
     ZW) + ";PD;" : ZX = SX: ZY = SY: ZZ = .5: R3#(2,2) = 1 + 2*SY: R3#(2,1) = 0
5190 FOR L1% = BN% + 1 TO BN% + 4: LPRINT "PA" + STR$(BX(L1%,1)*SZ + SW) +
     "," + STR$(BX(L1%,2)*SZ + ZW) + ";": NEXT L1% : LPRINT "PU;"
5192 FOR L1% = 2 TO 6: BX(L1%,1) = (BX(L1%,1) - .5)*.995 + (BX(L1%,3) - .5)*.1 + .5:
     NEXT L1%: SM = 0: SA = 0
5200 REM plot contours
5220 FOR L1% = 1 TO I3%
5230 R3#(1,2) = R3#(2,2) - 2*SY - 2*ZY: R3#(1,1) = R3#(2,1): N1% = (2*L1% - .5 +
     ZZ)*I0% + L1% + .5/ZZ: R3#(1,3) = S1(N1%): R3#(2,2) = R3#(2,2) - 2*SY - ZY:
     R3#(2,1) = .5 - ZZ: N2% = (2*L1% - .5 + ZZ)*I0% + L1%
5240 R3#(2,3) = S1(N2%): R3#(3,2) = R3#(2,2): R3#(3,1) = R3#(2,1) + ZX/2: N3% = N2%
     - .5/ZZ: R3#(3,3) = S1(N3%): AR# = AA#/2
5250 GOSUB 5820
5260 R3#(1,2) = R3#(1,2) + 2*ZY: N2% = (2*L1% - .5/ZZ - .5 - ZZ)*I0% + L1%: R3#(1,3) =
     S1(N2%)
5270 GOSUB 5820
```

```
5280 AR# = AA#
5290 FOR L2% = 1 TO GR
5300 R3#(2,2) = R3#(1,2): R3#(2,1) = R3#(1,1) + ZX: N4% = N2% + .5/ZZ: R3#(2,3) =
    S1(N4%)
5310 GOSUB 5820
5320 R3#(1,2) = R3#(3,2): R3#(1,1) = R3#(3,1) + ZX
5330 IF L2% = GR THEN R3#(1,1) = R3#(1,1) - ZX/2
5340 N6% = N3% - .5/ZZ: R3#(1,3) = S1(N6%): IF L2% = GR THEN AR# = AA#/2
5350 GOSUB 5820
5360 R3#(2,2) = R3#(2,2) - 2*ZY: N5% = N1% + .5/ZZ: R3#(2,3) = S1(N5%)
5370 GOSUB 5820
5380 R3#(1,2) = R3#(2,2): R3#(1,1) = R3#(2,1) - ZX: R3#(1,3) = S1(N1%): IF L2% = GR
    THEN AR# = AA#
5390 GOSUB 5820
5400 N1% = N5%: N2% = N4%: N3% = N6%: R3#(1,2) = R3#(1,2) + 2*ZY: R3#(1,1) =
    R3#(2,1) : R3#(1,3) = S1(N4%): R3#(3,1) = R3#(3,1) + ZX: R3#(3,3) = S1(N6%) 5410
    NEXT L2%
5420 ZY = - ZY: ZX = - ZX: ZZ = - ZZ
5430 NEXT L1%: REM end plotting contours
5450 REM plot data points
5460 IF PT<1 THEN GOTO 5770
5480 DZ% = DZ: FOR L1% = 1 TO ND%
5490 IF P1#(L1%,1)< - .0001 OR P1#(L1%,1)>1.0001 OR P1#(L1%,2)< - .0001 OR
    P1#(L1%,2)>1.0001 THEN GOTO 5610
5500 PP1# = P1#(L1%,1): PP2# = P1#(L1%,2): PP3# = P1#(L1%,3): IF DP<1 THEN
    GOTO 5530
5510 YY = (PP2# - .5)*R8(1,1) + (PP1# - .5)*R8(1,2) + .5: PP1# = (PP1# - .5)*R8(1,1) -
    (PP2# - .5)*R8(1,2) + .5
5520 ZZ3 = (PP3# - .5)*R8(2,1) + (YY - .5)*R8(2,2) + .5: PP2# = (YY - .5)*R8(2,1) - (PP3#
    - .5)*R8(2,2) + .5: PP3# = ZZ3: IF SR<2 THEN GOTO 5528
5525 PP1# = (PP1# - .5)*.99875 - (PP3# - .5)*.05 + .5
5528 IF SR<>1 THEN GOTO 5530
5529 PP1# = (PP1# - .5)*.99875 + (PP3# - .5)*.05 + .5
5530 S3(1,1) = PP1#*SZ + SW: S3(1,2) = PP2#*SZ + ZW
5540 LPRINT "PA" + STR$(S3(1,1) - DZ%) + "," + STR$(S3(1,2) + DZ%) + ";": LPRINT
    "PD;PA" + STR$(S3(1,1) + DZ%) + "," + STR$(S3(1,2) + DZ%) + ";": LPRINT "PA" +
    STR$(S3(1,1) + DZ%) + "," + STR$(S3(1,2) - DZ%) + ";"
5550 LPRINT "PA" + STR$(S3(1,1) - DZ%) + "," + STR$(S3(1,2) - DZ%) + ";": LPRINT
    "PA" + STR$(S3(1,1) - DZ%) + "," + STR$(S3(1,2) + DZ%) + ";": LPRINT "PA" +
    STR$(S3(1,1) + DZ%) + "," + STR$(S3(1,2) - DZ%) + ";PU;"
5610 NEXT L1%
5770 SR = SR - 1: SW = SV: IF SR = 1 THEN GOTO 5180
5775 LPRINT "PU;PA" + STR$(SU) + "," + STR$(LW - 500) + ";SI0.2,0.3;SL0.2;"
5777 LPRINT "LBFigure 1.0.1 - Example Isolines" + CHR$(3): LPRINT "PU;PA0,6800;":
    REM plot close
5790 SOUND 123,5: SOUND 185,5: SOUND 208,5: SOUND 165,5: TM$ = TIME$
5800 SM = SM/SA: PRINT " AVERAGE HEIGHT OF SURFACE IS" SM: PRINT " AT "
    TM$ " RUN COMPLETE": END
5810 REM curvemaker
5820 ST = R3#(1,3)
5830 IF ST>R3#(2,3) THEN ST = R3#(2,3)
5840 IF ST>R3#(3,3) THEN ST = R3#(3,3)
5850 IF ST< = - 999 THEN RETURN
5860 SM = SM + AR#*((R3#(1,3) + R3#(2,3) + R3#(3,3))/3*VS + P9(2,3)): SA = SA + AR#
6210 L8% = 0: IF DP<2 THEN GOTO 6310
6220 L8% = 1
6230 REM rotate triangle for orthogonal profiles
6240 FOR L3% = 1 TO 3
6250 Z# = R3#(L3%,3): R3#(L3%,3) = 1 - R3#(L3%,L8%): R3#(L3%,L8%) = Z#
6260 NEXT L3%
6270 ST = R3#(1,3)
6280 IF ST>R3#(2,3) THEN ST = R3#(2,3)
6290 IF ST>R3#(3,3) THEN ST = R3#(3,3)
6300 IF ST < = - 999 THEN RETURN
6310 TP = R3#(1,3)
6320 IF TP<R3#(2,3) THEN TP = R3#(2,3)
```

Appendix

```
6330 IF TP<R3#(3,3) THEN TP = R3#(3,3)
6340 IF TP = ST THEN RETURN
6350 FOR L0 = 1 TO NC%
6360 REM slice the triangle
6370 IF P8(L0)>TP OR P8(L0)<ST THEN GOTO 6610
6380 L5% = 1: L4% = 0
6390 L4% = L4% + 1
6400 L7% = 1: IF L7% = L4% THEN L7% = L7% + 1
6410 L3% = L7% + 1: IF L3% = L4% THEN L3% = L3% + 1
6420 IF R3#(L7%,3) = R3#(L3%,3) THEN GOTO 6460
6430 F = (P8(L0) - R3#(L7%,3))/(R3#(L3%,3) - R3#(L7%,3)): IF F<0 OR F>1 THEN GOTO 6460
6440 S3(L5%,1) = R3#(L7%,1) + (R3#(L3%,1) - R3#(L7%,1))*F: S3(L5%,2) = R3#(L7%,2) + (R3#(L3%,2) - R3#(L7%,2))*F
6450 L5% = L5% + 1
6460 IF L5%<3 THEN GOTO 6390
6470 S3(1,3) = P8(L0): S3(2,3) = P8(L0): IF DP<2 THEN GOTO 6500
6480 REM reverse rotate the intersection trace
6490 FOR L3% = 1 TO 2: Z# = S3(L3%,3): S3(L3%,3) = S3(L3%,L8%): S3(L3%,L8%) = 1 - Z#: NEXT L3%
6500 IF DP = 0 THEN GOTO 6570
6510 REM apply perspective
6520 FOR L3% = 1 TO 2
6530 YY = (S3(L3%,2) - .5)*R8(1,1) + (S3(L3%,1) - .5)*R8(1,2): S3(L3%,1) = (S3(L3%,1) - .5)*R8(1,1) - (S3(L3%,2) - .5)*R8(1,2) + .5
6540 S33 = (S3(L3%,3) - .5)*R8(2,1) + YY*R8(2,2) + .5: S3(L3%,2) = YY*R8(2,1) - (S3(L3%,3) - .5)*R8(2,2) + .5: S3(L3%,3) = S33
6550 NEXT L3%
6552 IF SR<2 THEN GOTO 6560
6554 S3(1,1) = (S3(1,1) - .5)*.99875 - (S3(1,3) - .5)*.05 + .5: S3(2,1) = (S3(2,1) - .5)*.99875 - (S3(2,3) - .5)*.05 + .5
6560 IF SR<>1 THEN GOTO 6570
6565 S3(1,1) = (S3(1,1) - .5)*.99875 + (S3(1,3) - .5)*.05 + .5: S3(2,1) = (S3(2,1) - .5)*.99875 + (S3(2,3) - .5)*.05 + .5
6567 REM draw a line from (S3(1,1),S3(1,2)) to (S3(2,1),S3(2,2))
6570 LPRINT "PA" + STR$(S3(1,1)*SZ + SW) + "," + STR$(S3(1,2)*SZ + ZW) + ";":
     LPRINT "PD;PA" + STR$(S3(2,1)*SZ + SW) + "," + STR$(S3(2,2)*SZ + ZW) + ";PU;"
6610 NEXT L0
6620 IF L8%<1 THEN RETURN
6630 REM reverse rotate the triangle
6640 FOR L3% = 1 TO 3
6650 Z# = R3#(L3%,3): R3#(L3%,3) = R3#(L3%,L8%): R3#(L3%,L8%) = 1 - Z#
6660 NEXT L3%
6670 L8% = L8% + 1: IF L8%<3 THEN GOTO 6240 ELSE RETURN
6680 REM solve R3#(LL%,LL%)xB#(LL%) = R9#(LL%)
6690 FOR J2% = 1 TO LL%: P4%(J2%) = J2%: B#(J2%) = 0
6700 FOR J3% = 1 TO LL%: IF ABS(R3#(J2%,J3%))>= B#(J2%) THEN B#(J2%) = ABS(R3#(J2%,J3%))
6710 NEXT J3%: NEXT J2%: LM1% = LL% - 1: LP1% = LL% + 1
6720 FOR J2% = 1 TO LM1%: X3# = 0
6730 FOR J3% = J2% TO LL%: X4# = ABS(R3#(P4%(J3%),J2%))/B#(P4%(J3%)): IF X4#< = X3# THEN GOTO 6750
6740 X3# = X4#: K3% = J3%
6750 NEXT J3%: IF K3% = J2% GOTO 6770
6760 K2% = P4%(J2%): P4%(J2%) = P4%(K3%): P4%(K3%) = K2%
6770 X3# = R3#(P4%(J2%),J2%): K4% = J2% + 1
6780 FOR J3% = K4% TO LL%: X4# = - R3#(P4%(J3%),J2%)/X3#: R3#(P4%(J3%),J2%) = - X4#
6790 FOR J4 = K4% TO LL%: R3#(P4%(J3%),J4) = R3#(P4%(J3%),J4) + X4#*R3#(P4%(J2%),J4): NEXT J4: NEXT J3%: NEXT J2%
6800 B#(1) = R9#(P4%(1))
6810 FOR J1% = 2 TO LL%
6820 I8% = J1% - 1: X3# = 0
6830 FOR J2% = 1 TO I8%: X3# = X3# + R3#(P4%(J1%),J2%)*B#(J2%): NEXT J2%
6840 B#(J1%) = R9#(P4%(J1%)) - X3#: NEXT J1%
6850 B#(LL%) = B#(LL%)/R3#(P4%(LL%),LL%)
```

```
6860 FOR J1% = 2 TO LL%: I8% = LP1% - J1%: I9% = I8% + 1: X3# = 0
6870 FOR J2% = I9% TO LL%: X3# = X3# + R3#(P4%(I8%),J2%)*B#(J2%): NEXT J2%
6880 B#(I8%) = (B#(I8%) - X3#)/R3#(P4%(I8%),I8%): NEXT J1%: RETURN


1000 REM PLOT\TRIANGLE.BAS TRIANGLE - BASED INTERPOLATION
     Copyright (c) 1988 D.F.Watson
1010 DIM P1#(55,6), P2#(237,3), P5#(55), P8(80), P9(2,3), S1(3038), S2(3), S4(55),
     P0(4,2)
1020 DIM R1#(3,3), R3#(55,55), R4#(3,2), R5#(3,3), R7#(3), R8(2,2), S3(2,3), BX(6,3),
     R9#(55), B#(55)
1030 DIM P3%(237,3), P4%(237), P6%(55), P7%(25,2), R2%(2,2), R6%(3,2)
1050 DATA 1,0,0,1,0,0,1,0,0,0,1,0,0,1,0,1, - 1, - 1,0,1,2,5, - 1,0,2,3, - 1,5,1E37,3,1
1060 DEF FNCV(RT#) = RT#^2*LOG(RT#): REM minimum curvature: REM DEF
     FNCV(RT#) = RT#^2*(LOG(RT#) - 1): REM biharmonic spline: REM DEF FNCV(RT#)
     = RT#: REM dummy function
1070 REM DEF FNCV(RT#) = NG + SI*(1 - EXP( - RT#/RG)): NG = .25: SI = 1: RG = 1/3:
     REM exponential semivariogram
1080 REM DEF FNCV(RT#) = NG + SI*(1.5*RT#/RG - (RT#/RG)^3/2): NG = .25: SI = 1:
     RG = 1: REM spherical semivariogram
1090 P9(1,1) = - 1E37: P9(1,2) = - 1E37: P9(1,3) = - 1E37: P9(2,1) = 1E37: P9(2,2) =
     1E37: P9(2,3) = 1E37
1100 SL = 0: SR = 0: SZ = 5100: GR = 19: DP = 0: CL = 1: PT = 1: DZ = 15: TT = 0: WD
     = 10000: HI = 6800: AZ = - 25: TL = - 15: RD = 2: BI = 1.5: BJ = 7: IF DP = 0 THEN
     SR = 0
1140 OPEN "A: \HILL.DAT" FOR INPUT AS #1
1150 REM OPEN "A: \JDAVIS.DAT" FOR INPUT AS #1
1180 INPUT#1,ND%,NC%: FOR I1% = 1 TO NC%: INPUT#1,P8(I1%): NEXT I1%: IF
     CR>0 AND NC%>31 THEN NC% = 31
1185 FOR I1% = 1 TO ND%: P4%(I1%) = I1%: NEXT I1%: FOR I1% = 1 TO ND%: I2% =
     INT(RND*ND% + .5): IF I2%<1 THEN I2% = 1
1187 IF I2%>ND% THEN I2% = ND%
1188 I3% = P4%(I1%): P4%(I1%) = P4%(I2%): P4%(I2%) = I3%: NEXT I1%
1190 INPUT#1,XS,YS,DT: I2% = 1
1220 I1% = P4%(I2%)
1222 INPUT#1,P1#(I1%,1),P1#(I1%,2),P1#(I1%,3): P1#(I1%,1) = (P1#(I1%,1) - XS)/DT:
     P1#(I1%,2) = (P1#(I1%,2) - YS)/DT: IF (P1#(I1%,1) - .5)^2 + (P1#(I1%,2) - .5)^2<RD
     THEN GOTO 1270
1230 ND% = ND% - 1: IF I1%>ND% THEN GOTO 1360
1260 GOTO 1222
1270 IF P9(1,1)<P1#(I1%,1) THEN P9(1,1) = P1#(I1%,1)
1280 IF P9(2,1)>P1#(I1%,1) THEN P9(2,1) = P1#(I1%,1)
1290 IF P9(1,2)<P1#(I1%,2) THEN P9(1,2) = P1#(I1%,2)
1300 IF P9(2,2)>P1#(I1%,2) THEN P9(2,2) = P1#(I1%,2)
1310 IF P9(1,3)<P1#(I1%,3) THEN P9(1,3) = P1#(I1%,3)
1320 IF P9(2,3)>P1#(I1%,3) THEN P9(2,3) = P1#(I1%,3)
1350 I2% = I2% + 1: IF I2%< = ND% THEN GOTO 1220
1360 IF ND%<4 THEN STOP
1370 XF = P9(1,1) - P9(2,1): YF = P9(1,2) - P9(2,2): VS = P9(1,3) - P9(2,3): R8(2,1) = AZ:
     R8(2,2) = TL
1380 FOR I1% = 1 TO 6: READ BX(I1%,1),BX(I1%,2),BX(I1%,3): NEXT I1%
1390 BN% = 1: IF DP<1 THEN GOTO 1450
1400 BN% = 2: BX(2,1) = 0: BX(2,3) = 1: IF R8(2,1)>0 THEN GOTO 1420
1410 BX(2,1) = 1: BX(3,1) = 1: BX(4,1) = 1: BX(5,1) = 0: BX(6,1) = 0
1420 IF DP<2 OR CR>0 THEN GOTO 1450
1430 NC1% = INT(SZ/100): NC% = NC1% + 1: U0 = .9999/NC1%: U1 = .0000511 - U0
1440 FOR I1% = 1 TO NC%: U1 = U1 + U0: P8(I1%) = U1: NEXT I1%: GOTO 1460
1450 FOR I1% = 1 TO NC%: P8(I1%) = (P8(I1%) - P9(2,3))/VS + .00001*(RND - .5):
     NEXT I1%
1460 FOR I1% = 1 TO ND%: P1#(I1%,1) = P1#(I1%,1) + .0001*(RND - .5): P1#(I1%,2) =
     P1#(I1%,2) + .0001*(RND - .5) : P1#(I1%,3) = (P1#(I1%,3) - P9(2,3))/VS: NEXT I1%
1470 FOR I1% = 1 TO 3: READ R1#(I1%,1), R1#(I1%,2), R1#(I1%,3), R6%(I1%,1),
     R6%(I1%,2)
1480 P1#(I1% + ND%,1) = R1#(I1%,1)*XF + P9(2,1): P1#(I1% + ND%,2) = R1#(I1%,2)*YF
     + P9(2,2): P1#(I1% + ND%,3) = 0: NEXT I1%
1490 IF DP<1 THEN GOTO 1550
```

289

Appendix

```
1500 FOR I1% = 1 TO 2: R8(I1%,1) = COS(R8(2,I1%)/57.29578): R8(I1%,2) =
     SIN(R8(2,I1%)/57.29578): NEXT I1%
1510 FOR I1% = 2 TO 6
1520 YY = (BX(I1%,2) - .5)*R8(1,1) + (BX(I1%,1) - .5)*R8(1,2) + .5: BX(I1%,1) =
     (BX(I1%,1) - .5)*R8(1,1) - (BX(I1%,2) - .5)*R8(1,2) + .5
1530 BX3 = (BX(I1%,3) - .5)*R8(2,1) + (YY - .5)*R8(2,2) + .5: BX(I1%,2) = (YY -
     .5)*R8(2,1) - (BX(I1%,3) - .5)*R8(2,2) + .5: BX(I1%,3) = BX3: NEXT I1%
1540 REM end data input, begin gradient estimation
1550 SOUND 208,5: SOUND 165,5: SOUND 185,5: SOUND 123,5
1560 IF CL<1 THEN GOTO 3110
1570 PRINT " AT " TIME$ " BEGIN GRADIENT ESTIMATION  -  PLEASE WAIT"
1580 FOR I0% = 1 TO ND%
1590 FOR I1% = 1 TO 3: P3%(1,I1%) = ND% + I1%: P3%(2,I1%) = ND% + I1%:
     P2#(1,I1%) = R1#(I1%,3): P2#(2,I1%) = R1#(I1%,3) : NEXT I1%
1600 FOR I1% = 1 TO 237: P4%(I1%) = I1%: NEXT I1%: R2%(1,1) = 1: R2%(1,2) = 1:
     R2%(2,1) = 3: R2%(2,2) = 4: K1% = I0% : K2% = I0%: K3% = ND% + 4: K4 = - .5
1610 FOR I1% = K1% TO K2%
1620 IF I1% = K3% THEN GOTO 2000
1630 K5 = K4
1640 K6 = 1.5 + K5: IF K3%>ND% THEN K5 = - K5
1650 U0 = 0: K7% = 0: L6% = K6 - 2
1660 FOR I2% = 1 TO R2%(1,K6)
1670 L6% = L6% + 2: IF P3%(L6%,1) = 0 THEN GOTO 1670
1680 U1# = P2#(L6%,3) - (P1#(I1%,1) - P2#(L6%,1))^2: IF U1#<0 THEN GOTO 1840
1690 U1# = U1# - (P1#(I1%,2) - P2#(L6%,2))^2: IF U1#<0 THEN GOTO 1840
1700 R2%(2,K6) = R2%(2,K6) - 2: P4%(R2%(2,K6)) = L6%: U0 = U0 - 1
1710 FOR I3% = 1 TO 3
1720 L1% = 1: IF L1% = I3% THEN L1% = L1% + 1
1730 L2% = L1% + 1: IF L2% = I3% THEN L2% = L2% + 1
1740 IF K7%<1 THEN GOTO 1810
1750 K9% = K7%
1760 FOR I4% = 1 TO K9%
1770 IF P3%(L6%,L1%)<>P7%(I4%,1) THEN GOTO 1800
1780 IF P3%(L6%,L2%)<>P7%(I4%,2) THEN GOTO 1800
1790 P7%(I4%,1) = P7%(K9%,1): P7%(I4%,2) = P7%(K9%,2): K7% = K7% - 1: GOTO
     1820
1800 NEXT I4%
1810 K7% = K7% + 1: P7%(K7%,1) = P3%(L6%,L1%): P7%(K7%,2) = P3%(L6%,L2%)
1820 NEXT I3%
1830 P3%(L6%,1) = 0
1840 NEXT I2%
1850 IF K7%<1 THEN GOTO 2000
1860 REM make new triangles
1870 FOR I2% = 1 TO K7%
1880 IF K5>0 THEN GOTO 1900
1890 IF P7%(I2%,1)<>I0% AND P7%(I2%,2)<>I0% THEN GOTO 1970
1900 FOR I3% = 1 TO 2
1910 R3#(I3%,1) = P1#(P7%(I2%,I3%),1) - P1#(I1%,1): R3#(I3%,2) =
     P1#(P7%(I2%,I3%),2) - P1#(I1%,2)
1920 R3#(I3%,3) = R3#(I3%,1)*(P1#(P7%(I2%,I3%),1) + P1#(I1%,1))/2 + R3#(I3%,2)
     *(P1#(P7%(I2%,I3%),2) + P1#(I1%,2))/2
1930 NEXT I3%
1940 U1# = R3#(1,1)*R3#(2,2) - R3#(2,1)*R3#(1,2): P2#(P4%(R2%(2,K6)),1) =
     (R3#(1,3)*R3#(2,2) - R3#(2,3)*R3#(1,2))/U1#: P2#(P4%(R2%(2,K6)),2) =
     (R3#(1,1)*R3#(2,3) - R3#(2,1)*R3#(1,3))/U1#
1950 P2#(P4%(R2%(2,K6)),3) = (P1#(I1%,1) - P2#(P4%(R2%(2,K6)),1))^2 + (P1#(I1%,2) -
     P2#(P4%(R2%(2,K6)),2))^2
1960 P3%(P4%(R2%(2,K6)),1) = P7%(I2%,1): P3%(P4%(R2%(2,K6)),2) = P7%(I2%,2) :
     P3%(P4%(R2%(2,K6)),3) = I1%: R2%(2,K6) = R2%(2,K6) + 2: U0 = U0 + 1
1970 NEXT I2%
1980 R2%(1,K6) = R2%(1,K6) + U0: IF K5>0 THEN GOTO 2000
1990 K5 = - K5: GOTO 1640
2000 NEXT I1%
2010 IF K2%>K1% GOTO 2040
2020 K1% = 1: K2% = ND%: K3% = I0%: GOTO 1610
2030 REM order triangles positively
```

```
2040 FOR I1% = 1 TO 2
2050 L8% = I1% - 2
2060 FOR I2% = 1 TO R2%(1,I1%)
2070 L8% = L8% + 2: IF P3%(L8%,1) = 0 THEN GOTO 2070
2080 IF P3%(L8%,1)>ND% THEN GOTO 2140
2090 FOR I3% = 1 TO 2
2100 R3#(1,I3%) = P1#(P3%(L8%,2),I3%) - P1#(P3%(L8%,1),I3%): R3#(2,I3%) =
    P1#(P3%(L8%,3),I3%) - P1#(P3%(L8%,1),I3%)
2110 NEXT I3%
2120 IF R3#(1,1)*R3#(2,2) - R3#(1,2)*R3#(2,1)> = 0 THEN GOTO 2140
2130 K8 = P3%(L8%,3): P3%(L8%,3) = P3%(L8%,2): P3%(L8%,2) = K8
2140 NEXT I2%
2150 NEXT I1%
2160 IF SL>0 THEN GOTO 2470
2170 REM calculate natural neighbor coordinates
2180 R5#(1,1) = P1#(I0%,1): R5#(1,2) = P1#(I0%,2): R5#(2,1) = R5#(1,1) + .0001 :
    R5#(2,2) = R5#(1,2): R5#(3,1) = R5#(1,1): R5#(3,2) = R5#(1,2) + .0001
2190 FOR I1% = 1 TO 3
2200 L4% = 0: L5% = 0: L6% = 0
2210 FOR I2% = 1 TO R2%(1,2)
2220 L6% = L6% + 2: IF P3%(L6%,1) = 0 THEN GOTO 2220
2230 IF P3%(L6%,1)>ND% THEN GOTO 2410
2240 U1# = P2#(L6%,3) - (R5#(I1%,1) - P2#(L6%,1))^2: IF U1#<0 THEN GOTO 2410
2250 U1# = U1# - (R5#(I1%,2) - P2#(L6%,2))^2: IF U1#<0 THEN GOTO 2410
2260 FOR I3% = 1 TO 3
2270 FOR I4% = 1 TO 2
2280 R3#(I4%,1) = P1#(P3%(L6%,R6%(I3%,I4%)),1) - R5#(I1%,1): R3#(I4%,2) =
    P1#(P3%(L6%,R6%(I3%,I4%)),2) - R5#(I1%,2)
2290 R3#(I4%,3) = R3#(I4%,1)*(P1#(P3%(L6%,R6%(I3%,I4%)),1) + R5#(I1%,1))/2 +
    R3#(I4%,2)*(P1#(P3%(L6%,R6%(I3%,I4%)),2) + R5#(I1%,2))/2: NEXT I4%
2300 U1# = R3#(1,1)*R3#(2,2) - R3#(1,2)*R3#(2,1)
2310 R4#(R6%(I3%,2),1) = (R3#(1,3)*R3#(2,2) - R3#(2,3)*R3#(1,2))/U1#:
    R4#(R6%(I3%,2),2) = (R3#(1,1)*R3#(2,3) - R3#(2,1)*R3#(1,3))/U1#: NEXT I3%: L3%=0
2320 FOR I3% = 1 TO 3: R3#(3,I3%) = ((R4#(R6%(I3%,1),1) -
    P2#(L6%,1))*(R4#(R6%(I3%,2),2) - P2#(L6%,2)) - (R4#(R6%(I3%,2),1) -
    P2#(L6%,1))*(R4#(R6%(I3%,1),2) - P2#(L6%,2)))/2: IF R3#(3,I3%)>0 THEN L3% =
    L3% + 1
2330 NEXT I3%
2340 IF L3%>2 THEN L4% = 1
2350 FOR I3% = 1 TO 3: IF L5%<1 THEN GOTO 2390
2360 FOR I4% = 1 TO L5%: IF P3%(L6%,I3%)<>P6%(I4%) THEN GOTO 2380
2370 P5#(I4%) = P5#(I4%) + R3#(3,I3%): GOTO 2400
2380 NEXT I4%
2390 L5% = L5% + 1: P6%(L5%) = P3%(L6%,I3%): P5#(L5%) = R3#(3,I3%)
2400 NEXT I3%
2410 NEXT I2%: IF L4%<1 THEN GOTO 2470
2420 U4 = 0
2430 FOR I2% = 1 TO L5%: U4 = U4 + P5#(I2%): NEXT I2%: R7#(I1%) = 0
2440 FOR I2% = 1 TO L5%: P5#(I2%) = P5#(I2%)/U4: R7#(I1%) = R7#(I1%) +
    P5#(I2%)*P1#(P6%(I2%),3): NEXT I2%: NEXT I1%
2450 P1#(I0%,6) = P1#(I0%,3) - R7#(1): P1#(I0%,4) = (R7#(1) - R7#(2))/.0001:
    P1#(I0%,5) = (R7#(1) - R7#(3))/.0001 : GOTO 3080
2460 REM cross product gradients
2470 IF SL>1 THEN GOTO 2610
2480 P1#(I0%,4) = 0: P1#(I0%,5) = 0: P1#(I0%,6) = 0: P7# = 0: L7% = - 1
2490 FOR I1% = 1 TO R2%(1,1)
2500 L7% = L7% + 2: IF P3%(L7%,1) = 0 THEN GOTO 2500
2510 IF P3%(L7%,1)>ND% THEN GOTO 2570
2520 FOR I2% = 1 TO 2: FOR I3% = 1 TO 3: R3#(I2%,I3%) = P1#(P3%(L7%,1),I3%) -
    P1#(P3%(L7%,I2% + 1),I3%): NEXT I3% : NEXT I2%
2530 R3#(3,1) = R3#(1,2)*R3#(2,3) - R3#(2,2)*R3#(1,3): R3#(3,2) = R3#(1,3)*R3#(2,1) -
    R3#(2,3)*R3#(1,1): R3#(3,3) = R3#(1,1)*R3#(2,2) - R3#(2,1)*R3#(1,2): U3 = 1
2540 IF R3#(3,3)<0 THEN U3 = - 1
2550 U2 = (R3#(3,1)^2 + R3#(3,2)^2 + R3#(3,3)^2)^.5: P7# = P7# + U2
2560 FOR I2% = 1 TO 3: P1#(I0%,I2% + 3) = P1#(I0%,I2% + 3) + R3#(3,I2%)*U3: NEXT
    I2%
```

Appendix

```
2570 NEXT I1%
2580 U2 = P1#(I0%,4)^2 + P1#(I0%,5)^2 + P1#(I0%,6)^2: P7# = 1 - U2^.5/P7#
2590 P1#(I0%,4) = P1#(I0%,4)/P1#(I0%,6): P1#(I0%,5) = P1#(I0%,5)/P1#(I0%,6):
     P1#(I0%,6) = P7#
2600 GOTO 3080
2610 IF SL>2 THEN GOTO 2810 : REM least squares gradients
2620 P1#(I0%,4) = 0: P1#(I0%,5) = 0: P1#(I0%,6) = .2: L7% = - 1: P6%(1) = I0%: L5% =1
2630 R4#(1,1) = P1#(I0%,3): R4#(2,1) = P1#(I0%,1)*P1#(I0%,3): R4#(3,1) =
     P1#(I0%,2)*P1#(I0%,3)
2640 R3#(2,1) = P1#(I0%,1): R3#(2,2) = P1#(I0%,1)^2: R3#(3,1) = P1#(I0%,2) : R3#(3,2)
     = P1#(I0%,1)*P1#(I0%,2): R3#(3,3) = P1#(I0%,2)^2
2650 FOR I1% = 1 TO R2%(1,1)
2660 L7% = L7% + 2
2670 IF P3%(L7%,1) = 0 THEN GOTO 2660
2680 IF P3%(L7%,1)>ND% THEN GOTO 2760
2690 FOR I2% = 1 TO 3
2700 FOR I3% = 1 TO L5%
2710 IF P3%(L7%,I2%) = P6%(I3%) THEN GOTO 2750
2720 NEXT I3%
2730 L5% = L5% + 1: P6%(L5%) = P3%(L7%,I2%): R4#(1,1) = R4#(1,1) +
     P1#(P6%(L5%),3) : R4#(2,1) = R4#(2,1) + P1#(P6%(L5%),1)*P1#(P6%(L5%),3):
     R4#(3,1) = R4#(3,1) + P1#(P6%(L5%),2)*P1#(P6%(L5%),3)
2740 R3#(2,1) = R3#(2,1) + P1#(P6%(L5%),1): R3#(2,2) = R3#(2,2) +
     P1#(P6%(L5%),1)^2: R3#(3,1) = R3#(3,1) + P1#(P6%(L5%),2): R3#(3,2) = R3#(3,2) +
     P1#(P6%(L5%),1)*P1#(P6%(L5%),2): R3#(3,3) = R3#(3,3) + P1#(P6%(L5%),2)^2
2750 NEXT I2%
2760 NEXT I1%
2770 U1# = L5%*(R3#(2,2)*R3#(3,3) - R3#(3,2)^2) - R3#(2,1)*(R3#(2,1)*R3#(3,3) -
     R3#(3,1)*R3#(3,2)) + R3#(3,1)*(R3#(2,1)*R3#(3,2) - R3#(3,1)*R3#(2,2))
2780 P1#(I0%,4) = - (L5%*(R4#(2,1)*R3#(3,3) - R3#(3,2)*R4#(3,1)) -
     R4#(1,1)*(R3#(2,1)*R3#(3,3) - R3#(3,1)*R3#(3,2)) + R3#(3,1)*(R3#(2,1)*R4#(3,1) -
     R3#(3,1)*R4#(2,1)))/U1#
2790 P1#(I0%,5) = - (L5%*(R3#(2,2)*R4#(3,1) - R3#(3,2)*R4#(2,1)) -
     R3#(2,1)*(R3#(2,1)*R4#(3,1) - R3#(3,1)*R4#(2,1)) + R4#(1,1)*(R3#(2,1)*R3#(3,2) -
     R3#(3,1)*R3#(2,2)))/U1#
2800 GOTO 3080 : REM spline gradients
2810 IF SL>3 THEN GOTO 3035
2815 P1#(I0%,4) = 0: P1#(I0%,5) = 0: P1#(I0%,6) = .2: L7% = - 1: L5% = 1: P6%(1) =
     I0%
2820 R5#(1,1) = P1#(I0%,1): R5#(1,2) = P1#(I0%,2): R5#(2,1) = R5#(1,1) + .0001 :
     R5#(2,2) = R5#(1,2): R5#(3,1) = R5#(1,1): R5#(3,2) = R5#(1,2) + .0001
2830 FOR I1% = 1 TO R2%(1,1) 2840 L7% = L7% + 2: IF P3%(L7%,1) = 0 THEN GOTO
     2840
2850 IF P3%(L7%,1)>ND% THEN GOTO 2890
2860 FOR I2% = 1 TO 3: FOR I3% = 1 TO L5%: IF P3%(L7%,I2%) = P6%(I3%) THEN
     GOTO 2880
2870 NEXT I3%: L5% = L5% + 1: P6%(L5%) = P3%(L7%,I2%)
2880 NEXT I2%
2890 NEXT I1%
2900 LL% = L5% + 3
2910 FOR I1% = 1 TO L5%
2920 R3#(I1%,1) = 1: R3#(LL%,I1% + 3) = 1: R3#(I1%,2) = P1#(P6%(I1%),1): R3#(L5% +
     1,I1% + 3) = R3#(I1%,2)
2930 R3#(I1%,3) = P1#(P6%(I1%),2): R3#(L5% + 2,I1% + 3) = R3#(I1%,3): R9#(I1%) =
     P1#(P6%(I1%),3): NEXT I1%
2940 FOR I1% = 1 TO 3: FOR I2% = 1 TO 3: R3#(I1% + L5%,I2%) = 0: NEXT I2%:
     R9#(I1% + L5%) = 0: NEXT I1%
2950 FOR I1% = 1 TO L5%: R3#(I1%,I1% + 3) = 0: IJ = I1% + 1: IF IJ>L5% THEN GOTO
     2970
2960 FOR I2% = IJ TO L5%: RT# = ((R3#(I1%,2) - R3#(I2%,2))^2 + (R3#(I1%,3) -
     R3#(I2%,3))^2)^.5: R3#(I2%,I1% + 3) = FNCV(RT#): R3#(I1%,I2% + 3) = R3#(I2%,I1%
     + 3): NEXT I2%
2970 NEXT I1%: GOSUB 6690
2980 TR# = 0: TS# = 0
2990 FOR I1% = 4 TO LL%
```

```
3000 RT# = ((P1#(P6%(I1% - 3),1) - R5#(2,1))^2 + (P1#(P6%(I1% - 3),2) -
     R5#(2,2))^2)^.5: TR# = TR# + B#(I1%)*FNCV(RT#)
3010 RS# = ((P1#(P6%(I1% - 3),1) - R5#(3,1))^2 + (P1#(P6%(I1% - 3),2) -
     R5#(3,2))^2)^.5: TS# = TS# + B#(I1%)*FNCV(RS#) : NEXT I1%
3020 R5#(2,3) = B#(1) + B#(2)*R5#(2,1) + B#(3)*R5#(2,2) + TR#: P1#(I0%,4) =
     (P1#(I0%,3) - R5#(2,3))/.0001
3030 R5#(3,3) = B#(1) + B#(2)*R5#(3,1) + B#(3)*R5#(3,2) + TS#: P1#(I0%,5) =
     (P1#(I0%,3) - R5#(3,3))/.0001: GOTO 3080
3033 REM hyperboloid gradients
3035 R3#(5,1) = 1: FOR I1% = 1 TO 4: R9#(I1%) = 0: FOR I2% = 1 TO 4: R3#(I1%,I2%) =
     0: NEXT I2%: NEXT I1%: LL% = 4
3036 P1#(I0%,4) = 0: P1#(I0%,5) = 0: P1#(I0%,6) = .2: L7% = - 1: L5% = 0: FOR I1% = 1
     TO R2%(1,1)
3037 L7% = L7% + 2: IF P3%(L7%,1) = 0 THEN GOTO 3037
3038 IF P3%(L7%,1)>ND% THEN GOTO 3070
3039 FOR I2% = 1 TO 3: IF L5%<1 THEN GOTO 3044
3041 FOR I3% = 1 TO L5%: IF P3%(L7%,I2%) = P6%(I3%) THEN GOTO 3060
3042 NEXT I3%
3044 L5% = L5% + 1: P6%(L5%) = P3%(L7%,I2%)
3050 R3#(5,2) = P1#(P6%(L5%),1): R3#(5,3) = P1#(P6%(L5%),2): R3#(5,4) =
     P1#(P6%(L5%),1)*P1#(P6%(L5%),2)
3055 FOR I3% = 1 TO 4: FOR I4% = 1 TO 4: R3#(I3%,I4%) = R3#(I3%,I4%) +
     R3#(5,I3%)*R3#(5,I4%): NEXT I4%: R9#(I3%) = R9#(I3%) +
     R3#(5,I3%)*P1#(P6%(L5%),3): NEXT I3%
3060 NEXT I2%
3070 NEXT I1%: GOSUB 6690: P1#(I0%,4) = - B#(2) - B#(4)*P1#(I0%,2): P1#(I0%,5) = -
     B#(3) - B#(4)*P1#(I0%,1): P1#(I0%,6) = .2
3080 NEXT I0%
3100 REM natural neighbor sort
3110 FOR I0% = 1 TO 3: P3%(1,I0%) = ND% + I0%: P2#(1,I0%) = R1#(I0%,3): NEXT I0%
3120 FOR I0% = 1 TO 237: P4%(I0%) = I0%: NEXT I0%: NT% = 1: L0% = 2
3130 FOR I0% = 1 TO ND%: L3% = 0
3140 FOR I1% = 1 TO NT%
3150 U1# = P2#(I1%,3) - (P1#(I0%,1) - P2#(I1%,1))^2: IF U1#<0 THEN GOTO 3300
3160 U1# = U1# - (P1#(I0%,2) - P2#(I1%,2))^2: IF U1#<0 THEN GOTO 3300
3170 L0% = L0% - 1: P4%(L0%) = I1%
3180 FOR I2% = 1 TO 3
3190 L1% = 1: IF L1% = I2% THEN L1% = L1% + 1
3200 L2% = L1% + 1: IF L2% = I2% THEN L2% = L2% + 1
3210 IF L3%<1 THEN GOTO 3280
3220 L5% = L3%
3230 FOR I3% = 1 TO L5%
3240 IF P3%(I1%,L1%)<>P7%(I3%,1) THEN GOTO 3270
3250 IF P3%(I1%,L2%)<>P7%(I3%,2) THEN GOTO 3270
3260 P7%(I3%,1) = P7%(L5%,1): P7%(I3%,2) = P7%(L5%,2): L3% = L3% - 1: GOTO
     3290
3270 NEXT I3%
3280 L3% = L3% + 1: P7%(L3%,1) = P3%(I1%,L1%): P7%(L3%,2) = P3%(I1%,L2%)
3290 NEXT I2%
3300 NEXT I1%
3310 FOR I1% = 1 TO L3%
3320 FOR I2% = 1 TO 2
3330 R3#(I2%,1) = P1#(P7%(I1%,I2%),1) - P1#(I0%,1): R3#(I2%,2) =
     P1#(P7%(I1%,I2%),2) - P1#(I0%,2): R3#(I2%,3) = R3#(I2%,1)*(P1#(P7%(I1%,I2%),1) +
     P1#(I0%,1))/2
3340 R3#(I2%,3) = R3#(I2%,3) + R3#(I2%,2)*(P1#(P7%(I1%,I2%),2) + P1#(I0%,2))/2 3350
     NEXT I2%
3360 U1# = R3#(1,1)*R3#(2,2) - R3#(2,1)*R3#(1,2): P2#(P4%(L0%),1) =
     (R3#(1,3)*R3#(2,2) - R3#(2,3)*R3#(1,2))/U1#: P2#(P4%(L0%),2) = (R3#(1,1)*R3#(2,3)
     - R3#(2,1)*R3#(1,3))/U1#
3370 P2#(P4%(L0%),3) = (P1#(I0%,1) - P2#(P4%(L0%),1))^2 + (P1#(I0%,2) -
     P2#(P4%(L0%),2))^2
3380 P3%(P4%(L0%),1) = P7%(I1%,1): P3%(P4%(L0%),2) = P7%(I1%,2):
     P3%(P4%(L0%),3) = I0%: L0% = L0% + 1
3390 NEXT I1%
3400 NT% = NT% + 2: NEXT I0%
```

293

Appendix

```
3410 FOR I0% = 1 TO NT%: IF P3%(I0%,1)>ND% THEN GOTO 3460
3420 FOR I1% = 1 TO 2
3430 R3#(1,I1%) = P1#(P3%(I0%,2),I1%) - P1#(P3%(I0%,1),I1%): R3#(2,I1%) =
     P1#(P3%(I0%,3),I1%) - P1#(P3%(I0%,1),I1%): NEXT I1%
3440 IF R3#(1,1)*R3#(2,2) - R3#(1,2)*R3#(2,1)> = 0 THEN GOTO 3460
3450 L4% = P3%(I0%,3): P3%(I0%,3) = P3%(I0%,2): P3%(I0%,2) = L4%
3460 NEXT I0%
3820 REM make triangular grid
3830 PRINT " AT " TIME$ " BEGIN SURFACE INTERPOLATION  -  PLEASE WAIT":
     SOUND 123,5: SOUND 185,5: SOUND 208,5 : SOUND 165,5
3840 I0% = GR + 1: IS = 1: I2% = 2*I0% + 1: I3% = 1 + INT(GR/3^.5): I4% = I3%*I2% +
     I0%: SX = 1/GR: SY = 1/(2*I3%): S2(3) = .000001
3850 WX# = S2(3): S2(1) = 1.00001: AA# = SX*SY*.5*DT*DT: WY# = 1: U8 = 1
3860 FOR L1% = 1 TO I4%
3870 S1(L1%) = - 999: WXO# = WX#: WYO# = WY#
4320 REM .....calculate barycentric coordinates of WX#,WY# **********
4340 K5 = 0
4350 FOR L2% = 1 TO NT%
4360 IF P3%(L2%,1)>ND% THEN GOTO 4460
4370 U1# = P2#(L2%,3) - (WX# - P2#(L2%,1))^2: IF U1#<0 THEN GOTO 4460
4380 U1# = U1# - (WY# - P2#(L2%,2))^2: IF U1#<0 THEN GOTO 4460
4390 K3% = 0
4400 FOR L3% = 1 TO 3
4410 P5#(L3%) = ((P1#(P3%(L2%,R6%(L3%,2)),1) -
     P1#(P3%(L2%,R6%(L3%,1)),1))*(WY# - P1#(P3%(L2%,R6%(L3%,1)),2)) - (WX# -
     P1#(P3%(L2%,R6%(L3%,1)),1))*(P1#(P3%(L2%,R6%(L3%,2)),2) -
     P1#(P3%(L2%,R6%(L3%,1)),2)))
4420 P6%(L3%) = P3%(L2%,(6 - R6%(L3%,1) - R6%(L3%,2))): IF P5#(L3%)>0 THEN
     K3% = K3% + 1
4430 NEXT L3%
4440 IF K3%<3 THEN GOTO 4460
4450 K5 = 3: GOTO 4560
4460 NEXT L2%
4470 IF K5<1 GOTO 4940
4560 S1(L1%) = 0: U4 = 0
4570 FOR L2% = 1 TO K5: U4 = U4 + P5#(L2%): NEXT L2%
4580 FOR L2% = 1 TO K5: P5#(L2%) = P5#(L2%)/U4: S1(L1%) = S1(L1%) +
     P5#(L2%)*P1#(P6%(L2%),3): NEXT L2%
4590 IF CL<1 THEN GOTO 4940
4620 FOR L2% = 1 TO K5: S4(L2%) = 0
4630 IF P5#(L2%)<.00001 OR P5#(L2%)>1 THEN GOTO 4690
4640 IF ABS(P1#(P6%(L2%),6))<.00001 THEN GOTO 4690
4650 RS# = ABS(P1#(P6%(L2%),6)) + BI: RT# = RS#*BJ: RB# = 1/RT#: BD# =
     P5#(L2%)^RT#: BB# = BD#*2: IF BD#>.5 THEN BB# = (1 - BD#)*2
4660 BB# = BB#^RS#/2: IF BD#>.5 THEN BB# = 1 - BB#
4670 HP# = BB#^RB#
4680 S4(L2%) = ((P1#(P6%(L2%),4)*P1#(P6%(L2%),1) +
     P1#(P6%(L2%),5)*P1#(P6%(L2%),2) + P1#(P6%(L2%),3) - P1#(P6%(L2%),4)*WX# -
     P1#(P6%(L2%),5)*WY#) - S1(L1%))*HP#
4690 NEXT L2%
4700 FOR L2% = 1 TO K5: S1(L1%) = S1(L1%) + S4(L2%): NEXT L2%
4940 IS = IS + SX*U8: IF IS>0 THEN GOTO 4960
4950 WXO# = WXO# + SX/2: IS = 1.001 - SX
4960 WX# = WXO# + SX*U8: WY# = WYO#: IF U8*WX#<S2(2 - U8) THEN GOTO 4980
4970 U8 = - U8: WX# = S2(2 + U8): WY# = WY# - SY: IS = U8
4980 NEXT L1%
5060 REM end interpolation to triangular grid and begin display construction
5070 SOUND 208,5: SOUND 165,5: SOUND 185,5: SOUND 123,5: PRINT " AT " TIME$;
5071 INPUT " PREPARE PLOTTER AND PRESS ENTER",PM$
5090 LW = (HI - SZ)/2: IF DP>0 THEN SZ = INT(SZ*2/3)
5091 IF DP>0 AND SZ>HI/2 THEN SZ = HI/2
5100 SW = (WD - SZ)/2: ZW = (HI - SZ)/2
5110 IF SR>0 THEN SW = (WD - 2*SZ)/3
5120 SW = SW + 1: SV = 2*SW + SZ: SR = SR*2: SZ = SZ - 2: SZ = SZ - 2: ZW = ZW -
     1: SU = SW: IF SR>0 THEN SU = (2*SW + SV)/3
5160 REM plot open
```

```
5170 LPRINT "IN;IP0,0," + STR$(WD) + "," + STR$(HI) + ";SP1;": IF SR<2 THEN GOTO
     5180
5175 FOR L1% = 2 TO 6: BX1 = (BX(L1%,1) - .5)*.99875 - (BX(L1%,3) - .5)*.05 + .5:
     BX(L1%,3) = (BX(L1%,3) - .5)*.99875 + (BX(L1%,1) - .5)*.05 + .5: BX(L1%,1) = BX1:
     NEXT L1%: LPRINT "PU;PA0,6800;"
5180 LPRINT "PU;PA" + STR$(BX(BN%,1)*SZ + SW) + "," + STR$(BX(BN%,2)*SZ +
     ZW) + ";PD;" : ZX = SX: ZY = SY: ZZ = .5: R3#(2,2) = 1 + 2*SY: R3#(2,1) = 0
5190 FOR L1% = BN% + 1 TO BN% + 4: LPRINT "PA" + STR$(BX(L1%,1)*SZ + SW) +
     "," + STR$(BX(L1%,2)*SZ + ZW) + ";": NEXT L1% : LPRINT "PU;"
5192 FOR L1% = 2 TO 6: BX(L1%,1) = (BX(L1%,1) - .5)*.995 + (BX(L1%,3) - .5)*.1 + .5:
     NEXT L1%: SM = 0: SA = 0
5200 REM plot contours
5220 FOR L1% = 1 TO I3%
5230 R3#(1,2) = R3#(2,2) - 2*SY - 2*ZY: R3#(1,1) = R3#(2,1): N1% = (2*L1% - .5 +
     ZZ)*I0% + L1% + .5/ZZ: R3#(1,3) = S1(N1%): R3#(2,2) = R3#(2,2) - 2*SY - ZY:
     R3#(2,1) = .5 - ZZ: N2% = (2*L1% - .5 + ZZ)*I0% + L1%
5240 R3#(2,3) = S1(N2%): R3#(3,2) = R3#(2,2): R3#(3,1) = R3#(2,1) + ZX/2: N3% = N2%
     - .5/ZZ: R3#(3,3) = S1(N3%): AR# = AA#/2
5250 GOSUB 5820
5260 R3#(1,2) = R3#(1,2) + 2*ZY: N2% = (2*L1% - .5/ZZ - .5 - ZZ)*I0% + L1%: R3#(1,3) =
     S1(N2%)
5270 GOSUB 5820
5280 AR# = AA#
5290 FOR L2% = 1 TO GR
5300 R3#(2,2) = R3#(1,2): R3#(2,1) = R3#(1,1) + ZX: N4% = N2% + .5/ZZ: R3#(2,3) =
     S1(N4%)
5310 GOSUB 5820
5320 R3#(1,2) = R3#(3,2): R3#(1,1) = R3#(3,1) + ZX
5330 IF L2% = GR THEN R3#(1,1) = R3#(1,1) - ZX/2
5340 N6% = N3% - .5/ZZ: R3#(1,3) = S1(N6%): IF L2% = GR THEN AR# = AA#/2
5350 GOSUB 5820
5360 R3#(2,2) = R3#(2,2) - 2*ZY: N5% = N1% + .5/ZZ: R3#(2,3) = S1(N5%)
5370 GOSUB 5820
5380 R3#(1,2) = R3#(2,2): R3#(1,1) = R3#(2,1) - ZX: R3#(1,3) = S1(N1%): IF L2% = GR
     THEN AR# = AA#
5390 GOSUB 5820
5400 N1% = N5%: N2% = N4%: N3% = N6%: R3#(1,2) = R3#(1,2) + 2*ZY: R3#(1,1) =
     R3#(2,1) : R3#(1,3) = S1(N4%): R3#(3,1) = R3#(3,1) + ZX: R3#(3,3) = S1(N6%) 5410
     NEXT L2%
5420 ZY = - ZY: ZX = - ZX: ZZ = - ZZ
5430 NEXT L1%: REM end plotting contours
5450 REM plot data points
5460 IF PT<1 THEN GOTO 5620
5480 DZ% = DZ: FOR L1% = 1 TO ND%
5490 IF P1#(L1%,1)< - .0001 OR P1#(L1%,1)>1.0001 OR P1#(L1%,2)< - .0001 OR
     P1#(L1%,2)>1.0001 THEN GOTO 5610
5500 PP1# = P1#(L1%,1): PP2# = P1#(L1%,2): PP3# = P1#(L1%,3): IF DP<1 THEN
     GOTO 5530
5510 YY = (PP2# - .5)*R8(1,1) + (PP1# - .5)*R8(1,2) + .5: PP1# = (PP1# - .5)*R8(1,1) -
     (PP2# - .5)*R8(1,2) + .5
5520 ZZ3 = (PP3# - .5)*R8(2,1) + (YY - .5)*R8(2,2) + .5: PP2# = (YY - .5)*R8(2,1) - (PP3#
     - .5)*R8(2,2) + .5: PP3# = ZZ3: IF SR<2 THEN GOTO 5528
5525 PP1# = (PP1# - .5)*.99875 - (PP3# - .5)*.05 + .5
5528 IF SR<>1 THEN GOTO 5530
5529 PP1# = (PP1# - .5)*.99875 + (PP3# - .5)*.05 + .5
5530 S3(1,1) = PP1#*SZ + SW: S3(1,2) = PP2#*SZ + ZW
5540 LPRINT "PA" + STR$(S3(1,1) - DZ%) + "," + STR$(S3(1,2) + DZ%) + ";": LPRINT
     "PD;PA" + STR$(S3(1,1) + DZ%) + "," + STR$(S3(1,2) + DZ%) + ";": LPRINT "PA" +
     STR$(S3(1,1) - DZ%) + "," + STR$(S3(1,2) - DZ%) + ";"
5550 LPRINT "PA" + STR$(S3(1,1) - DZ%) + "," + STR$(S3(1,2) - DZ%) + ";": LPRINT
     "PA" + STR$(S3(1,1) - DZ%) + "," + STR$(S3(1,2) + DZ%) + ";": LPRINT "PA" +
     STR$(S3(1,1) + DZ%) + "," + STR$(S3(1,2) - DZ%) + ";PU;"
5610 NEXT L1%
5620 IF TT<1 THEN GOTO 5770
5630 REM draw triangles
5640 FOR L1% = 1 TO NT%
```

Appendix

```
5650 IF P3%(L1%,1)>ND% THEN GOTO 5750
5651 FOR L2% = 1 TO 3: IF P1#(P3%(L1%,L2%),1)< - .01 OR
     P1#(P3%(L1%,L2%),1)>1.01 OR P1#(P3%(L1%,L2%),2)< - .01 OR
     P1#(P3%(L1%,L2%),2)>1.01 THEN GOTO 5750
5652 NEXT L2%
5653 PP1# = P1#(P3%(L1%,3),1): PP2# = P1#(P3%(L1%,3),2): PP3# =
     P1#(P3%(L1%,3),3): IF DP<1 THEN GOTO 5708
5701 YY = (PP2# - .5)*R8(1,1) + (PP1# - .5)*R8(1,2) + .5: PP1# = (PP1# - .5)*R8(1,1) -
     (PP2# - .5)*R8(1,2) + .5
5703 ZZ3 = (PP3# - .5)*R8(2,1) + (YY - .5)*R8(2,2) + .5: PP2# = (YY - .5)*R8(2,1) - (PP3#
     - .5)*R8(2,2) + .5: PP3# = ZZ3: IF SR<2 THEN GOTO 5706
5705 PP1# = (PP1# - .5)*.99875 - (PP3# - .5)*.05 + .5
5706 IF SR<>1 THEN GOTO 5708
5707 PP1# = (PP1# - .5)*.99875 + (PP3# - .5)*.05 + .5
5708 S3(1,1) = PP1#*SZ + SW: S3(1,2) = PP2#*SZ + ZW: LPRINT "PA" + STR$(S3(1,1))
     + "," + STR$(S3(1,2)) + ";": LPRINT "PD;"
5710 FOR L2% = 1 TO 3: PP1# = P1#(P3%(L1%,L2%),1): PP2# =
     P1#(P3%(L1%,L2%),2): PP3# = P1#(P3%(L1%,L2%),3): IF DP<1 THEN GOTO 5718
5711 YY = (PP2# - .5)*R8(1,1) + (PP1# - .5)*R8(1,2) + .5: PP1# = (PP1# - .5)*R8(1,1) -
     (PP2# - .5)*R8(1,2) + .5
5713 ZZ3 = (PP3# - .5)*R8(2,1) + (YY - .5)*R8(2,2) + .5: PP2# = (YY - .5)*R8(2,1) - (PP3#
     - .5)*R8(2,2) + .5: PP3# = ZZ3: IF SR<2 THEN GOTO 5716
5715 PP1# = (PP1# - .5)*.99875 - (PP3# - .5)*.05 + .5
5716 IF SR<>1 THEN GOTO 5718
5717 PP1# = (PP1# - .5)*.99875 + (PP3# - .5)*.05 + .5
5718 S3(1,1) = PP1#*SZ + SW: S3(1,2) = PP2#*SZ + ZW: LPRINT "PA" + STR$(S3(1,1))
     + "," + STR$(S3(1,2)) + ";": NEXT L2% : LPRINT "PU;"
5750 NEXT L1%
5770 SR = SR - 1: SW = SV: IF SR = 1 THEN GOTO 5180
5775 LPRINT "PU;PA" + STR$(SU) + "," + STR$(LW - 500) + ";SI0.2,0.3;SL0.2;" 5777
     LPRINT "LBFigure 1.0.1 - Example Isolines" + CHR$(3): LPRINT "PU;PA0,6800;":
     REM plot close
5790 SOUND 123,5: SOUND 185,5: SOUND 208,5: SOUND 165,5: TM$ = TIME$
5800 SM = SM/SA: PRINT " AVERAGE HEIGHT OF SURFACE IS" SM: PRINT " AT "
     TM$ " RUN COMPLETE": END
5810 REM curvemaker
5820 ST = R3#(1,3)
5830 IF ST>R3#(2,3) THEN ST = R3#(2,3)
5840 IF ST>R3#(3,3) THEN ST = R3#(3,3)
5850 IF ST< = - 999 THEN RETURN
5860 SM = SM + AR#*((R3#(1,3) + R3#(2,3) + R3#(3,3))/3*VS + P9(2,3)): SA = SA + AR#
6210 L8% = 0: IF DP<2 THEN GOTO 6310
6220 L8% = 1
6230 REM rotate triangle for orthogonal profiles
6240 FOR L3% = 1 TO 3
6250 Z# = R3#(L3%,3): R3#(L3%,3) = 1 - R3#(L3%,L8%): R3#(L3%,L8%) = Z#
6260 NEXT L3%
6270 ST = R3#(1,3)
6280 IF ST>R3#(2,3) THEN ST = R3#(2,3)
6290 IF ST>R3#(3,3) THEN ST = R3#(3,3)
6300 IF ST < = - 999 THEN RETURN
6310 TP = R3#(1,3)
6320 IF TP<R3#(2,3) THEN TP = R3#(2,3)
6330 IF TP<R3#(3,3) THEN TP = R3#(3,3)
6340 IF TP = ST THEN RETURN
6350 FOR L0 = 1 TO NC%
6360 REM slice the triangle
6370 IF P8(L0)>TP OR P8(L0)<ST THEN GOTO 6610
6380 L5% = 1: L4% = 0
6390 L4% = L4% + 1
6400 L7% = 1: IF L7% = L4% THEN L7% = L7% + 1
6410 L3% = L7% + 1: IF L3% = L4% THEN L3% = L3% + 1
6420 IF R3#(L7%,3) = R3#(L3%,3) THEN GOTO 6460
6430 F = (P8(L0) - R3#(L7%,3))/(R3#(L3%,3) - R3#(L7%,3)): IF F<0 OR F>1 THEN GOTO
     6460
```

```
6440 S3(L5%,1) = R3#(L7%,1) + (R3#(L3%,1) - R3#(L7%,1))*F: S3(L5%,2) = R3#(L7%,2)
     + (R3#(L3%,2) - R3#(L7%,2))*F
6450 L5% = L5% + 1
6460 IF L5%<3 THEN GOTO 6390
6470 S3(1,3) = P8(L0): S3(2,3) = P8(L0): IF DP<2 THEN GOTO 6500
6480 REM reverse rotate the intersection trace
6490 FOR L3% = 1 TO 2: Z# = S3(L3%,3): S3(L3%,3) = S3(L3%,L8%): S3(L3%,L8%) = 1
     - Z#: NEXT L3%
6500 IF DP = 0 THEN GOTO 6570
6510 REM apply perspective
6520 FOR L3% = 1 TO 2
6530 YY = (S3(L3%,2) - .5)*R8(1,1) + (S3(L3%,1) - .5)*R8(1,2): S3(L3%,1) = (S3(L3%,1) -
     .5)*R8(1,1) - (S3(L3%,2) - .5)*R8(1,2) + .5
6540 S33 = (S3(L3%,3) - .5)*R8(2,1) + YY*R8(2,2) + .5: S3(L3%,2) = YY*R8(2,1) -
     (S3(L3%,3) - .5)*R8(2,2) + .5: S3(L3%,3) = S33
6550 NEXT L3%
6552 IF SR<2 THEN GOTO 6560
6554 S3(1,1) = (S3(1,1) - .5)*.99875 - (S3(1,3) - .5)*.05 + .5: S3(2,1) = (S3(2,1) -
     .5)*.99875 - (S3(2,3) - .5)*.05 + .5
6560 IF SR<>1 THEN GOTO 6570
6565 S3(1,1) = (S3(1,1) - .5)*.99875 + (S3(1,3) - .5)*.05 + .5: S3(2,1) = (S3(2,1) -
     .5)*.99875 + (S3(2,3) - .5)*.05 + .5
6567 REM draw a line from (S3(1,1),S3(1,2)) to (S3(2,1),S3(2,2))
6570 LPRINT "PA" + STR$(S3(1,1)*SZ + SW) + "," + STR$(S3(1,2)*SZ + ZW) + ";":
     LPRINT "PD;PA" + STR$(S3(2,1)*SZ + SW) + "," + STR$(S3(2,2)*SZ + ZW) + ";PU;"
6610 NEXT L0
6620 IF L8%<1 THEN RETURN
6630 REM reverse rotate the triangle
6640 FOR L3% = 1 TO 3
6650 Z# = R3#(L3%,3): R3#(L3%,3) = R3#(L3%,L8%): R3#(L3%,L8%) = 1 - Z#
6660 NEXT L3%
6670 L8% = L8% + 1: IF L8%<3 THEN GOTO 6240 ELSE RETURN
6680 REM solve R3#(LL%,LL%)xB#(LL%) = R9#(LL%)
6690 FOR J2% = 1 TO LL%: P4%(J2%) = J2%: B#(J2%) = 0
6700 FOR J3% = 1 TO LL%: IF ABS(R3#(J2%,J3%))> = B#(J2%) THEN B#(J2%) =
     ABS(R3#(J2%,J3%))
6710 NEXT J3%: NEXT J2%: LM1 = LL% - 1: LP1 = LL% + 1
6720 FOR J2% = 1 TO LM1: X3# = 0
6730 FOR J3% = J2% TO LL%: X4# = ABS(R3#(P4%(J3%),J2%))/B#(P4%(J3%)): IF
     X4#< = X3# THEN GOTO 6750
6740 X3# = X4#: K3 = J3%
6750 NEXT J3%: IF K3 = J2% GOTO 6770
6760 K2 = P4%(J2%): P4%(J2%) = P4%(K3): P4%(K3) = K2
6770 X3# = R3#(P4%(J2%),J2%): K4 = J2% + 1
6780 FOR J3% = K4 TO LL%: X4# = - R3#(P4%(J3%),J2%)/X3#: R3#(P4%(J3%),J2%) =
     - X4#
6790 FOR J4 = K4 TO LL%: R3#(P4%(J3%),J4) = R3#(P4%(J3%),J4) +
     X4#*R3#(P4%(J2%),J4): NEXT J4: NEXT J3%: NEXT J2%
6800 B#(1) = R9#(P4%(1))
6810 FOR J1% = 2 TO LL%
6820 I8% = J1% - 1: X3# = 0
6830 FOR J2% = 1 TO I8%: X3# = X3# + R3#(P4%(J1%),J2%)*B#(J2%): NEXT J2%
6840 B#(J1%) = R9#(P4%(J1%)) - X3#: NEXT J1%
6850 B#(LL%) = B#(LL%)/R3#(P4%(LL%),LL%)
6860 FOR J1% = 2 TO LL%: I8% = LP1 - J1%: I9% = I8% + 1: X3# = 0
6870 FOR J2% = I9% TO LL%: X3# = X3# + R3#(P4%(I8%),J2%)*B#(J2%): NEXT J2%
6880 B#(I8%) = (B#(I8%) - X3#)/R3#(P4%(I8%),I8%): NEXT J1%: RETURN


1000 REM PLOT\DISTANCE.BAS DISTANCE - BASED INTERPOLATION
     Copyright (c) 1988 D.F.Watson
1010 DIM P1#(63,6), P2#(237,3), P5#(50), P8(80), P9(2,3), S1(2033), S2(3), S4(50),
     PO(4,2)
1020 DIM R1#(3,3), R3#(55,55), R4#(3,2), R5#(3,3), R7#(3), R8(2,2), S3(2,3), BX(6,3),
     R9#(55), B#(55)
1030 DIM P3%(237,3), P4%(237), P6%(50), P7%(25,2), R2%(2,2), R6%(3,2)
```

Appendix

```
1050 DATA 1,0,0,1,1,0,0,1,0,0,0,0,1,0,0,1,0,1, - 1, - 1,0,1,2,5, - 1,0,2,3, - 1,5,1E37,3,1
1060 DEF FNCV(RT#) = RT#^2*LOG(RT#): REM minimum curvature: REM DEF
     FNCV(RT#) = RT#^2*(LOG(RT#) - 1): REM biharmonic spline: REM DEF FNCV(RT#)
     = RT#: REM dummy function
1070 REM DEF FNCV(RT#) = NG + SI*(1 -  EXP( - RT#/RG)): NG = .25: SI = 1: RG = 1/3:
     REM exponential semivariogram
1080 REM DEF FNCV(RT#) = NG + SI*(1.5*RT#/RG - (RT#/RG)^3/2): NG = .25: SI = 1:
     RG = 1: REM spherical semivariogram
1090 P9(1,1) = - 1E + 37: P9(1,2) = - 1E + 37: P9(1,3) = - 1E + 37: P9(2,1) = 1E + 37:
     P9(2,2) = 1E + 37: P9(2,3) = 1E + 37
1100 SL = 0: SR = 0: SZ = 5100: GR = 19: DP = 0: CL = 1: PT = 1: DZ = 15: PW = 3: WD
     = 10000: HI = 6800: AZ = - 25: TL = - 15: RD = 2: BI = 1.5: BJ = 7: IF DP = 0 THEN
     SR = 0
1140 OPEN "A: \HILL.DAT" FOR INPUT AS #1
1150 REM OPEN "A: \JDAVIS.DAT" FOR INPUT AS #1
1180 INPUT#1,ND%,NC%: FOR I1% = 1 TO NC%: INPUT#1,P8(I1%): NEXT I1%: IF
     CR>0 AND NC%>31 THEN NC% = 31
1190 INPUT#1,XS,YS,DT: I1% = 1
1220 INPUT#1,P1#(I1%,1),P1#(I1%,2),P1#(I1%,3): P1#(I1%,1) = (P1#(I1%,1) - XS)/DT:
     P1#(I1%,2) = (P1#(I1%,2) - YS)/DT: IF (P1#(I1%,1) - .5)^2 + (P1#(I1%,2) - .5)^2<RD
     THEN GOTO 1270
1230 ND% = ND% - 1: IF I1%>ND% THEN GOTO 1360
1260 GOTO 1220
1270 IF P9(1,1)<P1#(I1%,1) THEN P9(1,1) = P1#(I1%,1)
1280 IF P9(2,1)>P1#(I1%,1) THEN P9(2,1) = P1#(I1%,1)
1290 IF P9(1,2)<P1#(I1%,2) THEN P9(1,2) = P1#(I1%,2)
1300 IF P9(2,2)>P1#(I1%,2) THEN P9(2,2) = P1#(I1%,2)
1310 IF P9(1,3)<P1#(I1%,3) THEN P9(1,3) = P1#(I1%,3)
1320 IF P9(2,3)>P1#(I1%,3) THEN P9(2,3) = P1#(I1%,3)
1350 I1% = I1% + 1: IF I1%< = ND% THEN GOTO 1220
1360 IF ND%<4 THEN STOP
1370 XF = P9(1,1) - P9(2,1): YF = P9(1,2) - P9(2,2): VS = P9(1,3) - P9(2,3): R8(2,1) = AZ:
     R8(2,2) = TL
1380 FOR I1% = 1 TO 6: READ BX(I1%,1),BX(I1%,2),BX(I1%,3): NEXT I1%
1390 BN% = 1: IF DP<1 THEN GOTO 1450
1400 BN% = 2: BX(2,1) = 0: BX(2,3) = 1: IF R8(2,1)>0 THEN GOTO 1420
1410 BX(2,1) = 1: BX(3,1) = 1: BX(4,1) = 1: BX(5,1) = 0: BX(6,1) = 0
1420 IF DP<2 THEN GOTO 1450
1430 NC1 = INT(SZ/100): NC% = NC1 + 1: U0 = .9999/NC1: U1 = .0000511 - U0
1440 FOR I1% = 1 TO NC%: U1 = U1 + U0: P8(I1%) = U1: NEXT I1%: GOTO 1460
1450 FOR I1% = 1 TO NC%: P8(I1%) = (P8(I1%) - P9(2,3))/VS + .00001*(RND - .5):
     NEXT I1%
1460 FOR I1% = 1 TO ND%: P1#(I1%,1) = P1#(I1%,1) + .0001*(RND - .5): P1#(I1%,2) =
     P1#(I1%,2) + .0001*(RND - .5) : P1#(I1%,3) = (P1#(I1%,3) - P9(2,3))/VS: NEXT I1%
1470 FOR I1% = 1 TO 3: READ R1#(I1%,1), R1#(I1%,2), R1#(I1%,3), R6%(I1%,1),
     R6%(I1%,2)
1480 P1#(I1% + ND%,1) = R1#(I1%,1)*XF + P9(2,1): P1#(I1% + ND%,2) = R1#(I1%,2)*YF
     + P9(2,2): P1#(I1% + ND%,3) = 0: NEXT I1%
1490 IF DP<1 THEN GOTO 1550
1500 FOR I1% = 1 TO 2: R8(I1%,1) = COS(R8(2,I1%)/57.29578): R8(I1%,2) =
     SIN(R8(2,I1%)/57.29578): NEXT I1%
1510 FOR I1% = 2 TO 6
1520 YY = (BX(I1%,2) - .5)*R8(1,1) + (BX(I1%,1) - .5)*R8(1,2) + .5: BX(I1%,1) =
     (BX(I1%,1) - .5)*R8(1,1) - (BX(I1%,2) - .5)*R8(1,2) + .5
1530 BX3 = (BX(I1%,3) - .5)*R8(2,1) + (YY - .5)*R8(2,2) + .5: BX(I1%,2) = (YY -
     .5)*R8(2,1) - (BX(I1%,3) - .5)*R8(2,2) + .5: BX(I1%,3) = BX3: NEXT I1%
1540 REM end data input, begin gradient estimation
1550 SOUND 208,5: SOUND 165,5: SOUND 185,5: SOUND 123,5
1560 IF CL<1 THEN GOTO 3830
1570 PRINT " AT " TIME$ " BEGIN GRADIENT ESTIMATION - PLEASE WAIT"
1580 FOR I0% = 1 TO ND%
1590 FOR I1% = 1 TO 3: P3%(1,I1%) = ND% + I1%: P3%(2,I1%) = ND% + I1%:
     P2#(1,I1%) = R1#(I1%,3): P2#(2,I1%) = R1#(I1%,3) : NEXT I1%
1600 FOR I1% = 1 TO 237: P4%(I1%) = I1%: NEXT I1%: R2%(1,1) = 1: R2%(1,2) = 1:
     R2%(2,1) = 3: R2%(2,2) = 4: K1% = I0%: K2% = I0% : K3% = ND% + 4: K4 = - .5
1610 FOR I1% = K1% TO K2%
```

BASIC PROGRAMS

```
1620 IF I1% = K3% THEN GOTO 2000
1630 K5 = K4
1640 K6 = 1.5 + K5: IF K3%>ND% THEN K5 = - K5
1650 U0 = 0: K7% = 0: L6% = K6 - 2
1660 FOR I2% = 1 TO R2%(1,K6)
1670 L6% = L6% + 2: IF P3%(L6%,1) = 0 THEN GOTO 1670
1680 U1# = P2#(L6%,3) - (P1#(I1%,1) - P2#(L6%,1))^2: IF U1#<0 THEN GOTO 1840
1690 U1# = U1# - (P1#(I1%,2) - P2#(L6%,2))^2: IF U1#<0 THEN GOTO 1840
1700 R2%(2,K6) = R2%(2,K6) - 2: P4%(R2%(2,K6)) = L6%: U0 = U0 - 1
1710 FOR I3% = 1 TO 3
1720 L1% = 1: IF L1% = I3% THEN L1% = L1% + 1
1730 L2% = L1% + 1: IF L2% = I3% THEN L2% = L2% + 1
1740 IF K7%<1 THEN GOTO 1810
1750 K9% = K7%
1760 FOR I4% = 1 TO K9%
1770 IF P3%(L6%,L1%)<>P7%(I4%,1) THEN GOTO 1800
1780 IF P3%(L6%,L2%)<>P7%(I4%,2) THEN GOTO 1800
1790 P7%(I4%,1) = P7%(K9%,1): P7%(I4%,2) = P7%(K9%,2): K7% = K7% - 1: GOTO 1820
1800 NEXT I4%
1810 K7% = K7% + 1: P7%(K7%,1) = P3%(L6%,L1%): P7%(K7%,2) = P3%(L6%,L2%)
1820 NEXT I3%
1830 P3%(L6%,1) = 0
1840 NEXT I2%
1850 IF K7%<1 THEN GOTO 2000
1860 REM make new triangles
1870 FOR I2% = 1 TO K7%
1880 IF K5>0 THEN GOTO 1900
1890 IF P7%(I2%,1)<>I0% AND P7%(I2%,2)<>I0% THEN GOTO 1970
1900 FOR I3% = 1 TO 2
1910 R3#(I3%,1) = P1#(P7%(I2%,I3%),1) - P1#(I1%,1): R3#(I3%,2) =
     P1#(P7%(I2%,I3%),2) - P1#(I1%,2)
1920 R3#(I3%,3) = R3#(I3%,1)*(P1#(P7%(I2%,I3%),1) + P1#(I1%,1))/2 + R3#(I3%,2)
     *(P1#(P7%(I2%,I3%),2) + P1#(I1%,2))/2
1930 NEXT I3%
1940 U1# = R3#(1,1)*R3#(2,2) - R3#(2,1)*R3#(1,2): P2#(P4%(R2%(2,K6)),1) =
     (R3#(1,3)*R3#(2,2) - R3#(2,3)*R3#(1,2))/U1#: P2#(P4%(R2%(2,K6)),2) =
     (R3#(1,1)*R3#(2,3) - R3#(2,1)*R3#(1,3))/U1#
1950 P2#(P4%(R2%(2,K6)),3) = (P1#(I1%,1) - P2#(P4%(R2%(2,K6)),1))^2 + (P1#(I1%,2) -
     P2#(P4%(R2%(2,K6)),2))^2
1960 P3%(P4%(R2%(2,K6)),1) = P7%(I2%,1): P3%(P4%(R2%(2,K6)),2) = P7%(I2%,2) :
     P3%(P4%(R2%(2,K6)),3) = I1%: R2%(2,K6) = R2%(2,K6) + 2: U0 = U0 + 1
1970 NEXT I2%
1980 R2%(1,K6) = R2%(1,K6) + U0: IF K5>0 THEN GOTO 2000
1990 K5 = - K5: GOTO 1640
2000 NEXT I1%
2010 IF K2%>K1% GOTO 2040
2020 K1% = 1: K2% = ND%: K3% = I0%: GOTO 1610
2030 REM order triangles positively
2040 FOR I1% = 1 TO 2
2050 L8% = I1% - 2
2060 FOR I2% = 1 TO R2%(1,I1%)
2070 L8% = L8% + 2: IF P3%(L8%,1) = 0 THEN GOTO 2070
2080 IF P3%(L8%,1)>ND% THEN GOTO 2140
2090 FOR I3% = 1 TO 2
2100 R3#(1,I3%) = P1#(P3%(L8%,2),I3%) - P1#(P3%(L8%,1),I3%): R3#(2,I3%) =
     P1#(P3%(L8%,3),I3%) - P1#(P3%(L8%,1),I3%)
2110 NEXT I3%
2120 IF R3#(1,1)*R3#(2,2) - R3#(1,2)*R3#(2,1)> = 0 THEN GOTO 2140
2130 K8% = P3%(L8%,3): P3%(L8%,3) = P3%(L8%,2): P3%(L8%,2) = K8% 2140 NEXT I2%
2150 NEXT I1%
2160 IF SL>0 THEN GOTO 2470
2170 REM calculate natural neighbor coordinates
2180 R5#(1,1) = P1#(I0%,1): R5#(1,2) = P1#(I0%,2): R5#(2,1) = R5#(1,1) + .0001 :
     R5#(2,2) = R5#(1,2): R5#(3,1) = R5#(1,1): R5#(3,2) = R5#(1,2) + .0001
```

299

Appendix

```
2190 FOR I1% = 1 TO 3 2200 L4% = 0: L5% = 0: L6% = 0
2210 FOR I2% = 1 TO R2%(1,2)
2220 L6% = L6% + 2: IF P3%(L6%,1) = 0 THEN GOTO 2220
2230 IF P3%(L6%,1)>ND% THEN GOTO 2410
2240 U1# = P2#(L6%,3) - (R5#(I1%,1) - P2#(L6%,1))^2: IF U1#<0 THEN GOTO 2410
2250 U1# = U1# - (R5#(I1%,2) - P2#(L6%,2))^2: IF U1#<0 THEN GOTO 2410
2260 FOR I3% = 1 TO 3
2270 FOR I4% = 1 TO 2
2280 R3#(I4%,1) = P1#(P3%(L6%,R6%(I3%,I4%)),1) - R5#(I1%,1): R3#(I4%,2) =
    P1#(P3%(L6%,R6%(I3%,I4%)),2) - R5#(I1%,2)
2290 R3#(I4%,3) = R3#(I4%,1)*(P1#(P3%(L6%,R6%(I3%,I4%)),1) + R5#(I1%,1))/2 +
    R3#(I4%,2) *(P1#(P3%(L6%,R6%(I3%,I4%)),2) + R5#(I1%,2))/2: NEXT I4%
2300 U1# = R3#(1,1)*R3#(2,2) - R3#(1,2)*R3#(2,1)
2310 R4#(R6%(I3%,2),1) = (R3#(1,3)*R3#(2,2) - R3#(2,3)*R3#(1,2))/U1#:
    R4#(R6%(I3%,2),2) = (R3#(1,1)*R3#(2,3) - R3#(2,1)*R3#(1,3))/U1#: NEXT I3%: L3%=0
2320 FOR I3% = 1 TO 3: R3#(3,I3%) = ((R4#(R6%(I3%,1),1) -
    P2#(L6%,1))*(R4#(R6%(I3%,2),2) - P2#(L6%,2)) - (R4#(R6%(I3%,2),1) -
    P2#(L6%,1))*(R4#(R6%(I3%,1),2) - P2#(L6%,2)))/2: IF R3#(3,I3%)>0 THEN L3% =
    L3% + 1
2330 NEXT I3%
2340 IF L3%>2 THEN L4% = 1
2350 FOR I3% = 1 TO 3: IF L5%<1 THEN GOTO 2390
2360 FOR I4% = 1 TO L5%: IF P3%(L6%,I3%)<>P6%(I4%) THEN GOTO 2380 2370
    P5#(I4%) = P5#(I4%) + R3#(3,I3%): GOTO 2400
2380 NEXT I4%
2390 L5% = L5% + 1: P6%(L5%) = P3%(L6%,I3%): P5#(L5%) = R3#(3,I3%) 2400 NEXT
    I3%
2410 NEXT I2%: IF L4%<1 THEN GOTO 2470
2420 U4 = 0
2430 FOR I2% = 1 TO L5%: U4 = U4 + P5#(I2%): NEXT I2%: R7#(I1%) = 0
2440 FOR I2% = 1 TO L5%: P5#(I2%) = P5#(I2%)/U4: R7#(I1%) = R7#(I1%) +
    P5#(I2%)*P1#(P6%(I2%),3): NEXT I2%: NEXT I1%
2450 P1#(I0%,6) = P1#(I0%,3) - R7#(1): P1#(I0%,4) = (R7#(1) - R7#(2))/.0001:
    P1#(I0%,5) = (R7#(1) - R7#(3))/.0001 : GOTO 3080
2460 REM cross product gradients
2470 IF SL>1 THEN GOTO 2610
2480 P1#(I0%,4) = 0: P1#(I0%,5) = 0: P1#(I0%,6) = 0: P7# = 0: L7% = - 1
2490 FOR I1% = 1 TO R2%(1,1)
2500 L7% = L7% + 2: IF P3%(L7%,1) = 0 THEN GOTO 2500
2510 IF P3%(L7%,1)>ND% THEN GOTO 2570
2520 FOR I2% = 1 TO 2: FOR I3% = 1 TO 3: R3#(I2%,I3%) = P1#(P3%(L7%,1),I3%) -
    P1#(P3%(L7%,I2% + 1),I3%): NEXT I3% : NEXT I2%
2530 R3#(3,1) = R3#(1,2)*R3#(2,3) - R3#(2,2)*R3#(1,3): R3#(3,2) = R3#(1,3)*R3#(2,1) -
    R3#(2,3)*R3#(1,1): R3#(3,3) = R3#(1,1)*R3#(2,2) - R3#(2,1)*R3#(1,2): U3 = 1
2540 IF R3#(3,3)<0 THEN U3 = - 1
2550 U2 = (R3#(3,1)^2 + R3#(3,2)^2 + R3#(3,3)^2)^.5: P7# = P7# + U2
2560 FOR I2% = 1 TO 3: P1#(I0%,I2% + 3) = P1#(I0%,I2% + 3) + R3#(3,I2%)*U3: NEXT
    I2%
2570 NEXT I1%
2580 U2 = P1#(I0%,4)^2 + P1#(I0%,5)^2 + P1#(I0%,6)^2: P7# = 1 - U2^.5/P7#
2590 P1#(I0%,4) = P1#(I0%,4)/P1#(I0%,6): P1#(I0%,5) = P1#(I0%,5)/P1#(I0%,6):
    P1#(I0%,6) = P7#
2600 GOTO 3080
2610 IF SL>2 THEN GOTO 2810 : REM least squares gradients
2620 P1#(I0%,4) = 0: P1#(I0%,5) = 0: P1#(I0%,6) = .2: L7% = - 1: P6%(1) = I0%: L5% =1
2630 R4#(1,1) = P1#(I0%,3): R4#(2,1) = P1#(I0%,1)*P1#(I0%,3): R4#(3,1) =
    P1#(I0%,2)*P1#(I0%,3)
2640 R3#(2,1) = P1#(I0%,1): R3#(2,2) = P1#(I0%,1)^2: R3#(3,1) = P1#(I0%,2) : R3#(3,2)
    = P1#(I0%,1)*P1#(I0%,2): R3#(3,3) = P1#(I0%,2)^2
2650 FOR I1% = 1 TO R2%(1,1)
2660 L7% = L7% + 2
2670 IF P3%(L7%,1) = 0 THEN GOTO 2660
2680 IF P3%(L7%,1)>ND% THEN GOTO 2760
2690 FOR I2% = 1 TO 3
2700 FOR I3% = 1 TO L5%
2710 IF P3%(L7%,I2%) = P6%(I3%) THEN GOTO 2750
```

```
2720 NEXT I3%
2730 L5% = L5% + 1: P6%(L5%) = P3%(L7%,I2%): R4#(1,1) = R4#(1,1) +
     P1#(P6%(L5%),3) : R4#(2,1) = R4#(2,1) + P1#(P6%(L5%),1)*P1#(P6%(L5%),3):
     R4#(3,1) = R4#(3,1) + P1#(P6%(L5%),2)*P1#(P6%(L5%),3)
2740 R3#(2,1) = R3#(2,1) + P1#(P6%(L5%),1): R3#(2,2) = R3#(2,2) +
     P1#(P6%(L5%),1)^2: R3#(3,1) = R3#(3,1) + P1#(P6%(L5%),2): R3#(3,2) = R3#(3,2) +
     P1#(P6%(L5%),1)*P1#(P6%(L5%),2): R3#(3,3) = R3#(3,3) + P1#(P6%(L5%),2)^2
2750 NEXT I2%
2760 NEXT I1%
2770 U1# = L5%*(R3#(2,2)*R3#(3,3) - R3#(3,2)^2) - R3#(2,1)*(R3#(2,1)*R3#(3,3) -
     R3#(3,1)*R3#(3,2)) + R3#(3,1)*(R3#(2,1)*R3#(3,2) - R3#(3,1)*R3#(2,2))
2780 P1#(I0%,4) = - (L5%*(R4#(2,1)*R3#(3,3) - R3#(3,2)*R4#(3,1)) -
     R4#(1,1)*(R3#(2,1)*R3#(3,3) - R3#(3,1)*R3#(3,2)) + R3#(3,1)*(R3#(2,1)*R4#(3,1) -
     R3#(3,1)*R4#(2,1)))/U1#
2790 P1#(I0%,5) = - (L5%*(R3#(2,2)*R4#(3,1) - R3#(3,2)*R4#(2,1)) -
     R3#(2,1)*(R3#(2,1)*R4#(3,1) - R3#(3,1)*R4#(2,1)) + R4#(1,1)*(R3#(2,1)*R3#(3,2) -
     R3#(3,1)*R3#(2,2)))/U1#
2800 GOTO 3080 : REM spline gradients
2810 IF SL>3 THEN GOTO 3035
2815 P1#(I0%,4) = 0: P1#(I0%,5) = 0: P1#(I0%,6) = .2: L7% = - 1: L5% = 1: P6%(1) =
     I0%
2820 R5#(1,1) = P1#(I0%,1): R5#(1,2) = P1#(I0%,2): R5#(2,1) = R5#(1,1) + .0001 :
     R5#(2,2) = R5#(1,2): R5#(3,1) = R5#(1,1): R5#(3,2) = R5#(1,2) + .0001
2830 FOR I1% = 1 TO R2%(1,1)
2840 L7% = L7% + 2: IF P3%(L7%,1) = 0 THEN GOTO 2840
2850 IF P3%(L7%,1)>ND% THEN GOTO 2890
2860 FOR I2% = 1 TO 3: FOR I3% = 1 TO L5%: IF P3%(L7%,I2%) = P6%(I3%) THEN
     GOTO 2880
2870 NEXT I3%: L5% = L5% + 1: P6%(L5%) = P3%(L7%,I2%)
2880 NEXT I2%
2890 NEXT I1%
2900 LL% = L5% + 3
2910 FOR I1% = 1 TO L5%
2920 R3#(I1%,1) = 1: R3#(LL%,I1% + 3) = 1: R3#(I1%,2) = P1#(P6%(I1%),1): R3#(L5% +
     1,I1% + 3) = R3#(I1%,2)
2930 R3#(I1%,3) = P1#(P6%(I1%),2): R3#(L5% + 2,I1% + 3) = R3#(I1%,3): R9#(I1%) =
     P1#(P6%(I1%),3): NEXT I1%
2940 FOR I1% = 1 TO 3: FOR I2% = 1 TO 3: R3#(I1% + L5%,I2%) = 0: NEXT I2%:
     R9#(I1% + L5%) = 0: NEXT I1%
2950 FOR I1% = 1 TO L5%: R3#(I1%,I1% + 3) = 0: IJ = I1% + 1: IF IJ>L5% THEN GOTO
     2970
2960 FOR I2% = IJ TO L5%: RT# = ((R3#(I1%,2) - R3#(I2%,2))^2 + (R3#(I1%,3) -
     R3#(I2%,3))^2)^.5: R3#(I2%,I1% + 3) = FNCV(RT#): R3#(I1%,I2% + 3) = R3#(I2%,I1%
     + 3): NEXT I2%
2970 NEXT I1%: GOSUB 6690
2980 TR# = 0: TS# = 0
2990 FOR I1% = 4 TO LL%
3000 RT# = ((P1#(P6%(I1% - 3),1) - R5#(2,1))^2 + (P1#(P6%(I1% - 3),2) -
     R5#(2,2))^2)^.5: TR# = TR# + B#(I1%)*FNCV(RT#)
3010 RS# = ((P1#(P6%(I1% - 3),1) - R5#(3,1))^2 + (P1#(P6%(I1% - 3),2) -
     R5#(3,2))^2)^.5: TS# = TS# + B#(I1%)*FNCV(RS#) : NEXT I1%
3020 R5#(2,3) = B#(1) + B#(2)*R5#(2,1) + B#(3)*R5#(2,2) + TR#: P1#(I0%,4) =
     (P1#(I0%,3) - R5#(2,3))/.0001
3030 R5#(3,3) = B#(1) + B#(2)*R5#(3,1) + B#(3)*R5#(3,2) + TS#: P1#(I0%,5) =
     (P1#(I0%,3) - R5#(3,3))/.0001: GOTO 3080
3033 REM hyperboloid gradients
3035 R3#(5,1) = 1: FOR I1% = 1 TO 4: R9#(I1%) = 0: FOR I2% = 1 TO 4: R3#(I1%,I2%) =
     0: NEXT I2%: NEXT I1%: LL% = 4
3036 P1#(I0%,4) = 0: P1#(I0%,5) = 0: P1#(I0%,6) = .2: L7% = - 1: L5% = 0: FOR I1% = 1
     TO R2%(1,1)
3037 L7% = L7% + 2: IF P3%(L7%,1) = 0 THEN GOTO 3037
3038 IF P3%(L7%,1)>ND% THEN GOTO 3070
3039 FOR I2% = 1 TO 3: IF L5%<1 THEN GOTO 3044
3041 FOR I3% = 1 TO L5%: IF P3%(L7%,I2%) = P6%(I3%) THEN GOTO 3060
3042 NEXT I3%
3044 L5% = L5% + 1: P6%(L5%) = P3%(L7%,I2%)
```

Appendix

```
3050 R3#(5,2) = P1#(P6%(L5%),1): R3#(5,3) = P1#(P6%(L5%),2): R3#(5,4) =
     P1#(P6%(L5%),1)*P1#(P6%(L5%),2)
3055 FOR I3% = 1 TO 4: FOR I4% = 1 TO 4: R3#(I3%,I4%) = R3#(I3%,I4%) +
     R3#(5,I3%)*R3#(5,I4%): NEXT I4% : R9#(I3%) = R9#(I3%) +
     R3#(5,I3%)*P1#(P6%(L5%),3): NEXT I3%
3060 NEXT I2%
3070 NEXT I1%: GOSUB 6690: P1#(I0%,4) =  - B#(2) - B#(4)*P1#(I0%,2): P1#(I0%,5) =  -
     B#(3) - B#(4)*P1#(I0%,1): P1#(I0%,6) = .2
3080 NEXT I0%
3820 REM make triangular grid
3830 PRINT " AT " TIME$ " BEGIN SURFACE INTERPOLATION - PLEASE WAIT":
     SOUND 123,5: SOUND 185,5: SOUND 208,5 : SOUND 165,5
3840 I0% = GR + 1: IS = 1: I2% = 2*I0% + 1: I3% = 1 + INT(GR/3^.5): I4% = I3%*I2% +
     I0%: SX = 1/GR: SY = 1/(2*I3%): S2(3) = .000001
3850 WX# = S2(3): S2(1) = 1.00001: AA# = SX*SY*.5*DT*DT: SM# = 0: SA# = 0: WY# =
     1: U8 = 1
3860 FOR L1% = 1 TO I4%
3870 S1(L1%) =  - 999: WXO# = WX#: WYO# = WY#
4490 REM .....calculate distance weighted coordinates of WX#,WY# *********
4500 IF PW<8 THEN GOTO 4550
4510 P5#(1) = 1E37
4520 FOR L2% = 1 TO ND%: U4 = (P1#(L2%,1) - WX#)^2 + (P1#(L2%,2) - WY#)^2: IF
     U4>P5#(1) THEN GOTO 4540
4530 P6%(1) = L2%: P5#(1) = U4
4540 NEXT L2%: S1(L1%) = P1#(P6%(1),3): GOTO 4590
4550 FOR L2% = 1 TO ND%: P5#(L2%) = 1/(((P1#(L2%,1) - WX#)^2 + (P1#(L2%,2) -
     WY#)^2)^.5)^PW: P6%(L2%) = L2%: NEXT L2%: K5 = ND%
4560 S1(L1%) = 0: U4 = 0
4570 FOR L2% = 1 TO K5: U4 = U4 + P5#(L2%): NEXT L2%
4580 FOR L2% = 1 TO K5: P5#(L2%) = P5#(L2%)/U4: S1(L1%) = S1(L1%) +
     P5#(L2%)*P1#(P6%(L2%),3): NEXT L2%
4590 IF CL<1 THEN GOTO 4940
4600 IF PW<8 THEN GOTO 4620
4610 S1(L1%) = (P1#(P6%(1),4)*P1#(P6%(1),1) + P1#(P6%(1),5)*P1#(P6%(1),2) +
     P1#(P6%(1),3) - P1#(P6%(1),4)*WX# - P1#(P6%(1),5)*WY#): GOTO 4940
4620 FOR L2% = 1 TO K5: S4(L2%) = 0
4630 IF P5#(L2%)<.00001 OR P5#(L2%)>1 THEN GOTO 4690
4640 IF ABS(P1#(P6%(L2%),6))<.00001 THEN GOTO 4690
4650 RS# = ABS(P1#(P6%(L2%),6)) + BI: RT# = RS#*BJ: RB# = 1/RT#: BD# =
     P5#(L2%)^RT#: BB# = BD#*2: IF BD#>.5 THEN BB# = (1 - BD#)*2
4660 BB# = BB#^RS#/2: IF BD#>.5 THEN BB# = 1 - BB#
4670 HP# = BB#^RB#
4680 S4(L2%) = ((P1#(P6%(L2%),4)*P1#(P6%(L2%),1) +
     P1#(P6%(L2%),5)*P1#(P6%(L2%),2) + P1#(P6%(L2%),3) - P1#(P6%(L2%),4)*WX# -
     P1#(P6%(L2%),5)*WY#) - S1(L1%))*HP#
4690 NEXT L2%
4700 FOR L2% = 1 TO K5: S1(L1%) = S1(L1%) + S4(L2%): NEXT L2%
4940 IS = IS + SX*U8: IF IS>0 THEN GOTO 4960
4950 WXO# = WXO# + SX/2: IS = 1.001 - SX
4960 WX# = WXO# + SX*U8: WY# = WYO#: IF U8*WX#<S2(2 - U8) THEN GOTO 4980
4970 U8 =  - U8: WX# = S2(2 + U8): WY# = WY# - SY: IS = U8
4980 NEXT L1%
5060 REM end interpolation to triangular grid and begin display construction
5070 SOUND 208,5: SOUND 165,5: SOUND 185,5: SOUND 123,5: PRINT " AT " TIME$;
5071 INPUT " PREPARE PLOTTER AND PRESS ENTER",PM$
5090 LW = (HI - SZ)/2: IF DP>0 THEN SZ = INT(SZ*2/3)
5091 IF DP>0 AND SZ>HI/2 THEN SZ = HI/2
5100 SW = (WD - SZ)/2: ZW = (HI - SZ)/2
5110 IF SR>0 THEN SW = (WD - 2*SZ)/3
5120 SW = SW + 1: SV = 2*SW + SZ: SR = SR*2: SZ = SZ - 2: SZ = SZ - 2: ZW = ZW -
     1: SU = SW: IF SR>0 THEN SU = (2*SW + SV)/3
5160 REM plot open
5170 LPRINT "IN;IP0,0," + STR$(WD) + "," + STR$(HI) + ";SP1;": IF SR<2 THEN GOTO
     5180
```

```
5175 FOR L1% = 2 TO 6: BX1 = (BX(L1%,1) - .5)*.99875 - (BX(L1%,3) - .5)*.05 + .5:
     BX(L1%,3) = (BX(L1%,3) - .5)*.99875 + (BX(L1%,1) - .5)*.05 + .5: BX(L1%,1) = BX1:
     NEXT L1%: LPRINT "PU;PA0,6800;"
5180 LPRINT "PU;PA" + STR$(BX(BN%,1)*SZ + SW) + "," + STR$(BX(BN%,2)*SZ +
     ZW) + ";PD;" : ZX = SX: ZY = SY: ZZ = .5: R3#(2,2) = 1 + 2*SY: R3#(2,1) = 0
5190 FOR L1% = BN% + 1 TO BN% + 4: LPRINT "PA" + STR$(BX(L1%,1)*SZ + SW) +
     "," + STR$(BX(L1%,2)*SZ + ZW) + ";": NEXT L1% : LPRINT "PU;"
5192 FOR L1% = 2 TO 6: BX(L1%,1) = (BX(L1%,1) - .5)*.995 + (BX(L1%,3) - .5)*.1 + .5:
     NEXT L1%: SM = 0: SA = 0
5200 REM plot contours
5220 FOR L1% = 1 TO I3%
5230 R3#(1,2) = R3#(2,2) - 2*SY - 2*ZY: R3#(1,1) = R3#(2,1): N1% = (2*L1% - .5 +
     ZZ)*I0% + L1% + .5/ZZ: R3#(1,3) = S1(N1%): R3#(2,2) = R3#(2,2) - 2*SY - ZY:
     R3#(2,1) = .5 - ZZ: N2% = (2*L1% - .5 + ZZ)*I0% + L1%
5240 R3#(2,3) = S1(N2%): R3#(3,2) = R3#(2,2): R3#(3,1) = R3#(2,1) + ZX/2: N3% = N2%
     - .5/ZZ: R3#(3,3) = S1(N3%): AR# = AA#/2
5250 GOSUB 5820
5260 R3#(1,2) = R3#(1,2) + 2*ZY: N2% = (2*L1% - .5/ZZ - .5 - ZZ)*I0% + L1%: R3#(1,3) =
     S1(N2%)
5270 GOSUB 5820
5280 AR# = AA#
5290 FOR L2% = 1 TO GR
5300 R3#(2,2) = R3#(1,2): R3#(2,1) = R3#(1,1) + ZX: N4% = N2% + .5/ZZ: R3#(2,3) =
     S1(N4%)
5310 GOSUB 5820
5320 R3#(1,2) = R3#(3,2): R3#(1,1) = R3#(3,1) + ZX
5330 IF L2% = GR THEN R3#(1,1) = R3#(1,1) - ZX/2
5340 N6% = N3% - .5/ZZ: R3#(1,3) = S1(N6%): IF L2% = GR THEN AR# = AA#/2
5350 GOSUB 5820
5360 R3#(2,2) = R3#(2,2) - 2*ZY: N5% = N1% + .5/ZZ: R3#(2,3) = S1(N5%)
5370 GOSUB 5820
5380 R3#(1,2) = R3#(2,2): R3#(1,1) = R3#(2,1) - ZX: R3#(1,3) = S1(N1%): IF L2% = GR
     THEN AR# = AA#
5390 GOSUB 5820
5400 N1% = N5%: N2% = N4%: N3% = N6%: R3#(1,2) = R3#(1,2) + 2*ZY: R3#(1,1) =
     R3#(2,1) : R3#(1,3) = S1(N4%): R3#(3,1) = R3#(3,1) + ZX: R3#(3,3) = S1(N6%)
5410 NEXT L2%
5420 ZY = - ZY: ZX = - ZX: ZZ = - ZZ
5430 NEXT L1%: REM end plotting contours
5450 REM plot data points
5460 IF PT<1 THEN GOTO 5770
5480 DZ% = DZ: FOR L1% = 1 TO ND%
5490 IF P1#(L1%,1)< - .0001 OR P1#(L1%,1)>1.0001 OR P1#(L1%,2)< - .0001 OR
     P1#(L1%,2)>1.0001 THEN GOTO 5610
5500 PP1# = P1#(L1%,1): PP2# = P1#(L1%,2): PP3# = P1#(L1%,3): IF DP<1 THEN
     GOTO 5530
5510 YY = (PP2# - .5)*R8(1,1) + (PP1# - .5)*R8(1,2) + .5: PP1# = (PP1# - .5)*R8(1,1) -
     (PP2# - .5)*R8(1,2) + .5
5520 ZZ3 = (PP3# - .5)*R8(2,1) + (YY - .5)*R8(2,2) + .5: PP2# = (YY - .5)*R8(2,1) - (PP3#
     - .5)*R8(2,2) + .5: PP3# = ZZ3: IF SR<2 THEN GOTO 5528
5525 PP1# = (PP1# - .5)*.99875 - (PP3# - .5)*.05 + .5
5528 IF SR<>1 THEN GOTO 5530
5529 PP1# = (PP1# - .5)*.99875 + (PP3# - .5)*.05 + .5
5530 S3(1,1) = PP1#*SZ + SW: S3(1,2) = PP2#*SZ + ZW
5540 LPRINT "PA" + STR$(S3(1,1) - DZ%) + "," + STR$(S3(1,2) + DZ%) + ";": LPRINT
     "PD;PA" + STR$(S3(1,1) + DZ%) + "," + STR$(S3(1,2) + DZ%) + ";": LPRINT "PA" +
     STR$(S3(1,1) + DZ%) + "," + STR$(S3(1,2) - DZ%) + ";"
5550 LPRINT "PA" + STR$(S3(1,1) - DZ%) + "," + STR$(S3(1,2) - DZ%) + ";": LPRINT
     "PA" + STR$(S3(1,1) - DZ%) + "," + STR$(S3(1,2) + DZ%) + ";": LPRINT "PA" +
     STR$(S3(1,1) + DZ%) + "," + STR$(S3(1,2) - DZ%) + ";PU;"
5610 NEXT L1%
5770 SR = SR - 1: SW = SV: IF SR = 1 THEN GOTO 5180
5775 LPRINT "PU;PA" + STR$(SU) + "," + STR$(LW - 500) + ";SI0.2,0.3;SL0.2;" 5777
     LPRINT "LBFigure 1.0.1 - Example Isolines" + CHR$(3): LPRINT "PU;PA0,6800;":
     REM plot close
5790 SOUND 123,5: SOUND 185,5: SOUND 208,5: SOUND 165,5: TM$ = TIME$
```

Appendix

```
5800 SM = SM/SA: PRINT " AVERAGE HEIGHT OF SURFACE IS" SM: PRINT " AT "
    TM$ " RUN COMPLETE": END
5810 REM curvemaker
5820 ST = R3#(1,3)
5830 IF ST>R3#(2,3) THEN ST = R3#(2,3)
5840 IF ST>R3#(3,3) THEN ST = R3#(3,3)
5850 IF ST< = - 999 THEN RETURN
5860 SM = SM + AR#*((R3#(1,3) + R3#(2,3) + R3#(3,3))/3*VS + P9(2,3)): SA = SA + AR#
6210 L8% = 0: IF DP<2 THEN GOTO 6310
6220 L8% = 1
6230 REM rotate triangle for orthogonal profiles
6240 FOR L3% = 1 TO 3
6250 Z# = R3#(L3%,3): R3#(L3%,3) = 1 - R3#(L3%,L8%): R3#(L3%,L8%) = Z#
6260 NEXT L3%
6270 ST = R3#(1,3)
6280 IF ST>R3#(2,3) THEN ST = R3#(2,3)
6290 IF ST>R3#(3,3) THEN ST = R3#(3,3)
6300 IF ST < = - 999 THEN RETURN
6310 TP = R3#(1,3)
6320 IF TP<R3#(2,3) THEN TP = R3#(2,3)
6330 IF TP<R3#(3,3) THEN TP = R3#(3,3)
6340 IF TP = ST THEN RETURN
6350 FOR L0% = 1 TO NC%
6360 REM slice the triangle
6370 IF P8(L0%)>TP OR P8(L0%)<ST THEN GOTO 6610
6380 L5% = 1: L4% = 0
6390 L4% = L4% + 1
6400 L7% = 1: IF L7% = L4% THEN L7% = L7% + 1
6410 L3% = L7% + 1: IF L3% = L4% THEN L3% = L3% + 1
6420 IF R3#(L7%,3) = R3#(L3%,3) THEN GOTO 6460
6430 F = (P8(L0%) - R3#(L7%,3))/(R3#(L3%,3) - R3#(L7%,3)): IF F<0 OR F>1 THEN
    GOTO 6460
6440 S3(L5%,1) = R3#(L7%,1) + (R3#(L3%,1) - R3#(L7%,1))*F: S3(L5%,2) - R3#(L7%,2)
    + (R3#(L3%,2) - R3#(L7%,2))*F
6450 L5% = L5% + 1
6460 IF L5%<3 THEN GOTO 6390
6470 S3(1,3) = P8(L0%): S3(2,3) = P8(L0%): IF DP<2 THEN GOTO 6500
6480 REM reverse rotate the intersection trace
6490 FOR L3% = 1 TO 2: Z# = S3(L3%,3): S3(L3%,3) = S3(L3%,L8%): S3(L3%,L8%) = 1
    - Z#: NEXT L3%
6500 IF DP = 0 THEN GOTO 6570
6510 REM apply perspective
6520 FOR L3% = 1 TO 2
6530 YY = (S3(L3%,2) - .5)*R8(1,1) + (S3(L3%,1) - .5)*R8(1,2): S3(L3%,1) = (S3(L3%,1) -
    .5)*R8(1,1) - (S3(L3%,2) - .5)*R8(1,2) + .5
6540 S33 = (S3(L3%,3) - .5)*R8(2,1) + YY*R8(2,2) + .5: S3(L3%,2) = YY*R8(2,1) -
    (S3(L3%,3) - .5)*R8(2,2) + .5: S3(L3%,3) = S33
6550 NEXT L3%
6552 IF SR<2 THEN GOTO 6560
6554 S3(1,1) = (S3(1,1) - .5)*.99875 - (S3(1,3) - .5)*.05 + .5: S3(2,1) = (S3(2,1) -
    .5)*.99875 - (S3(2,3) - .5)*.05 + .5
6560 IF SR<>1 THEN GOTO 6570
6565 S3(1,1) = (S3(1,1) - .5)*.99875 + (S3(1,3) - .5)*.05 + .5: S3(2,1) = (S3(2,1) -
    .5)*.99875 + (S3(2,3) - .5)*.05 + .5
6567 REM draw a line from (S3(1,1),S3(1,2)) to (S3(2,1),S3(2,2))
6570 LPRINT "PA" + STR$(S3(1,1)*SZ + SW) + "," + STR$(S3(1,2)*SZ + ZW) + ";":
    LPRINT "PD;PA" + STR$(S3(2,1)*SZ + SW) + "," + STR$(S3(2,2)*SZ + ZW) + ";PU;"
6610 NEXT L0%
6620 IF L8%<1 THEN RETURN
6630 REM reverse rotate the triangle
6640 FOR L3% = 1 TO 3
6650 Z# = R3#(L3%,3): R3#(L3%,3) = R3#(L3%,L8%): R3#(L3%,L8%) = 1 - Z#
6660 NEXT L3%
6670 L8% = L8% + 1: IF L8%<3 THEN GOTO 6240 ELSE RETURN
6680 REM solve R3#(LL%,LL%)xB#(LL%) = R9#(LL%)
6690 FOR J2% = 1 TO LL%: P4%(J2%) = J2%: B#(J2%) = 0
```

```
6700 FOR J3% = 1 TO LL%: IF ABS(R3#(J2%,J3%))> = B#(J2%) THEN B#(J2%) =
     ABS(R3#(J2%,J3%))
6710 NEXT J3%: NEXT J2%: LM1% = LL% - 1: LP1% = LL% + 1
6720 FOR J2% = 1 TO LM1%: X3# = 0
6730 FOR J3% = J2% TO LL%: X4# = ABS(R3#(P4%(J3%),J2%))/B#(P4%(J3%)): IF
     X4#< = X3# THEN GOTO 6750
6740 X3# = X4#: K3% = J3%
6750 NEXT J3%: IF K3% = J2% GOTO 6770
6760 K2% = P4%(J2%): P4%(J2%) = P4%(K3%): P4%(K3%) = K2%
6770 X3# = R3#(P4%(J2%),J2%): K4 = J2% + 1
6780 FOR J3% = K4 TO LL%: X4# = - R3#(P4%(J3%),J2%)/X3#: R3#(P4%(J3%),J2%) =
     - X4#
6790 FOR J4 = K4 TO LL%: R3#(P4%(J3%),J4) = R3#(P4%(J3%),J4) +
     X4#*R3#(P4%(J2%),J4): NEXT J4: NEXT J3%: NEXT J2%
6800 B#(1) = R9#(P4%(1))
6810 FOR J1% = 2 TO LL%
6820 I8% = J1% - 1: X3# = 0
6830 FOR J2% = 1 TO I8%: X3# = X3# + R3#(P4%(J1%),J2%)*B#(J2%): NEXT J2%
6840 B#(J1%) = R9#(P4%(J1%)) - X3#: NEXT J1%
6850 B#(LL%) = B#(LL%)/R3#(P4%(LL%),LL%)
6860 FOR J1% = 2 TO LL%: I8% = LP1% - J1%: I9% = I8% + 1: X3# = 0
6870 FOR J2% = I9% TO LL%: X3# = X3# + R3#(P4%(I8%),J2%)*B#(J2%): NEXT J2%
6880 B#(I8%) = (B#(I8%) - X3#)/R3#(P4%(I8%),I8%): NEXT J1%: RETURN


1000 REM PLOT\FITFUNCT.BAS FITTED FUNCTION INTERPOLATION
     Copyright (c) 1988 D.F.Watson
1010 DIM P1#(55,3),P8(80),P9(2,3),S1(5934),S2(3),RO(55),PO(4,2)
1020 DIM R1#(3,3), R3#(55,55), R8(2,2), S3(2,3), BX(6,3), R9#(55), B#(55)
1030 DIM P4%(55),R6%(3,2)
1050 DATA 1,0,0,1,1,0,0,1,0,0,0,0,1,0,0,1,0,1, - 1, - 1,0,1,2,5, - 1,0,2,3, - 1,5,1E37,3,1
1060 DEF FNCV(RT#) = RT#^2*LOG(RT#): REM minimum curvature: REM DEF
     FNCV(RT#) = RT#^2*(LOG(RT#) - 1): REM biharmonic spline: REM DEF FNCV(RT#)
     = RT#: REM dummy function
1070 REM DEF FNCV(RT#) = NG + SI*(1 -  EXP( - RT#/RG)): NG = .25: SI = 1: RG = 1/3:
     REM exponential semivariogram
1080 REM DEF FNCV(RT#) = NG + SI*(1.5*RT#/RG - (RT#/RG)^3/2): NG = .25: SI = 1:
     RG = 1: REM spherical semivariogram
1090 P9(1,1) = - 1E + 37: P9(1,2) = - 1E + 37: P9(1,3) = - 1E + 37: P9(2,1) = 1E + 37:
     P9(2,2) = 1E + 37: P9(2,3) = 1E + 37
1100 SR = 0: SZ = 5100: GR = 19: DP = 0: TB = 3: PT = 1: DZ = 15: DG = 3: WD =
     10000: HI = 6800: AZ = - 25: TL = - 15: RD = 2: BI = 1.5: BJ = 7: IF DP = 0 THEN
     SR= 0
1140 OPEN "A: \HILL.DAT" FOR INPUT AS #1
1150 REM OPEN "A: \JDAVIS.DAT" FOR INPUT AS #1
1180 INPUT#1,ND%,NC%: FOR I1% = 1 TO NC%: INPUT#1,P8(I1%): NEXT I1%: IF
     CR>0 AND NC%>31 THEN NC% = 31
1190 INPUT#1,XS,YS,DT: I1% = 1
1220 INPUT#1,P1#(I1%,1), P1#(I1%,2), P1#(I1%,3): RO(I1%) = .1*RND: P1#(I1%,1) =
     (P1#(I1%,1) - XS)/DT: P1#(I1%,2) = (P1#(I1%,2) - YS)/DT: IF (P1#(I1%,1) - .5)^2 +
     (P1#(I1%,2) - .5)^2<RD THEN GOTO 1270
1230 ND% = ND% - 1: IF I1%>ND% THEN GOTO 1360
1260 GOTO 1220
1270 IF P9(1,1)<P1#(I1%,1) THEN P9(1,1) = P1#(I1%,1)
1280 IF P9(2,1)>P1#(I1%,1) THEN P9(2,1) = P1#(I1%,1)
1290 IF P9(1,2)<P1#(I1%,2) THEN P9(1,2) = P1#(I1%,2)
1300 IF P9(2,2)>P1#(I1%,2) THEN P9(2,2) = P1#(I1%,2)
1310 IF P9(1,3)<P1#(I1%,3) THEN P9(1,3) = P1#(I1%,3)
1320 IF P9(2,3)>P1#(I1%,3) THEN P9(2,3) = P1#(I1%,3)
1350 I1% = I1% + 1: IF I1%< = ND% THEN GOTO 1220
1360 IF ND%<4 THEN STOP
1370 XF = P9(1,1) - P9(2,1): YF = P9(1,2) - P9(2,2): VS = P9(1,3) - P9(2,3): R8(2,1) = AZ:
     R8(2,2) = TL
1380 FOR I1% = 1 TO 6: READ BX(I1%,1), BX(I1%,2), BX(I1%,3): NEXT I1%
1390 BN% = 1: IF DP<1 THEN GOTO 1450
1400 BN% = 2: BX(2,1) = 0: BX(2,3) = 1: IF R8(2,1)>0 THEN GOTO 1420
```

Appendix

```
1410 BX(2,1) = 1: BX(3,1) = 1: BX(4,1) = 1: BX(5,1) = 0: BX(6,1) = 0
1420 IF DP<2 THEN GOTO 1450
1430 NC1 = INT(SZ/100): NC% = NC1 + 1: U0 = .9999/NC1: U1 = .0000511 - U0
1440 FOR I1% = 1 TO NC%: U1 = U1 + U0: P8(I1%) = U1: NEXT I1%: GOTO 1460
1450 FOR I1% = 1 TO NC%: P8(I1%) = (P8(I1%) - P9(2,3))/VS + .00001*(RND - .5):
     NEXT I1%
1460 FOR I1% = 1 TO ND%: P1#(I1%,1) = P1#(I1%,1) + .0001*(RND - .5): P1#(I1%,2) =
     P1#(I1%,2) + .0001*(RND - .5) : P1#(I1%,3) = (P1#(I1%,3) - P9(2,3))/VS: NEXT I1%
1470 FOR I1% = 1 TO 3: READ R1#(I1%,1), R1#(I1%,2), R1#(I1%,3), R6%(I1%,1),
     R6%(I1%,2)
1480 P1#(I1% + ND%,1) = R1#(I1%,1)*XF + P9(2,1): P1#(I1% + ND%,2) = R1#(I1%,2)*YF
     + P9(2,2): P1#(I1% + ND%,3) = 0: NEXT I1%
1490 IF DP<1 THEN GOTO 1550
1500 FOR I1% = 1 TO 2: R8(I1%,1) = COS(R8(2,I1%)/57.29578): R8(I1%,2) =
     SIN(R8(2,I1%)/57.29578): NEXT I1%
1510 FOR I1% = 2 TO 6
1520 YY = (BX(I1%,2) - .5)*R8(1,1) + (BX(I1%,1) - .5)*R8(1,2) + .5: BX(I1%,1) =
     (BX(I1%,1) - .5)*R8(1,1) - (BX(I1%,2) - .5)*R8(1,2) + .5
1530 BX3 = (BX(I1%,3) - .5)*R8(2,1) + (YY - .5)*R8(2,2) + .5: BX(I1%,2) = (YY -
     .5)*R8(2,1) - (BX(I1%,3) - .5)*R8(2,2) + .5: BX(I1%,3) = BX3: NEXT I1%
1540 REM end data input
1550 SOUND 208,5: SOUND 165,5: SOUND 185,5: SOUND 123,5
3480 PRINT " AT " TIME$ " BEGIN SURFACE DEFINITION  -  PLEASE WAIT"
3490 REM minimum curvature splines
3500 IF TB>3 THEN GOTO 3600
3510 LL% = ND% + 3
3520 FOR I1% = 1 TO ND%
3530 R3#(I1%,1) = 1: R3#(LL%,I1% + 3) = 1: R3#(I1%,2) = P1#(I1%,1): R3#(ND% +
     1,I1% + 3) = R3#(I1%,2)
3540 R3#(I1%,3) = P1#(I1%,2): R3#(ND% + 2,I1% + 3) = R3#(I1%,3): R9#(I1%) =
     P1#(I1%,3): NEXT I1%
3550 FOR I1% = 1 TO 3: FOR I2% = 1 TO 3: R3#(I1% + ND%,I2%) = 0: NEXT I2%:
     R9#(I1% + ND%) = 0: NEXT I1%
3560 FOR I1% = 1 TO ND%: R3#(I1%,I1% + 3) = RO(I1%): IJ = I1% + 1: IF IJ>ND%
     THEN GOTO 3580
3570 FOR I2% = IJ TO ND%: RR# = ((R3#(I1%,2) - R3#(I2%,2))^2 + (R3#(I1%,3) -
     R3#(I2%,3))^2)^.5: R3#(I2%,I1% + 3) = FNCV(RR#): R3#(I2%,I2% + 3) = R3#(I2%,I1%
     + 3): NEXT I2%
3580 NEXT I1%: GOSUB 6690: GOTO 3830
3590 REM polynomial trend surface
3600 IF TB>4 THEN GOTO 3750
3610 B#(1) = 1: LL% = (DG + 1)*(DG + 2)/2: FOR I1% = 1 TO LL%: FOR I2% = 1 TO
     LL%: R3#(I1%,I2%) = 0: NEXT I2%: R9#(I1%) = 0 : NEXT I1%
3620 FOR I1% = 1 TO ND%: I5% = 1
3630 FOR I2% = 1 TO DG
3640 FOR I3% = 1 TO DG: I5% = I5% + 1: I6% = I5% - I2%: B#(I5%) =
     B#(I6%)*P1#(I1%,1): NEXT I3%
3650 I5% = I5% + 1: B#(I5%) = B#(I6%)*P1#(I1%,2): NEXT I2%
3660 FOR I2% = 1 TO LL%: R9#(I2%) = R9#(I2%) + B#(I2%)*P1#(I1%,3)
3670 FOR I3% = 1 TO LL%: R3#(I2%,I3%) = R3#(I2%,I3%) + B#(I2%)*B#(I3%): NEXT
     I3%
3680 NEXT I2%
3690 NEXT I1%: GOSUB 6690
3700 FOR L2% = 1 TO ND%: R9#(1) = 1: L5% = 1: RO(L2%) = 0
3710 FOR L3% = 1 TO DG
3720 FOR L4% = 1 TO L3%: L5% = L5% + 1: L6% = L5% - L3%: R9#(L5%) =
     R9#(L6%)*P1#(L2%,1): NEXT L4% : L5% = L5% + 1: R9#(L5%) =
     R9#(L6%)*P1#(L2%,2): NEXT L3%
3730 FOR L3% = 1 TO LL%: RO(L2%) = RO(L2%) + B#(L3%)*R9#(L3%): NEXT L3%:
     RO(L2%) = (RO(L2%) - P1#(L2%,3))*VS: NEXT L2% : GOTO 3830
3740 REM multiquadric collocation
3750 LL% = ND%
3760 FOR I1% = 1 TO LL%
3770 FOR I2% = 1 TO LL%
3780 IF I1% = I2% THEN GOTO 3800
```

```
3790 R3#(I1%,I2%) = ((P1#(I1%,1) - P1#(I2%,1))^2 + (P1#(I1%,2) - P1#(I2%,2))^2 +
     RO(I2%)^2)^.5: IF TB<8 THEN GOTO 3810
3795 R3#(I1%,I2%) = 2*((P1#(I1%,1) - P1#(I2%,1))^2 + (P1#(I1%,2) - P1#(I2%,2))^2)^.5 +
     RO(I2%) - R3#(I1%,I2%) : GOTO 3810
3800 R3#(I1%,I2%) = RO(I1%): IF TB = 8 THEN R3#(I1%,I2%) = 0
3810 NEXT I2%: R9#(I1%) = P1#(I1%,3): NEXT I1%: GOSUB 6690
3820 REM make triangular grid
3830 PRINT " AT " TIME$ " BEGIN SURFACE INTERPOLATION - PLEASE WAIT":
     SOUND 123,5: SOUND 185,5: SOUND 208,5 : SOUND 165,5
3840 I0% = GR + 1: IS = 1: I2% = 2*I0% + 1: I3% = 1 + INT(GR/3^.5): I4% = I3%*I2% +
     I0%: SX = 1/GR: SY = 1/(2*I3%): S2(3) = .000001
3850 WX# = S2(3): S2(1) = 1.00001: AA# = SX*SY*.5*DT*DT: SM# = 0: SA# = 0: WY# =
     1: U8 = 1
3860 FOR L1% = 1 TO I4%
3870 S1(L1%) = - 999: WXO# = WX#: WYO# = WY#
4710 REM minimum curvature splines
4720 S1(L1%) = 0: IF TB>3 THEN GOTO 4760
4730 FOR L2% = 4 TO LL%: RR# = ((P1#(L2% - 3,1) - WX#)^2 + (P1#(L2% - 3,2) -
     WY#)^2)^.5: S1(L1%) = S1(L1%) + B#(L2%)*FNCV(RR#): NEXT L2%
4740 S1(L1%) = S1(L1%) + B#(1) + B#(2)*WX# + B#(3)*WY#: GOTO 4940
4750 REM polynomial trend surface
4760 IF TB>4 THEN GOTO 4830
4770 R9#(1) = 1: L5% = 1
4780 FOR L2% = 1 TO DG
4790 FOR L3% = 1 TO L2%: L5% = L5% + 1: L6% = L5% - L2%: R9#(L5%) =
     R9#(L6%)*WX#: NEXT L3%
4800 L5% = L5% + 1: R9#(L5%) = R9#(L6%)*WY#: NEXT L2%
4810 FOR L2% = 1 TO LL%: S1(L1%) = S1(L1%) + B#(L2%)*R9#(L2%): NEXT L2%:
     GOTO 4940
4830 REM multiquadric collocation
4840 FOR L2% = 1 TO LL%: IF TB = 8 THEN GOTO 4844
4842 S1(L1%) = S1(L1%) + B#(L2%)*((P1#(L2%,1) - WX#)^2 + (P1#(L2%,2) - WY#)^2 +
     RO(L2%)^2)^.5: GOTO 4848
4844 S1(L1%) = S1(L1%) + B#(L2%)*(2*((P1#(L2%,1) - WX#)^2 + (P1#(L2%,2) -
     WY#)^2)^.5 + RO(L2%) - ((P1#(L2%,1) - WX#)^2 + (P1#(L2%,2) - WY#)^2 +
     RO(L2%)^2)^.5)
4848 NEXT L2%
4940 IS = IS + SX*U8: IF IS>0 THEN GOTO 4960
4950 WXO# = WXO# + SX/2: IS = 1.001 - SX
4960 WX# = WXO# + SX*U8: WY# = WYO#: IF U8*WX#<S2(2 - U8) THEN GOTO 4980
4970 U8 = - U8: WX# = S2(2 + U8): WY# = WY# - SY: IS = U8
4980 NEXT L1%
5060 REM end interpolation to triangular grid and begin display construction
5070 SOUND 208,5: SOUND 165,5: SOUND 185,5: SOUND 123,5: PRINT " AT " TIME$;
5071 INPUT " PREPARE PLOTTER AND PRESS ENTER",PM$
5090 LW = (HI - SZ)/2: IF DP>0 THEN SZ = INT(SZ*2/3)
5091 IF DP>0 AND SZ>HI/2 THEN SZ = HI/2
5100 SW = (WD - SZ)/2: ZW = (HI - SZ)/2
5110 IF SR>0 THEN SW = (WD - 2*SZ)/3 5120 SW = SW + 1: SV = 2*SW + SZ: SR =
     SR*SZ: SZ = SZ - 2: SZ = SZ - 2: ZW = ZW - 1: SU = SW: IF SR>0 THEN SU = (2*SW
     + SV)/3
5160 REM plot open
5170 LPRINT "IN;IP0,0," + STR$(WD) + "," + STR$(HI) + ";SP1;": IF SR<2 THEN GOTO
     5180
5175 FOR L1% = 2 TO 6: BX1 = (BX(L1%,1) - .5)*.99875 - (BX(L1%,3) - .5)*.05 + .5:
     BX(L1%,3) = (BX(L1%,3) - .5)*.99875 + (BX(L1%,1) - .5)*.05 + .5: BX(L1%,1) = BX1:
     NEXT L1%: LPRINT "PU;PA0,6800;"
5180 LPRINT "PU;PA" + STR$(BX(BN%,1)*SZ + SW) + "," + STR$(BX(BN%,2)*SZ +
     ZW) + ";PD;" : ZX = SX: ZY = SY: ZZ = .5: R3#(2,2) = 1 + 2*SY: R3#(2,1) = 0
5190 FOR L1% = BN% + 1 TO BN% + 4: LPRINT "PA" + STR$(BX(L1%,1)*SZ + SW) +
     "," + STR$(BX(L1%,2)*SZ + ZW) + ";": NEXT L1% : LPRINT "PU;" 5192 FOR L1% = 2
     TO 6: BX(L1%,1) = (BX(L1%,1) - .5)*.995 + (BX(L1%,3) - .5)*.1 + .5: NEXT L1%: SM =
     0: SA = 0
5200 REM plot contours
5220 FOR L1% = 1 TO I3%
```

307

Appendix

```
5230 R3#(1,2) = R3#(2,2) - 2*SY - 2*ZY: R3#(1,1) = R3#(2,1): N1% = (2*L1% - .5 +
     ZZ)*I0% + L1% + .5/ZZ: R3#(1,3) = S1(N1%): R3#(2,2) = R3#(2,2) - 2*SY - ZY:
     R3#(2,1) = .5 - ZZ: N2% = (2*L1% - .5 + ZZ)*I0% + L1%
5240 R3#(2,3) = S1(N2%): R3#(3,2) = R3#(2,2): R3#(3,1) = R3#(2,1) + ZX/2: N3% = N2%
     - .5/ZZ: R3#(3,3) = S1(N3%): AR# = AA#/2
5250 GOSUB 5820
5260 R3#(1,2) = R3#(1,2) + 2*ZY: N2% = (2*L1% - .5/ZZ - .5 - ZZ)*I0% + L1%: R3#(1,3) =
     S1(N2%)
5270 GOSUB 5820
5280 AR# = AA#
5290 FOR L2% = 1 TO GR
5300 R3#(2,2) = R3#(1,2): R3#(2,1) = R3#(1,1) + ZX: N4% = N2% + .5/ZZ: R3#(2,3) =
     S1(N4%)
5310 GOSUB 5820
5320 R3#(1,2) = R3#(3,2): R3#(1,1) = R3#(3,1) + ZX
5330 IF L2% = GR THEN R3#(1,1) = R3#(1,1) - ZX/2
5340 N6% = N3% - .5/ZZ: R3#(1,3) = S1(N6%): IF L2% = GR THEN AR# = AA#/2
5350 GOSUB 5820
5360 R3#(2,2) = R3#(2,2) - 2*ZY: N5% = N1% + .5/ZZ: R3#(2,3) = S1(N5%)
5370 GOSUB 5820
5380 R3#(1,2) = R3#(2,2): R3#(1,1) = R3#(2,1) - ZX: R3#(1,3) = S1(N1%): IF L2% = GR
     THEN AR# = AA#
5390 GOSUB 5820
5400 N1% = N5%: N2% = N4%: N3% = N6%: R3#(1,2) = R3#(1,2) + 2*ZY: R3#(1,1) =
     R3#(2,1) : R3#(1,3) = S1(N4%): R3#(3,1) = R3#(3,1) + ZX: R3#(3,3) = S1(N6%) 5410
     NEXT L2%
5420 ZY = - ZY: ZX = - ZX: ZZ = - ZZ
5430 NEXT L1%: REM end plotting contours
5450 REM plot data points
5460 IF PT<1 THEN GOTO 5770
5480 DZ% = DZ: FOR L1% = 1 TO ND%
5490 IF P1#(L1%,1)< - .0001 OR P1#(L1%,1)>1.0001 OR P1#(L1%,2)< - .0001 OR
     P1#(L1%,2)>1.0001 THEN GOTO 5610
5500 PP1# = P1#(L1%,1): PP2# = P1#(L1%,2): PP3# = P1#(L1%,3): IF DP<1 THEN
     GOTO 5530
5510 YY = (PP2# - .5)*R8(1,1) + (PP1# - .5)*R8(1,2) + .5: PP1# = (PP1# - .5)*R8(1,1) -
     (PP2# - .5)*R8(1,2) + .5
5520 ZZ3 = (PP3# - .5)*R8(2,1) + (YY - .5)*R8(2,2) + .5: PP2# = (YY - .5)*R8(2,1) - (PP3#
     - .5)*R8(2,2) + .5: PP3# = ZZ3: IF SR<2 THEN GOTO 5528
5525 PP1# = (PP1# - .5)*.99875 - (PP3# - .5)*.05 + .5
5528 IF SR<>1 THEN GOTO 5530
5529 PP1# = (PP1# - .5)*.99875 + (PP3# - .5)*.05 + .5
5530 S3(1,1) = PP1#*SZ + SW: S3(1,2) = PP2#*SZ + ZW
5540 LPRINT "PA" + STR$(S3(1,1) - DZ%) + "," + STR$(S3(1,2) + DZ%) + ";": LPRINT
     "PD;PA" + STR$(S3(1,1) + DZ%) + "," + STR$(S3(1,2) + DZ%) + ";": LPRINT "PA" +
     STR$(S3(1,1) + DZ%) + "," + STR$(S3(1,2) - DZ%) + ";"
5550 LPRINT "PA" + STR$(S3(1,1) - DZ%) + "," + STR$(S3(1,2) + DZ%) + ";": LPRINT
     "PA" + STR$(S3(1,1) - DZ%) + "," + STR$(S3(1,2) + DZ%) + ";": LPRINT "PA" +
     STR$(S3(1,1) + DZ%) + "," + STR$(S3(1,2) - DZ%) + ";PU;"
5610 NEXT L1%
5770 SR = SR - 1: SW = SV: IF SR = 1 THEN GOTO 5180
5775 LPRINT "PU;PA" + STR$(SU) + "," + STR$(LW - 500) + ";SI0.2,0.3;SL0.2;" 5777
     LPRINT "LBFigure 1.0.1 - Example Isolines" + CHR$(3); "PU;PA0,6800;":
     REM plot close
5790 SOUND 123,5: SOUND 185,5: SOUND 208,5: SOUND 165,5: TM$ = TIME$
5800 SM = SM/SA: PRINT " AVERAGE HEIGHT OF SURFACE IS" SM: PRINT " AT "
     TM$ " RUN COMPLETE": END
5810 REM curvemaker
5820 ST = R3#(1,3)
5830 IF ST>R3#(2,3) THEN ST = R3#(2,3)
5840 IF ST>R3#(3,3) THEN ST = R3#(3,3)
5850 IF ST< = - 999 THEN RETURN
5860 SM = SM + AR#*((R3#(1,3) + R3#(2,3) + R3#(3,3))/3*VS + P9(2,3)): SA = SA + AR#
6210 L8% = 0: IF DP<2 THEN GOTO 6310
6220 L8% = 1
6230 REM rotate triangle for orthogonal profiles
```

```
6240 FOR L3% = 1 TO 3
6250 Z# = R3#(L3%,3): R3#(L3%,3) = 1 - R3#(L3%,L8%): R3#(L3%,L8%) = Z#
6260 NEXT L3%
6270 ST = R3#(1,3)
6280 IF ST>R3#(2,3) THEN ST = R3#(2,3)
6290 IF ST>R3#(3,3) THEN ST = R3#(3,3)
6300 IF ST < = - 999 THEN RETURN
6310 TP = R3#(1,3)
6320 IF TP<R3#(2,3) THEN TP = R3#(2,3)
6330 IF TP<R3#(3,3) THEN TP = R3#(3,3)
6340 IF TP = ST THEN RETURN
6350 FOR L0% = 1 TO NC%
6360 REM slice the triangle
6370 IF P8(L0%)>TP OR P8(L0%)<ST THEN GOTO 6610
6380 L5% = 1: L4% = 0
6390 L4% = L4% + 1
6400 L7% = 1: IF L7% = L4% THEN L7% = L7% + 1
6410 L3% = L7% + 1: IF L3% = L4% THEN L3% = L3% + 1
6420 IF R3#(L7%,3) = R3#(L3%,3) THEN GOTO 6460
6430 F = (P8(L0%) - R3#(L7%,3))/(R3#(L3%,3) - R3#(L7%,3)): IF F<0 OR F>1 THEN
     GOTO 6460
6440 S3(L5%,1) = R3#(L7%,1) + (R3#(L3%,1) - R3#(L7%,1))*F: S3(L5%,2) = R3#(L7%,2)
     + (R3#(L3%,2) - R3#(L7%,2))*F
6450 L5% = L5% + 1
6460 IF L5%<3 THEN GOTO 6390
6470 S3(1,3) = P8(L0%): S3(2,3) = P8(L0%): IF DP<2 THEN GOTO 6500
6480 REM reverse rotate the intersection trace
6490 FOR L3% = 1 TO 2: Z# = S3(L3%,3): S3(L3%,3) = S3(L3%,L8%): S3(L3%,L8%) = 1
     - Z#: NEXT L3%
6500 IF DP = 0 THEN GOTO 6570
6510 REM apply perspective
6520 FOR L3% = 1 TO 2
6530 YY = (S3(L3%,2) - .5)*R8(1,1) + (S3(L3%,1) - .5)*R8(1,2): S3(L3%,1) = (S3(L3%,1) -
     .5)*R8(1,1) - (S3(L3%,2) - .5)*R8(1,2) + .5
6540 S33 = (S3(L3%,3) - .5)*R8(2,1) + YY*R8(2,2) + .5: S3(L3%,2) = YY*R8(2,1) -
     (S3(L3%,3) - .5)*R8(2,2) + .5: S3(L3%,3) = S33
6550 NEXT L3%
6552 IF SR<2 THEN GOTO 6560
6554 S3(1,1) = (S3(1,1) - .5)*.99875 - (S3(1,3) - .5)*.05 + .5: S3(2,1) = (S3(2,1) -
     .5)*.99875 - (S3(2,3) - .5)*.05 + .5
6560 IF SR<>1 THEN GOTO 6570
6565 S3(1,1) = (S3(1,1) - .5)*.99875 + (S3(1,3) - .5)*.05 + .5: S3(2,1) = (S3(2,1) -
     .5)*.99875 + (S3(2,3) - .5)*.05 + .5
6567 REM draw a line from (S3(1,1),S3(1,2)) to (S3(2,1),S3(2,2))
6570 LPRINT "PA" + STR$(S3(1,1)*SZ + SW) + "," + STR$(S3(1,2)*SZ + ZW) + ";":
     LPRINT "PD;PA" + STR$(S3(2,1)*SZ + SW) + "," + STR$(S3(2,2)*SZ + ZW) + ";PU;"
6610 NEXT L0%
6620 IF L8%<1 THEN RETURN
6630 REM reverse rotate the triangle
6640 FOR L3% = 1 TO 3
6650 Z# = R3#(L3%,3): R3#(L3%,3) = R3#(L3%,L8%): R3#(L3%,L8%) = 1 - Z#
6660 NEXT L3%
6670 L8% = L8% + 1: IF L8%<3 THEN GOTO 6240 ELSE RETURN
6680 REM solve R3#(LL%,LL%)xB#(LL%) = R9#(LL%)
6690 FOR J2% = 1 TO LL%: P4%(J2%) = J2%: B#(J2%) = 0
6700 FOR J3% = 1 TO LL%: IF ABS(R3#(J2%,J3%))> = B#(J2%) THEN B#(J2%) =
     ABS(R3#(J2%,J3%))
6710 NEXT J3%: NEXT J2%: LM1% = LL% - 1: LP1% = LL% + 1
6720 FOR J2% = 1 TO LM1%: X3# = 0
6730 FOR J3% = J2% TO LL%: X4# = ABS(R3#(P4%(J3%),J2%))/B#(P4%(J3%)): IF
     X4#< = X3# THEN GOTO 6750
6740 X3# = X4#: K3% = J3%
6750 NEXT J3%: IF K3% = J2% GOTO 6770
6760 K2% = P4%(J2%): P4%(J2%) = P4%(K3%): P4%(K3%) = K2%
6770 X3# = R3#(P4%(J2%),J2%): K4 = J2% + 1
```

Appendix

```
6780 FOR J3% = K4 TO LL%: X4# =  - R3#(P4%(J3%),J2%)/X3#: R3#(P4%(J3%),J2%) =
     - X4#
6790 FOR J4 = K4 TO LL%: R3#(P4%(J3%),J4) = R3#(P4%(J3%),J4) +
     X4#*R3#(P4%(J2%),J4): NEXT J4: NEXT J3%: NEXT J2%
6800 B#(1) = R9#(P4%(1))
6810 FOR J1% = 2 TO LL%
6820 I8% = J1% - 1: X3# = 0
6830 FOR J2% = 1 TO I8%: X3# = X3# + R3#(P4%(J1%),J2%)*B#(J2%): NEXT J2%
6840 B#(J1%) = R9#(P4%(J1%)) - X3#: NEXT J1%
6850 B#(LL%) = B#(LL%)/R3#(P4%(LL%),LL%)
6860 FOR J1% = 2 TO LL%: I8% = LP1% - J1%: I9% = I8% + 1: X3# = 0
6870 FOR J2% = I9% TO LL%: X3# = X3# + R3#(P4%(I8%),J2%)*B#(J2%): NEXT J2%
6880 B#(I8%) = (B#(I8%) - X3#)/R3#(P4%(I8%),I8%): NEXT J1%: RETURN


1000 REM PLOT\RECTANGL.BAS RECTANGLE - BASED INTERPOLATION
     Copyright (c) 1988 D.F.Watson
1010 DIM P1#(63,6), P5#(50), P8(80), P9(2,3),S1(2033),S2(3),S4(50),PO(4,2)
1020 DIM R1#(3,3), R3#(50,50), R4#(3,2), R5#(3,3), R8(2,2), S3(2,3), BX(6,3), R9#(50),
     B#(55)
1030 DIM P4%(50),P6%(50)
1050 DATA 1,0,0,1,1,0,0,1,0,0,0,0,1,0,0,1,0,1, - 1, - 1,0,5, - 1,0, - 1,5,1E37
1060 DEF FNCV(RT#) = RT#^2*LOG(RT#): REM minimum curvature: REM DEF
     FNCV(RT#) = RT#^2*(LOG(RT#) - 1): REM biharmonic spline: REM DEF FNCV(RT#)
     = RT#: REM dummy function
1070 REM DEF FNCV(RT#) = NG + SI*(1 -  EXP( - RT#/RG)): NG = .25: SI = 1: RG = 1/3:
     REM exponential semivariogram
1080 REM DEF FNCV(RT#) = NG + SI*(1.5*RT#/RG - (RT#/RG)^3/2): NG = .25: SI = 1:
     RG = 1: REM spherical semivariogram
1090 P9(1,1) =  - 1E37: P9(1,2) =  - 1E37: P9(1,3) =  - 1E37: P9(2,1) = 1E37: P9(2,2) =
     1E37: P9(2,3) = 1E37
1100 SL = 2: SR = 1: SZ = 5100: GR = 19: DP = 1: CL = 1: PT = 1: DZ = 15: TB = 6: WD
     = 10000: HI = 6800: AZ =  - 25: TL =  - 15: RD = 2: BI = 1.5: BJ = 7: IF DP = 0 THEN
     SR = 0
1125 IF TB>6 THEN CL = 0
1130 OPEN "A:\SHARPG.DAT" FOR INPUT AS #1
1180 INPUT#1,ND%,NC%: FOR I1% = 1 TO NC%: INPUT#1,P8(I1%): NEXT I1%: IF
     CR>0 AND NC%>31 THEN NC% = 31
1190 INPUT#1,XS,YS,DT: I1% = 1: JJ = 1: INPUT#1,FX,FY,DS,NS: P1#(I1%,1) = FX:
     P1#(I1%,2) = FY
1210 INPUT#1,P1#(I1%,3): P1#(I1%,1) = (P1#(I1%,1) - XS)/DT: P1#(I1%,2) = (P1#(I1%,2)
     - YS)/DT: IF ABS(P1#(I1%,1) - .5)<RD AND ABS(P1#(I1%,2) - .5)<RD THEN GOTO
     1270
1230 ND% = ND% - 1: IF I1%>ND% THEN GOTO 1360
1240 P1#(I1% + 1,1) = FX + JJ*DS: P1#(I1% + 1,2) = FY: JJ = JJ + 1: IF JJ< = NS THEN
     GOTO 1210
1250 FY = FY + DS: P1#(I1% + 1,2) = FY: JJ = 1: P1#(I1% + 1,2) = FY
1260 GOTO 1210
1270 IF P9(1,1)<P1#(I1%,1) THEN P9(1,1) = P1#(I1%,1)
1280 IF P9(2,1)>P1#(I1%,1) THEN P9(2,1) = P1#(I1%,1)
1290 IF P9(1,2)<P1#(I1%,2) THEN P9(1,2) = P1#(I1%,2)
1300 IF P9(2,2)>P1#(I1%,2) THEN P9(2,2) = P1#(I1%,2)
1310 IF P9(1,3)<P1#(I1%,3) THEN P9(1,3) = P1#(I1%,3)
1320 IF P9(2,3)>P1#(I1%,3) THEN P9(2,3) = P1#(I1%,3)
1330 P1#(I1% + 1,2) = FY + JJ*DS: P1#(I1% + 1,1) = FX: JJ = JJ + 1: IF JJ< = NS THEN
     GOTO 1350
1340 FX = FX + DS: P1#(I1% + 1,1) = FX: JJ = 1: P1#(I1% + 1,2) = FY
1350 I1% = I1% + 1: IF I1%< = ND% THEN GOTO 1210
1360 IF ND%<4 THEN STOP
1370 XF = P9(1,1) - P9(2,1): YF = P9(1,2) - P9(2,2): VS = P9(1,3) - P9(2,3): R8(2,1) = AZ:
     R8(2,2) = TL: IF INT(ND%^.5 + .1)<NS THEN NS = INT(ND%^.5)
1380 FOR I1% = 1 TO 6: READ BX(I1%,1),BX(I1%,2),BX(I1%,3): NEXT I1%
1390 BN% = 1: IF DP<1 THEN GOTO 1450
1400 BN% = 2: BX(2,1) = 0: BX(2,3) = 1: IF R8(2,1)>0 THEN GOTO 1420
1410 BX(2,1) = 1: BX(3,1) = 1: BX(4,1) = 1: BX(5,1) = 0: BX(6,1) = 0
1420 IF DP<2 OR CR>0 THEN GOTO 1450
```

BASIC PROGRAMS

```
1430 NC1% = INT(SZ/100): NC% = NC1% + 1: U0 = .9999/NC1%: U1 = .0000511 - U0
1440 FOR I1% = 1 TO NC%: U1 = U1 + U0: P8(I1%) = U1: NEXT I1%: GOTO 1460
1450 FOR I1% = 1 TO NC%: P8(I1%) = (P8(I1%) - P9(2,3))/VS + .00001*(RND - .5):
    NEXT I1%
1460 FOR I1% = 1 TO ND%: P1#(I1%,1) = P1#(I1%,1) + .0001*(RND - .5): P1#(I1%,2) =
    P1#(I1%,2) + .0001*(RND - .5) : P1#(I1%,3) = (P1#(I1%,3) - P9(2,3))/VS: NEXT I1%
1470 FOR I1% = 1 TO 3: READ R1#(I1%,1);R1#(I1%,2),R1#(I1%,3)
1480 P1#(I1% + ND%,1) = R1#(I1%,1)*XF + P9(2,1): P1#(I1% + ND%,2) = R1#(I1%,2)*YF
    + P9(2,2): P1#(I1% + ND%,3) = 0: NEXT I1%
1490 IF DP<1 THEN GOTO 1550
1500 FOR I1% = 1 TO 2: R8(I1%,1) = COS(R8(2,I1%)/57.29578): R8(I1%,2) =
    SIN(R8(2,I1%)/57.29578): NEXT I1%
1510 FOR I1% = 2 TO 6
1520 YY = (BX(I1%,2) - .5)*R8(1,1) + (BX(I1%,1) - .5)*R8(1,2) + .5: BX(I1%,1) =
    (BX(I1%,1) - .5)*R8(1,1) - (BX(I1%,2) - .5)*R8(1,2) + .5
1530 BX3 = (BX(I1%,3) - .5)*R8(2,1) + (YY - .5)*R8(2,2) + .5: BX(I1%,2) = (YY -
    .5)*R8(2,1) - (BX(I1%,3) - .5)*R8(2,2) + .5: BX(I1%,3) = BX3: NEXT I1%
1540 REM end data input, begin gradient estimation
1550 SOUND 208,5: SOUND 165,5: SOUND 185,5: SOUND 123,5
1560 IF CL<1 THEN GOTO 3830
1570 PRINT " AT " TIME$ " BEGIN GRADIENT ESTIMATION - PLEASE WAIT"
1580 FOR I0% = 1 TO ND%
2610 IF SL>2 THEN GOTO 2810 : REM least squares gradients
2620 P1#(I0%,4) = 0: P1#(I0%,5) = 0: P1#(I0%,6) = .2: L5% = 1
2641 R4#(1,1) = 0: R4#(2,1) = 0: R4#(3,1) = 0: R3#(2,1) = 0: R3#(2,2) = 0: R3#(3,1) = 0:
    R3#(3,2) = 0: R3#(3,3) = 0
2642 FOR I2% = - NS TO NS STEP NS: FOR I3% = - 1 TO 1 STEP 1: P6%(L5%) = I0%
    + I2% + I3%: IF P6%(L5%)<1 OR P6%(L5%)>ND% THEN GOTO 2745
2730 R4#(1,1) = R4#(1,1) + P1#(P6%(L5%),3): R4#(2,1) = R4#(2,1) + P1#(P6%(L5%),1)
    *P1#(P6%(L5%),3): R4#(3,1) = R4#(3,1) + P1#(P6%(L5%),2)*P1#(P6%(L5%),3)
2740 R3#(2,1) = R3#(2,1) + P1#(P6%(L5%),1): R3#(2,2) = R3#(2,2) +
    P1#(P6%(L5%),1)^2: R3#(3,1) = R3#(3,1) + P1#(P6%(L5%),2): R3#(3,2) = R3#(3,2) +
    P1#(P6%(L5%),1)*P1#(P6%(L5%),2): R3#(3,3) = R3#(3,3) + P1#(P6%(L5%),2)^2: L5%
    = L5% + 1
2745 NEXT I3%: NEXT I2%: L5% = L5% - 1
2770 U1# = L5%*(R3#(2,2)*R3#(3,3) - R3#(3,2)^2) - R3#(2,1)*(R3#(2,1)*R3#(3,3) -
    R3#(3,1)*R3#(3,2)) + R3#(3,1)*(R3#(2,1)*R3#(3,2) - R3#(3,1)*R3#(2,2))
2780 P1#(I0%,4) = - (L5%*(R4#(2,1)*R3#(3,3) - R3#(3,2)*R4#(3,1)) -
    R4#(1,1)*(R3#(2,1)*R3#(3,3) - R3#(3,1)*R3#(3,2)) + R3#(3,1)*(R3#(2,1)*R4#(3,1) -
    R3#(3,1)*R4#(2,1)))/U1#
2790 P1#(I0%,5) = - (L5%*(R3#(2,2)*R4#(3,1) - R3#(3,2)*R4#(2,1)) -
    R3#(2,1)*(R3#(2,1)*R4#(3,1) - R3#(3,1)*R4#(2,1)) + R4#(1,1)*(R3#(2,1)*R3#(3,2) -
    R3#(3,1)*R3#(2,2)))/U1#: GOTO 3080
2810 IF SL>3 THEN GOTO 3035 : REM spline gradients
2815 P1#(I0%,4) = 0: P1#(I0%,5) = 0: P1#(I0%,6) = .2: L5% = 1
2820 R5#(1,1) = P1#(I0%,1): R5#(1,2) = P1#(I0%,2): R5#(2,1) = R5#(1,1) + .0001 :
    R5#(2,2) = R5#(1,2): R5#(3,1) = R5#(1,1): R5#(3,2) = R5#(1,2) + .0001
2822 FOR I2% = - NS TO NS STEP NS: FOR I3% = - 1 TO 1 STEP 1: P6%(L5%) = I0%
    + I2% + I3%: IF P6%(L5%)<1 OR P6%(L5%)>ND% THEN GOTO 2826
2824 L5% = L5% + 1
2826 NEXT I3%: NEXT I2%: L5% = L5% - 1
2900 LL% = L5% + 3
2910 FOR I1% = 1 TO L5%
2920 R3#(I1%,1) = 1: R3#(LL%,I1% + 3) = 1: R3#(I1%,2) = P1#(P6%(I1%),1): R3#(L5% +
    1,I1% + 3) = R3#(I1%,2)
2930 R3#(I1%,3) = P1#(P6%(I1%),2): R3#(L5% + 2,I1% + 3) = R3#(I1%,3): R9#(I1%) =
    P1#(P6%(I1%),3): NEXT I1%
2940 FOR I1% = 1 TO 3: FOR I2% = 1 TO 3: R3#(I1% + L5%,I2%) = 0: NEXT I2%:
    R9#(I1% + L5%) = 0: NEXT I1%
2950 FOR I1% = 1 TO L5%: R3#(I1%,I1% + 3) = 0: IJ = I1% + 1: IF IJ>L5% THEN GOTO
    2970
2960 FOR I2% = IJ TO L5%: RT# = ((R3#(I1%,2) - R3#(I2%,2))^2 + (R3#(I1%,3) -
    R3#(I2%,3))^2)^.5: R3#(I2%,I1% + 3) = FNCV(RT#): R3#(I1%,I2% + 3) = R3#(I2%,I1%
    + 3): NEXT I2%
2970 NEXT I1%: GOSUB 6690
2980 TR# = 0: TS# = 0
```

Appendix

```
2990 FOR I1% = 4 TO LL% 3000 RT# = ((P1#(P6%(I1% - 3),1) - R5#(2,1))^2 +
     (P1#(P6%(I1% - 3),2) - R5#(2,2))^2)^.5: TR# = TR# + B#(I1%)*FNCV(RT#)
3010 RS# = ((P1#(P6%(I1% - 3),1) - R5#(3,1))^2 + (P1#(P6%(I1% - 3),2) -
     R5#(3,2))^2)^.5: TS# = TS# + B#(I1%)*FNCV(RS#) : NEXT I1%
3020 R5#(2,3) = B#(1) + B#(2)*R5#(2,1) + B#(3)*R5#(2,2) + TR#: P1#(I0%,4) =
     (P1#(I0%,3) - R5#(2,3))/.0001
3030 R5#(3,3) = B#(1) + B#(2)*R5#(3,1) + B#(3)*R5#(3,2) + TS#: P1#(I0%,5) =
     (P1#(I0%,3) - R5#(3,3))/.0001: GOTO 3080
3033 REM hyperboloid gradients
3035 R3#(5,1) = 1: FOR I1% = 1 TO 4: R9#(I1%) = 0: FOR I2% = 1 TO 4: R3#(I1%,I2%) =
     0: NEXT I2%: NEXT I1%: LL% = 1
3040 FOR I1% = - NS TO NS STEP NS: FOR I2% = - 1 TO 1 STEP 1: P6%(LL%) = I0%
     + I1% + I2%: IF P6%(LL%)<1 OR P6%(LL%)>ND% THEN GOTO 3060
3050 R3#(5,2) = P1#(P6%(LL%),1): R3#(5,3) = P1#(P6%(LL%),2): R3#(5,4) =
     P1#(P6%(LL%),1)*P1#(P6%(LL%),2): FOR I3% = 1 TO 4: FOR I4% = 1 TO 4
3055 R3#(I3%,I4%) = R3#(I3%,I4%) + R3#(5,I3%)*R3#(5,I4%): NEXT I4% : R9#(I3%) =
     R9#(I3%) + R3#(5,I3%)*P1#(P6%(LL%),3): NEXT I3%: LL% = LL% + 1
3060 NEXT I2%
3070 NEXT I1%: LL% = 4: GOSUB 6690: P1#(I0%,4) = - B#(2) - B#(4)*P1#(I0%,2):
     P1#(I0%,5) = - B#(3) - B#(4)*P1#(I0%,1): P1#(I0%,6) = .2
3080 NEXT I0%
3820 REM make triangular grid
3830 PRINT " AT " TIME$ " BEGIN SURFACE INTERPOLATION - PLEASE WAIT":
     SOUND 123,5: SOUND 185,5: SOUND 208,5 : SOUND 165,5
3831 IF TB<7 THEN GOTO 3840 : REM minimum curvature spline
3833 FOR I1% = 1 TO 3: FOR I2% = 1 TO 3: R3#(I1% + ND%,I2%) = 0: NEXT I2%:
     R9#(I1% + ND%) = 0: NEXT I1%: LL% = ND% + 3
3835 FOR I1% = 1 TO ND%: R3#(I1%,1) = 1: R3#(LL%,I1% + 3) = 1: R3#(I1%,2) =
     P1#(I1%,1) : R3#(ND% + 1,I1% + 3) = R3#(I1%,2): R3#(I1%,3) = P1#(I1%,2):
     R3#(ND% + 2,I1% + 3) = R3#(I1%,3): R9#(I1%) = P1#(I1%,3) : R3#(I1%,I1% + 3) = 0:
     IF I1% = 1 THEN GOTO 3838
3837 FOR I2% = I1% TO ND%: RT# = ((P1#(I2%,1) - R3#(I1% - 1,2))^2 + (P1#(I2%,2) -
     R3#(I1% - 1,3))^2)^.5 : R3#(I2%,I1% + 2) = FNCV(RT#): R3#(I1% - 1,I2% + 3) =
     R3#(I2%,I1% + 2): NEXT I2%
3838 NEXT I1%: GOSUB 6690
3840 I0% = GR + 1: IS = 1: I2% = 2*I0% + 1: I3% = 1 + INT(GR/3^.5): I4% = I3%*I2% +
     I0%: SX = 1/GR: SY = 1/(2*I3%): S2(3) = .000001
3850 WX# = S2(3): S2(1) = 1.00001: AA# = SX*SY*.5*DT*DT: WY# = 1: U8 = 1: K5% = 4:
     OX = 0: OY = 0
3860 FOR L1% = 1 TO I4%
3865 IF TB>6 THEN GOTO 4870
3870 S1(L1%) = - 999: RX# = 1 + (WX# - P9(2,1))*DT/DS: QX = INT(RX#): RY# = 1 +
     (WY# - P9(2,2))*DT/DS: QY = INT(RY#): IF QX<1 OR QX> = NS OR QY<1 OR QY> =
     NS THEN GOTO 4940
4558 S1(L1%) = 0: IF QX = OX AND QY = OY THEN GOTO 4562
4560 P6%(1) = (QX - 1)*NS + QY: P6%(2) = P6%(1) + NS: P6%(4) = P6%(1) + 1: P6%(3)
     = P6%(4) + NS
4562 RX# = RX# - QX: RY# = RY# - QY
4563 P5#(1) = (1 - RX#)*(1 - RY#): P5#(2) = RX#*(1 - RY#): P5#(3) = RX#*RY#: P5#(4) =
     RY#*(1 - RX#) 4580 FOR L2% = 1 TO K5%: S1(L1%) = S1(L1%) +
     P5#(L2%)*P1#(P6%(L2%),3): NEXT L2%: OX = QX: OY = QY
4590 SII = S1(L1%): IF CL<1 THEN GOTO 4940: REM blend bilinear and gradients 4620
     FOR L2% = 1 TO K5%: S4(L2%) = 0
4630 IF P5#(L2%)<.00001 OR P5#(L2%)>1 THEN GOTO 4690
4640 IF ABS(P1#(P6%(L2%),6))<.00001 THEN GOTO 4690
4650 RS# = ABS(P1#(P6%(L2%),6)) + BI: RT# = RS#*BJ: RB# = 1/RT#: BD# =
     P5#(L2%)^RT#: BB# = BD#*2: IF BD#>.5 THEN BB# = (1 - BD#)*2
4660 BB# = BB#^RS#/2: IF BD#>.5 THEN BB# = 1 - BB#
4670 HP# = BB#^RB#
4680 S4(L2%) = ((P1#(P6%(L2%),4)*P1#(P6%(L2%),1) +
     P1#(P6%(L2%),5)*P1#(P6%(L2%),2) + P1#(P6%(L2%),3) - P1#(P6%(L2%),4)*WX# -
     P1#(P6%(L2%),5)*WY#) - S1(L1%))*HP#
4690 NEXT L2%
4700 FOR L2% = 1 TO K5%: S1(L1%) = S1(L1%) + S4(L2%): NEXT L2%: GOTO 4940
4870 TR# = 0: REM minimum curvature splines
```

```
4880 FOR L2% = 4 TO LL%: L3% = L2% - 3: RT# = ((P1#(L3%,1) - WX#)^2 +
     (P1#(L3%,2) - WY#)^2)^.5: TR# = TR# + B#(L2%)*FNCV(RT#) : NEXT L2%: S1(L1%)
     = B#(1) + B#(2)*WX# + B#(3)*WY# + TR#
4940 IS = IS + SX*U8: IF IS>0 THEN GOTO 4960
4950 WX# = WX# + SX/2: IS = 1.001 - SX
4960 WX# = WX# + SX*U8: IF U8*WX#<S2(2 - U8) THEN GOTO 4980
4970 U8 = - U8: WX# = S2(2 + U8): WY# = WY# - SY: IS = U8
4980 NEXT L1%
5060 REM end interpolation to triangular grid and begin display construction
5070 SOUND 208,5: SOUND 165,5: SOUND 185,5: SOUND 123,5: PRINT " AT " TIME$;
5071 INPUT " PREPARE PLOTTER AND PRESS ENTER",PM$
5090 LW = (HI - SZ)/2: IF DP>0 THEN SZ = INT(SZ*2/3)
5091 IF DP>0 AND SZ>HI/2 THEN SZ = HI/2
5100 SW = (WD - SZ)/2: ZW = (HI - SZ)/2
5110 IF SR>0 THEN SW = (WD - 2*SZ)/3
5120 SW = SW + 1: SV = 2*SW + SZ: SR = SR*2: SZ = SZ - 2: SZ = SZ - 2: ZW = ZW -
     1: SU = SW: IF SR>0 THEN SU = (2*SW + SV)/3
5160 REM plot open
5170 LPRINT "IN;IP0,0," + STR$(WD) + "," + STR$(HI) + ";SP1;": IF SR<2 THEN GOTO
     5180
5175 FOR L1% = 2 TO 6: BX1 = (BX(L1%,1) - .5)*.99875 - (BX(L1%,3) - .5)*.05 + .5:
     BX(L1%,3) = (BX(L1%,3) - .5)*.99875 + (BX(L1%,1) - .5)*.05 + .5: BX(L1%,1) = BX1:
     NEXT L1%: LPRINT "PU;PA0,6800;"
5180 LPRINT "PU;PA" + STR$(BX(BN%,1)*SZ + SW) + "," + STR$(BX(BN%,2)*SZ +
     ZW) + ";PD;" : ZX = SX: ZY = SY: ZZ = .5: R3#(2,2) = 1 + 2*SY: R3#(2,1) = 0
5190 FOR L1% = BN% + 1 TO BN% + 4: LPRINT "PA" + STR$(BX(L1%,1)*SZ + SW) +
     "," + STR$(BX(L1%,2)*SZ + ZW) + ";": NEXT L1% : LPRINT "PU;"
5192 FOR L1% = 2 TO 6: BX(L1%,1) = (BX(L1%,1) - .5)*.995 + (BX(L1%,3) - .5)*.1 + .5:
     NEXT L1%: SM = 0: SA = 0
5200 REM plot contours
5220 FOR L1% = 1 TO I3%
5230 R3#(1,2) = R3#(2,2) - 2*SY - 2*ZY: R3#(1,1) = R3#(2,1): N1% = (2*L1% - .5 +
     ZZ)*I0% + L1% + .5/ZZ: R3#(1,3) = S1(N1%): R3#(2,2) = R3#(2,2) - 2*SY - ZY:
     R3#(2,1) = .5 - ZZ: N2% = (2*L1% - .5 + ZZ)*I0% + L1%
5240 R3#(2,3) = S1(N2%): R3#(3,2) = R3#(2,2) - 2*SY: R3#(3,1) = R3#(2,1) + ZX/2: N3% = N2%
     - .5/ZZ: R3#(3,3) = S1(N3%): AR# = AA#/2
5250 GOSUB 5820
5260 R3#(1,2) = R3#(1,2) + 2*ZY: N2% = (2*L1% - .5/ZZ - .5 - ZZ)*I0% + L1%: R3#(1,3) =
     S1(N2%)
5270 GOSUB 5820
5280 AR# = AA#
5290 FOR L2% = 1 TO GR
5300 R3#(2,2) = R3#(1,2): R3#(2,1) = R3#(1,1) + ZX: N4% = N2% + .5/ZZ: R3#(2,3) =
     S1(N4%)
5310 GOSUB 5820
5320 R3#(1,2) = R3#(3,2): R3#(1,1) = R3#(3,1) + ZX
5330 IF L2% = GR THEN R3#(1,1) = R3#(1,1) - ZX/2
5340 N6% = N3% - .5/ZZ: R3#(1,3) = S1(N6%): IF L2% = GR THEN AR# = AA#/2
5350 GOSUB 5820
5360 R3#(2,2) = R3#(2,2) - 2*ZY: N5% = N1% + .5/ZZ: R3#(2,3) = S1(N5%)
5370 GOSUB 5820
5380 R3#(1,2) = R3#(2,2): R3#(1,1) = R3#(2,1) - ZX: R3#(1,3) = S1(N1%): IF L2% = GR
     THEN AR# = AA#
5390 GOSUB 5820
5400 N1% = N5%: N2% = N4%: N3% = N6%: R3#(1,2) = R3#(1,2) + 2*ZY : R3#(1,1) =
     R3#(2,1): R3#(1,3) = S1(N4%): R3#(3,1) = R3#(3,1) + ZX: R3#(3,3) = S1(N6%) 5410
     NEXT L2%
5420 ZY = - ZY: ZX = - ZX: ZZ = - ZZ
5430 NEXT L1%: REM end plotting contours
5450 REM plot data points
5460 IF PT<1 THEN GOTO 5770
5480 DZ% = DZ: FOR L1% = 1 TO ND%
5490 IF P1#(L1%,1)< - .0001 OR P1#(L1%,1)>1.0001 OR P1#(L1%,2)< - .0001 OR
     P1#(L1%,2)>1.0001 THEN GOTO 5610
5500 PP1# = P1#(L1%,1): PP2# = P1#(L1%,2): PP3# = P1#(L1%,3): IF DP<1 THEN
     GOTO 5530
```

Appendix

```
5510 YY = (PP2# - .5)*R8(1,1) + (PP1# - .5)*R8(1,2) + .5: PP1# = (PP1# - .5)*R8(1,1) -
     (PP2# - .5)*R8(1,2) + .5
5520 ZZ3 = (PP3# - .5)*R8(2,1) + (YY - .5)*R8(2,2) + .5: PP2# = (YY - .5)*R8(2,1) - (PP3#
     - .5)*R8(2,2) + .5: PP3# = ZZ3: IF SR<2 THEN GOTO 5528
5525 PP1# = (PP1# - .5)*.99875 - (PP3# - .5)*.05 + .5
5528 IF SR<>1 THEN GOTO 5530
5529 PP1# = (PP1# - .5)*.99875 + (PP3# - .5)*.05 + .5
5530 S3(1,1) = PP1#*SZ + SW: S3(1,2) = PP2#*SZ + ZW
5540 LPRINT "PA" + STR$(S3(1,1) - DZ%) + "," + STR$(S3(1,2) + DZ%) + ";": LPRINT
     "PD;PA" + STR$(S3(1,1) + DZ%) + "," + STR$(S3(1,2) + DZ%) + ";": LPRINT "PA" +
     STR$(S3(1,1) + DZ%) + "," + STR$(S3(1,2) - DZ%) + ";"
5550 LPRINT "PA" + STR$(S3(1,1) - DZ%) + "," + STR$(S3(1,2) - DZ%) + ";": LPRINT
     "PA" + STR$(S3(1,1) - DZ%) + "," + STR$(S3(1,2) + DZ%) + ";": LPRINT "PA" +
     STR$(S3(1,1) + DZ%) + "," + STR$(S3(1,2) - DZ%) + ";PU;"
5610 NEXT L1%  5770 SR = SR - 1: SW = SV: IF SR = 1 THEN GOTO 5180
5775 LPRINT "PU;PA" + STR$(SU) + "," + STR$(LW - 500) + ";SI0.2,0.3;SL0.2;"
5777 LPRINT "LBFigure 1.0.1 - Example Isolines" + CHR$(3): LPRINT "PU;PA0,6800;":
     REM plot close
5790 SOUND 123,5: SOUND 185,5: SOUND 208,5: SOUND 165,5: TM$ = TIME$
5800 SM = SM/SA: PRINT " AVERAGE HEIGHT OF SURFACE IS" SM: PRINT " AT "
     TM$ " RUN COMPLETE": END
5810 REM curvemaker
5820 ST = R3#(1,3)
5830 IF ST>R3#(2,3) THEN ST = R3#(2,3)
5840 IF ST>R3#(3,3) THEN ST = R3#(3,3)
5850 IF ST< = - 999 THEN RETURN
5860 SM = SM + AR#*((R3#(1,3) + R3#(2,3) + R3#(3,3))/3*VS + P9(2,3)): SA = SA + AR#
6210 L8% = 0: IF DP<2 THEN GOTO 6310
6220 L8% = 1
6230 REM rotate triangle for orthogonal profiles
6240 FOR L3% = 1 TO 3
6250 Z# = R3#(L3%,3): R3#(L3%,3) = 1 - R3#(L3%,L8%): R3#(L3%,L8%) = Z#  6260
     NEXT L3%
6270 ST = R3#(1,3)
6280 IF ST>R3#(2,3) THEN ST = R3#(2,3)
6290 IF ST>R3#(3,3) THEN ST = R3#(3,3)
6300 IF ST < = - 999 THEN RETURN
6310 TP = R3#(1,3)
6320 IF TP<R3#(2,3) THEN TP = R3#(2,3)
6330 IF TP<R3#(3,3) THEN TP = R3#(3,3)
6340 IF TP = ST THEN RETURN
6350 FOR L0 = 1 TO NC%
6360 REM slice the triangle
6370 IF P8(L0)>TP OR P8(L0)<ST THEN GOTO 6610
6380 L5% = 1: L4% = 0
6390 L4% = L4% + 1
6400 L7% = 1: IF L7% = L4% THEN L7% = L7% + 1
6410 L3% = L7% + 1: IF L3% = L4% THEN L3% = L3% + 1
6420 IF R3#(L7%,3) = R3#(L3%,3) THEN GOTO 6460
6430 F = (P8(L0) - R3#(L7%,3))/(R3#(L3%,3) - R3#(L7%,3)): IF F<0 OR F>1 THEN GOTO
     6460
6440 S3(L5%,1) = R3#(L7%,1) + (R3#(L3%,1) - R3#(L7%,1))*F: S3(L5%,2) = R3#(L7%,2)
     + (R3#(L3%,2) - R3#(L7%,2))*F
6450 L5% = L5% + 1
6460 IF L5%<3 THEN GOTO 6390
6470 S3(1,3) = P8(L0): S3(2,3) = P8(L0): IF DP<2 THEN GOTO 6500
6480 REM reverse rotate the intersection trace
6490 FOR L3% = 1 TO 2: Z# = S3(L3%,3): S3(L3%,3) = S3(L3%,L8%): S3(L3%,L8%) = 1
     - Z#: NEXT L3%
6500 IF DP = 0 THEN GOTO 6570
6510 REM apply perspective
6520 FOR L3% = 1 TO 2
6530 YY = (S3(L3%,2) - .5)*R8(1,1) + (S3(L3%,1) - .5)*R8(1,2): S3(L3%,1) = (S3(L3%,1) -
     .5)*R8(1,1) - (S3(L3%,2) - .5)*R8(1,2) + .5
6540 S33 = (S3(L3%,3) - .5)*R8(2,1) + YY*R8(2,2) + .5: S3(L3%,2) = YY*R8(2,1) -
     (S3(L3%,3) - .5)*R8(2,2) + .5: S3(L3%,3) = S33
```

```
6550 NEXT L3%
6552 IF SR<2 THEN GOTO 6560
6554 S3(1,1) = (S3(1,1) - .5)*.99875 - (S3(1,3) - .5)*.05 + .5: S3(2,1) = (S3(2,1) -
     .5)*.99875 - (S3(2,3) - .5)*.05 + .5
6560 IF SR<>1 THEN GOTO 6570
6565 S3(1,1) = (S3(1,1) - .5)*.99875 + (S3(1,3) - .5)*.05 + .5: S3(2,1) = (S3(2,1) -
     .5)*.99875 + (S3(2,3) - .5)*.05 + .5
6567 REM draw a line from (S3(1,1),S3(1,2)) to (S3(2,1),S3(2,2))
6570 LPRINT "PA" + STR$(S3(1,1)*SZ + SW) + "," + STR$(S3(1,2)*SZ + ZW) + ";":
     LPRINT "PD;PA"  + STR$(S3(2,1)*SZ + SW) + "," + STR$(S3(2,2)*SZ + ZW) + ";PU;"
6610 NEXT L0
6620 IF L8%<1 THEN RETURN
6630 REM reverse rotate the triangle
6640 FOR L3% = 1 TO 3
6650 Z# = R3#(L3%,3): R3#(L3%,3) = R3#(L3%,L8%): R3#(L3%,L8%) = 1 - Z#
6660 NEXT L3%
6670 L8% = L8% + 1: IF L8%<3 THEN GOTO 6240 ELSE RETURN
6680 REM solve R3#(LL%,LL%)xB#(LL%) = R9#(LL%)
6690 FOR J2% = 1 TO LL%: P4%(J2%) = J2%: B#(J2%) = 0
6700 FOR J3% = 1 TO LL%: IF ABS(R3#(J2%,J3%))> = B#(J2%) THEN B#(J2%) =
     ABS(R3#(J2%,J3%))
6710 NEXT J3%: NEXT J2%: LM1% = LL% - 1: LP1% = LL% + 1
6720 FOR J2% = 1 TO LM1%: X3# = 0
6730 FOR J3% = J2% TO LL%: X4# = ABS(R3#(P4%(J3%),J2%))/B#(P4%(J3%)): IF
     X4#< = X3# THEN GOTO 6750
6740 X3# = X4#: K3 = J3%
6750 NEXT J3%: IF K3 = J2% GOTO 6770
6760 K2% = P4%(J2%): P4%(J2%) = P4%(K3): P4%(K3) = K2%
6770 X3# = R3#(P4%(J2%),J2%): K4% = J2% + 1
6780 FOR J3% = K4% TO LL%: X4# =  - R3#(P4%(J3%),J2%)/X3#: R3#(P4%(J3%),J2%)
     =  - X4#
6790 FOR J4 = K4% TO LL%: R3#(P4%(J3%),J4) = R3#(P4%(J3%),J4) +
     X4#*R3#(P4%(J2%),J4): NEXT J4: NEXT J3%: NEXT J2%
6800 B#(1) = R9#(P4%(1))
6810 FOR J1% = 2 TO LL%
6820 I8% = J1% - 1: X3# = 0
6830 FOR J2% = 1 TO I8%: X3# = X3# + R3#(P4%(J1%),J2%)*B#(J2%): NEXT J2%
6840 B#(J1%) = R9#(P4%(J1%)) - X3#: NEXT J1%
6850 B#(LL%) = B#(LL%)/R3#(P4%(LL%),LL%)
6860 FOR J1% = 2 TO LL%: I8% = LP1% - J1%: I9% = I8% + 1: X3# = 0
6870 FOR J2% = I9% TO LL%: X3# = X3# + R3#(P4%(I8%),J2%)*B#(J2%): NEXT J2%
6880 B#(I8%) = (B#(I8%) - X3#)/R3#(P4%(I8%),I8%): NEXT J1%: RETURN
```

Index

Approximation and interpolation, 102
  B-spline patches, 149
  minimum curvature, 128
  objective analysis, 126, 129
  residuals, 126
  response surface, 128
  surface defined, 126
  trend surface, 126
Area of a
  3 D triangle, 32
  parallelogram, 28
  triangle, 29
Area based interpolation, 113
Average grade, 187

B-spline patches, 149
Barycentric coordinates, 76
Basis functions
  B-spline, 149
  Bernstein, 147
  Hermite, 145
  polynomial, 126
  power, 164
  serendipity, 170
Bernstein polynomials, 147, 169
Beta-splines, 151
Bezier patches, 147
Bicubic patches, 143
Biharmonic splines, 121
Biharmonic surface, 125
Bilinear patches, 139
Binary coordinates, 76
Biquadratic patches, 141
Blended gradients, 116, 135, 157, 160
Blending functions, 163
  compound exponential, 170
  parametric, 169
Block averages, 43, 186
Boolean sum, 146, 157
Bounded surfaces
  distance based, 117
  neighborhood based, 157
  rectangle based, 141
  triangle based, 136
Brown's square, 147

Index

Cartesian product, 141
Centroid, 109
Chebyshev polynomials, 165
Choropleths, 50, 103, 182
Circumcenter, 31, 33, 109
  3 D triangle, 34
  tetrahedron, 33
  triangle, 31
Clothoidal splines, 151
Clough-Tocher, 132, 137
Collocation, 120
Color filling, 183
  efficiency, 192
CON2R.BAS, 226, 273
Continuity, 54
  topographical, 45
Contours
  defined, 47
  first computer, 49
  history, 48
Convex hull, 102
Convexity, 152
Coons' patch, 137, 146
Coordinates
  barycentric, 76, 131
binary, 76
natural neighbor, 81
parallelogram, 79
rectangular, 78
Cramer's rule, 32, 80, 88
Cross product, 33, 95, 171
Cross-eyed stereograms, 185
  Crosshatching, 184

Data
  choropleth, 50
  gridded, 54, 154, 178
  precision, 55
  punctulate, 49
  resolution, 55
  rotation of, 182
  scattered, 54
  sorting, 57

  subsets, 65
  topographical, 48
  traverse, 54
Degeneracies
  natural neighbor, 68
Delaunay triangulation, 59
Deleting natural neighbors, 67
Dirichlet, 59
Display efficiency, 192
Distance based, 113
DISTANCE.BAS, 116, 117, 253, 297
Dot density maps, 50
Drawing curves, 178

Exponential spline, 152
Extrapolation, 102

Fast Fourier FFT, 154
Fast Hartley FHT, 154
Ferguson patch, 146
Finite elements, 59, 133
FITFUNCT.BAS, 124, 127, 129, 261, 305
Fitted functions, 105, 118
Four-triangle patch, 138
Fourier surface, 154

Geometric algorithms, 59
Gradient estimation, 85
  efficiency, 190
Gradients
  cross products, 94
  inverse distance, 97
  least squares, 88
  maximum slope, 85
  neighborhood based, 97
  parametric mean, 97
  quadratics, 90
  spherical quadratics, 92
  splines, 92
  volume under, 87
Gregory's patch, 147

Hachures, 107, 184
Hermite patch, 143
Hermite polynomials, 145, 169
　Heron's formula, 33
Heteroscedasticity, 72
Histograms, 42
Histosplines,
Laplacian, 125
Hyperboloid, 111, 139

Incenter, 32, 109
Interpolation
　area based, 130
　bicubic, 144, 147, 149
　biharmonic, 122, 125
　bilinear, 140
　biquadratic, 141
　Clough-Tocher elements, 133
　collocation, 120
　constrained optimal, 129
　distance based, 113
　efficiency, 191
　globally based, 118
　Hermite, 143
　ideal, 103, 160
　Kergin, 151
　kriging, 123
　Lagrange, 119
　Laplacian, 124
　linear facets, 138
　manual, 106
　minimum curvature, 122
　multiquadric, 121
　neighborhood based, 155
　Pade, 154
　piecewise bicubic, 143
　piecewise bilinear, 139
　piecewise polynomials, 132
　rational, 153
　rectangle based, 137
　relaxation, 124
　slant top, 108
　transfinite, 146
　triangle based, 130

Inverse distance weighted gradients, 116
Inverse distance weighting, 76, 114
Isobaths, 49
Isochor maps, 50
Isogones, 49
Isolines, finding, 53, 180
Isometric views, 23, 181
Isted's formula, 35, 186

Kernel of influence, 104
Kernel functions, 165
Kriging, 123

L-splines, 152
Lagrange interpolation, 119
Laplacian histospline, 52
Laplacian surface, 124
Least squares, 127
Linear regression, 88, 113
Local coordinates, 75
　barycentric, 76
　binary, 76
　computing, 82
　natural neighbor, 81
　parallelogram, 79
　rectangular, 78
Local variability, 157
Lofted surfaces
　linear, 139
　non-linear, 146

Madison triangle, 137
Manual methods
　inverse distance, 112
　linear regression, 113
　proximal polygons, 107
　rectangular grids, 111
　statistical method, 106
　triangulations, 108
　Voronoi polygons, 108
Maps
　chorochromatic, 51

dot density, 50
first computer, 127
isochor, 50
isoline, 49
statistical, 107
Minimum curvature, 122
basis functions, 123
splines, 122
Monotonicity, 152, 153
Multiquadric interpolation, 121

Natural neighbor
circumcircles, 59
coordinates, 81
degeneracies, 68
proximal polygons, 108
Nautical origins, 49
NEIGHBOR.BAS, 67, 156, 158, 234, 280
Neighborhood based
gradients, 97
interpolation, 155
Nonnegativity, 153
Numerical estimates, 185

Objective analysis, 129
Ogives, 42, 185
Optimal interpolation, 129
Order
arbitrary, 58
greedy, 58
natural neighbor, 59
optimal, 58
proximal, 57
Orthogonal polynomials, 165
Orthogonal profiles, 23, 184
Outlier index, 173
Output database, 178

Perspective views, 181
Plausibility, 54
Plotter output, 178
Polynomial approximation, 126

Power functions, 118
Precision
computational, 55
data set, 55
display, 55
Prism map, 50
Projected inverse distance weighting, 116
Proximal order
arbitrary, 58
greedy, 58
natural neighbor, 59
optimal, 58
Proximal polygons, 107
Pseudoperspective, 181
Pycnophylactic, 125
Pycnophylactic methods, 52

Quadtrees, 180

Rational splines, 153
RECTANGL.BAS, 7, 141, 267, 310
Rectangle based, 139
Rectangular hyperboloid, 111, 139
Regression surface, 127
Regular grids, 178
Relaxation surface, 124
Residuals, 126, 129
Response surface, 128
Rotation of data, 182
Roughness index, 171
Ruled surfaces, 140

Semivariograms, 124
Serendipity functions, 120, 132, 170
Shape-preserving, 152
Slant-top proximal polygons, 108
Sorting
natural neighbor, 64
spatial, 57
Spatial order lists, 63
Splines bicubic, 143
Stable surfaces, 103

Statistical method, 106
Statistics
  average grade, 186
  block averages, 186
  parametric, 48
Steepest slope, 53
Stereograms, 24, 184
  construction, 193
  cross-eyed, 185
Subsets
  fixed area, 72
  fixed distance, 71
  fixed number, 71
  natural neighbor, 72
  selection, 69
  selection criteria, 71
Surface types, 86
Surfaces
  created or explored, 102
  lofted, 146
  nonnegative, 153
  ruled, 140
Systematic triangulations, 60

Tangent planes, 87
Tautness, 160
Taylor interpolants, 151
Tension patch, 152
Tensor product, 141

Tessellations, 58
  Delaunay, 62
  Voronoi, 62
Thiessen, 59
Trend surface, 126
Triangle based, 130
TRIANGLE.BAS, 67, 136, 244, 289
Triangular Coons' patch, 137
Triangular grids, 179
Triangulations, 58
Two-triangle patch, 138

Variability index, 157
  outlier, 173
  roughness, 171
Volume
  estimates, 185
  numerical integration, 31
  parallelepiped, 34
  prism, 29
  tetrahedron, 34
  under a surface, 30
Volume preserving, 125
Voronoi polygons, 59

Weighted averages, 105
Wigner-Seitz cells, 59
Wireframe views, 182